PATIENT
EARTH

WITH INVITED CONTRIBUTIONS BY
Herman E. Daly
Alice Taylor Day and Lincoln H. Day
Alfred W. Eipper
Richard A. Falk
Arthur W. Galston
Austin Heller and Edward Ferrand
Richard D. Lamm
Orie L. Loucks
Michael McCloskey and Albert Hill
Charlotte Alber Price
Jeremy A. Sabloff
Paul Sears
Kent Shifferd
H. Lyle Stotts

PATIENT EARTH

JOHN HARTE
ROBERT H. SOCOLOW
YALE UNIVERSITY

HOLT, RINEHART AND WINSTON, INC.
New York   Chicago   San Francisco
Atlanta   Dallas   Montreal   Toronto
London   Sydney

Library of Congress Catalog Card Number: 77-148032

ISBN: 0-03-085103-3 (Paper)
ISBN: 0-03-086571-9 (Cloth)

Printed in the United States of America
4 3 2 1   090   1 2 3 4 5 6 7 8 9

This book is printed on recycled paper

TO DAVID AND ALEXIS HARTE
AND DAVID SOCOLOW
WITH THE BELIEF THAT THE EARTH
WILL BE WORTH INHERITING

PREFACE

The diseases that afflict our planet may not be fatal, but by all appearances they are cruel and debilitating. The symptoms are all around us: in the air, in the water, in our cities, and in our cells.

But the purpose of this book is not to remind the reader of this—he knows it all too well. Rather, this is a book with a more constructive goal. It tells about the work of some people—we like to call them "geophysicians"—who are diagnosing and healing the earth, and it provides some of the information that will allow the reader to join in this activity.

Two lessons from the past begin our book and provide a sober perspective. One of the deans of the American conservation movement describes his own experiences in the 1930s with the Dust Bowl on the Great Plains. Then an anthropologist tells the story of the Classic Maya civilization, which ruined its land and vanished.

Ten case studies describing environmental problems in contemporary American society form the main section of the book. These are all stories of positive accomplishments, and they are told by authors who have been directly involved in the problems. We have selected the case study approach because it begins to do justice to the intellectual complexities of environmental problems and because it leads the reader into the real world.

There is a tendency for the two words "pollution" and "environment" to be almost interchangeable today. In part, this is because, quite naturally, all of us prefer to direct our attention toward those environmental problems that most affront the senses and have quite direct technical solutions. But in so doing we may fail to face those environmental problems whose solutions require a more substantial change in our system of values. In selecting the case studies, we have tried deliberately to enlarge the list of problems that are properly associated with environment. The case studies raise some of the tougher questions associated with urban blight, population control, resource management, the ecological impact of military activity, and alternative uses of land.

Because a scientific component underlies each of the case studies and needs to be grasped to appreciate both the problem and the possible solutions, the authors of the case studies have taken the time to present the relevant science. In a number of instances, we have amplified their discussions with our own remarks. Here and there throughout the book we show the reader how to do "back of the envelope" calculations, which give an approximate answer and point the way to a more refined calculation.

We have also written three extensive essays on water, energy, and radiation. We expect them to help the general reader comprehend the science that underlies the issues posed in the case studies. We are unaware of any other treatments of these subjects that have quite the same brevity and still "level" with the reader.

We assume that the reader likes science, but not that he knows more than is in a basic high school curriculum. Once in a rare while we do use some calculus, and there are a few structural formulas from organic chemistry. We expect the reader to recognize difficult mathematics and science when he sees it and to stay away from such passages if they are bad for him.

The authors of the case studies go beyond their descriptions of the scientific aspects of each environmental problem to involve the reader in the social and moral aspects as well. The authors leave the reader with at least a partial understanding of how the citizen, having become aware of an impending ecological catastrophe, starts doing something about it. The reader will learn what role the courts, the legislatures, and citizens' pressure groups can actually play in controlling the man-made forces that ravage our earth.

The harder we thought about the case studies, the more we realized that they were encapsulating a time span of years, and that their lessons did not immediately generalize to a time span of generations. The majority of our case studies describe an accommodation to growth, often an effective one. But longer-range considerations require that growth be confronted, not just accommodated.

Thus we decided to ask the reader to stop with us for a while to consider three speculative essays which attempt to describe the Equilibrium Society. One essay ponders a society with a constant population, another a society with a constant gross national product. The third essay of this section suggests what a world would be like in which international relations were adequate to the task of protecting the global environment. The issues of social philosophy raised in these essays, no less than scientific issues like those raised in the essays on water, energy, and radiation, elucidate the conflicts that ultimately surface as "case studies" in particular times and places.

There are three ecosystems today: the ecosystem of species, the ecosystem of intellectual disciplines, and the ecosystem of nations. The ecosystem of species is the direct concern of this book; it needs to be understood and cherished not only for itself but also because it provides models of interdependence which suggest how other systems might behave. The interactions of diverse intellectual disciplines throughout this book reveal some of the interdependence within the ecosystem of ideas. And the context of many of the essays—global scarcity and competition—suggests some of the interdependence in the ecosystem of nations.

If this book is successful, the reader will come to share our perspective that the task of restoring the world to health requires that mankind become more aware of interdependence in all these domains.   The three ecosystems are not even independent of one another. Each is ultimately dependent on the proper functioning of the other two for its suvival.

New Haven, Connecticut   John Harte
January 1971   Robert H. Socolow

ACKNOWLEDGMENTS

Our first thanks go to our contributors: for responding with unexpected enthusiasm to our initial proposal and for permitting us to collaborate aggressively in the construction of their essays.

The impetus for this book came out of our participation in the Everglades Jetport study organized by Marvin Goldberger and Gordon MacDonald through the Environmental Studies Board of the National Academy of Sciences and the National Academy of Engineering. Discussions with these two men and also with Joe Browder, James Fay, Harold Feiveson, Murray Gell-Mann, William Hall, and Milton Kolipinski greatly enhanced our contribution to that study and our essay on the Everglades here. The insights of Frank Egler, David Smith, and Harold Thomas have also been of benefit in the preparation of this essay.

Discussions with many others have helped us to clarify scientific matters throughout the book. We especially wish to thank Daniel Botkin, Min-Shih Chen, Gary Haller, Elliot Lapan, Richard Slansky, Ingrid Waldron, Charles Walker, James Walker, and Arthur Westing for their advice.

The presentation of material has been greatly helped by editorial suggestions from Grace Brush, Katherine Cole, David Dirks, Wendy Goble, Jerome Harte, Thomas Keller, Lucille Morowitz, Elizabeth Socolow, Joan Socolow, Frances Sussman, and Thomas Walker.

Kendall Getman, Richard Miller, and Donald Schumacher encouraged us at the outset and secured the support of our publisher. Michael Coe, Norman Cousins, Carol Harte, Franklin Long, Richard Miller, and Charles Remington each had the wisdom to identify an appropriate contributor. Suzanne Weiss was imaginative and resourceful in suggesting and locating many of the photographs that appear here. Brenda Preston did seemingly endless typing cheerfully, and Dorothy Garbose has been of enormous help in organizing the production of this book.

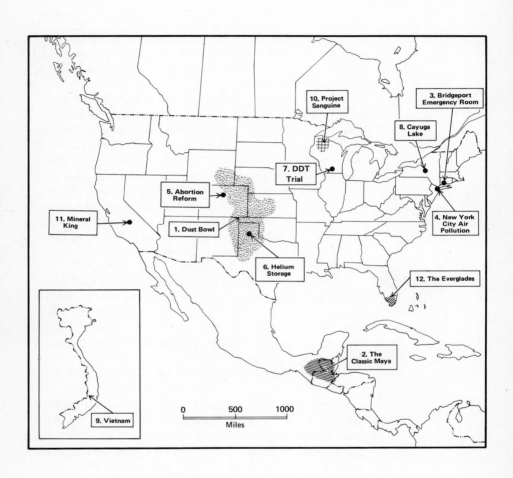

10. Project Sanguine

3. Bridgeport Emergency Room

8. Cayuga Lake

7. DDT Trial

5. Abortion Reform

11. Mineral King

1. Dust Bowl

6. Helium Storage

4. New York City Air Pollution

12. The Everglades

2. The Classic Maya

9. Vietnam

0   500   1000

Miles

CONTENTS

PATIENT
EARTH

I. LESSONS FROM THE PAST

The first of these essays is about being deaf; the second is about being blind. Professor Sears tells us about warnings unheeded: perennials hold the soil together where annuals will not; this was well known to the scientists, but no one listened.

The Mayas, Professor Sabloff suggests, did not see where they were going. No one knew how to make the metal tools which could have cut down the savanna and extended the land available for farming.

Every society has its blind spots, and from a distance one's reactions to them are instinctively charitable. But, to the deaf spots in a society, how should one respond?

1. AN EMPIRE OF DUST [1]

Paul Sears

Loss of soil through erosion by water began with European settlement. New England farmers seem to have been slow to recognize erosion of soil by water as a major cause of worn-out land, but from Virginia south its effects were unmistakable. There the combination of loose soils and the growing of row crops—corn, tobacco, and cotton—led to gully formation and serious concern in colonial times.

Curiously, however, it was erosion by wind with its spectacular dust storms following the introduction of dry farming in the semiarid shortgrass country of the West that finally aroused public opinion and led to official action.

The convergence of prolonged drought and severe economic depression during the 1930s, along with mass emigration from the distressed areas, brought home to the American people at long last the inseparable tie between the good earth and human destiny. One is reminded that the slow progress of disease is often brought into focus only when a sharp crisis develops.

Students of the living landscape were aware of what was happening long before the dramatic climax that made erosion more than an academic matter. Their warnings were received with indifference, even hostility. Because my own most vivid experience preceded the events of the 1930s by a decade, I shall begin with it. And because an understanding of the principles involved is important to the citizen, I shall devote some space to them before discussing the tragedy of the Dust Bowl.

With Figure 1 before him, a few very general reminders may help orient the reader. Recalling that natural vegetation is an indicator of environmental conditions, we observe that the great belts of forest and grassland run roughly north and south. Beginning at the Atlantic Coast, forest occupies a humid region where rainfall tends to exceed evaporation. Westward beyond the Great Lakes it tends to give way to the tallgrass prairie, characteristic of the subhumid climate and rich

[1] This title has been chosen out of respect for the book of that title by Lawrence Svobida, published by Caxton Printers, Ltd., Caldwell, Idaho, and now out of print.

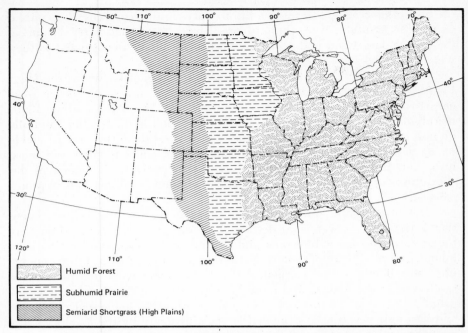

FIGURE 1 Major regions of climate and vegetation in the United States east of the Rockies.

soil that compose the granaries of this and other continents. Still farther west, where evaporation greatly exceeds rainfall, shortgrasses replace those of the subhumid prairie. This transition to semiarid conditions occurs— again roughly—along the 100th meridian. Transitions between these great plant formations and the climates they express are irregular, shifting with changes in climate, and are vulnerable even to short-term climatic episodes such as groups of drought years.

During Easter week of 1920, as a new junior member of the Botany Department at the University of Nebraska, it was my privilege to cross the Sandhills of that state with my chief, the late Raymond J. Pool. Dr. Pool had spent years of study in these same Sandhills, an area of grassland comprising some 11 million acres of dune and pan. Most of our 60-mile trip from south to north was on foot, with ample time to look and listen. At the time of our visit, the area was marked by great blowouts, scars of abortive plowing and overgrazing. A few years later, after these scars had begun to heal, I viewed it from horseback in company with Dr. Adolph Waller of Ohio State University. Seen thus in early summer the deeply undulating Sandhills, covered with waving grass and dotted with brilliantly colored flowers, afforded a scene of incomparable beauty.

What we saw were the final stages of a tragedy that had its beginning soon after 1900, when a congressman, the Honorable Moses Kincaid, with the best of motives got through a bill to open up the Sandhills, then public domain, to homesteads one mile square. I have heard that he tried to make the parcels larger but had no luck with his colleagues, accustomed as they were to smaller farms in the humid East.

The Sandhills are within the vast semiarid region known as the High Plains, or shortgrass country, that stretches north and south like a great carpet roughly 200 miles wide between the Rocky Mountains and the 100th meridian. Rainfall here is meager and evaporation high; but the Sandhills lie in a great rock bowl, and the sand, which gives them their name, yields up its water generously. Thus their native plant cover is taller and more lush than the grama and buffalo grasses that prevail over much of the shortgrass country. Small wonder that the verdant Sandhills looked promising.[2]

Not to everyone, however. Scientists (who were not consulted) knew that if the native cover were destroyed by the plow, wind would convert the surface into great blowouts and drifting sandy soil, even in the sub-humid prairie climate of eastern Nebraska. As a matter of fact, the geographer Nils Bengston had warned of this very danger when the plowing of sandy soil at the eastern and more humid portion of the Plains was advocated. Thereby he incurred the vengeful wrath of an influential politician, and was only saved from dismissal at the University by the loyalty of students and other friends.

Appropriately the homesteaders in the Sandhills were known as Kincaiders. Lacking both the space and capital to depend upon the grazing of livestock, they had no choice but to plow the land for gardens and grain. Many of their homes were built of blocks of turf, using a minimum of scarce timber and plastered inside with marl from the lakes (often alkaline) that occupied many of the depressions between hills.

By the end of World War I most of the settlers had given up and left in the face of crop failures and drifting sand. But our route in the spring of 1920 took us past a sod house whose owners were hanging on, subsisting on potatoes and white sauce, hoping against hope for better luck in the next growing season. Almost everywhere else the mile-square holdings were being bought up by ranchers. These buyers had sufficient working capital to acquire the large areas of several square miles which are an economic necessity where rainfall is deficient and to protect these areas from plowing and excessive grazing until a cover of native perennial grasses could be restored. The typical sandhill grass puts out runners over the loose bare soil which root and establish a kind of green quilt that holds the soil in place and affords nutritious forage for grazing animals, as it had done for the bison through millennia.

Meanwhile frustration and failure were developing in a much wider setting. Depression, which was to strike urban, fiscal, and industrial centers nine years later, began for the American family farm with the advent of the 1920s. By a curious perversity this took place at a time of great technical

[2] Frank Ellis Smith, *The Politics of Conservation.* New York: Pantheon, 1966.

advances in agriculture, encouraged by high official prices for cereals. Contracts for land and power machinery to produce these wartime necessities involved huge indebtedness that could not always be met when price supports were withdrawn.

The effect was much like that of supermarkets on the corner store, or of giant mergers on small business in general. Foreclosures drove many an average farmer and his family off the land. Luckily for these refugees there was a boom period elsewhere which gave considerable employment until the great crash of 1929 and the ensuing years.

Masking this tragic phase in American rural life was the apparent success of heavily mechanized, heavily capitalized large-scale farming on level land. Journals of the day ran feature articles on "factories in the fields," and the immense resources of government and industry fell in step. Instead of a few score acres, the unit of cultivation became one of several hundreds. Driving through Iowa, for example, during the mid-twenties one could see a "spread" consisting of a square mile of corn, another of wheat, a third of soybeans, and a fourth of legume-grass mixture for hay.

Among those who stayed afloat in the new agriculture and profited by it were large operators who practiced "dry farming" in the High Plains. Here the ground was open and either level or slightly rolling, perfectly adapted to mechanical tillage and harvest. The land was also cheap; repeated failure had shown the hopelessness of breaking it up into small family-sized holdings of 160 acres (a quarter section), which had been sufficient for the homesteads in the humid forest climate of the East and even in the fertile, subhumid, tallgrass region of the prairies. The only profit-making use of the land before modern technology was available was for livestock pasturage, requiring a high ratio of acreage to animal. But one steer per ten acres or more could not compete as a profit maker with wheat at ten bushels or even less per acre. Actually, under dry farming, wheat yields in the High Plains averaged 15 bushels to the acre during times of favorable rainfall, ranging down to zero on loose soils in times of drought.

What was happening made perfectly good sense from the standpoint of conventional economics where the rule is that what pays best *is* best and that if you can't make a thing pay, get out. (One of the notable exceptions seems to be armed conflict, where considerations of various kinds, including "honor" and "policy," prevail.) But justification of an economic operation cannot long ignore scrutiny that goes beyond what I have called conventional economics. The type of analysis known as accounting must be employed; so must another kind of analysis—the use of science to evaluate change in the landscape; and a third, still complicated by inherited legal principles, which would balance private benefit against impact on the general good.

Now accounting is not, as many may suppose, a mere matter of bookkeeping. It can be a very complex and difficult process; for example, how can a fair rate for electric power be determined? Should it be what the traffic will bear—that is, what the customer will pay before he decides to do without? Should a fair rate of return be based on the original investment plus cost of operating the facility in question? Or should it be based upon what it would cost to replace plant and installations at a particular time? As in modern mathematics (and much else) the answer depends upon what assumptions one starts with. Whatever his assumptions, the accountant must identify certain elements as he proceeds. Assets and liabilities, profit and loss are familiar enough terms. But so far as the usable environment is concerned, there exists the same problem that is faced by ownership of a factory or other working facility. What has happened to its value by the end of a given period of operation? Normally this is a matter of wear and tear and obsolescence, and is called depreciation.

One can, of course, merely collect his accounts due, pay his bills, subtract expenses from income and calculate the difference as profit (or loss). However, if his plant is worth less than it was a year before and he fails to add this lost value to expense before estimating his profit, he cheats himself.

To be quite direct about it, this kind of defective accounting has been the rule in land use throughout the history of the United States, as well as in older nations whose lands are now far less productive than they were. Self-deception has been easy, because our land area remains the same while our numbers have increased manyfold and the price per acre has risen accordingly. The inherent capacity of the land to produce food, fiber, and minerals has shown no such increase, but rather the reverse. The obvious answer to this is to point out that fewer farmers on fewer farms and fewer acres are producing more than before. This change, however, has been due to an enormous input in the way of research, education, nutrients, and equipment; it does not represent any fundamental increase in the quality of the land itself. Excepting those cases in which measures are taken to restore perennial, soil-building cover such as grass-legume pasture (or woodland), fertilizers are harvested with the crops or leached out by rain. And as a further charge (or input) there must be added the cost of sustaining workers who are no longer "needed" in the country and who have not been absorbed into the urban economy.

Thus any narrowly fiscal analysis must be expanded in the light of a broader perspective that takes account of the basic relationship of life to environment. And one of the first steps in such an analysis is to examine how that relationship made possible the existence of life—for not millions, but billions of years before the advent of man—while actually enhancing the capacity of the planet to sustain life. As proof that this process was a constructive one, we only need recall the proverbial richness of virgin lands. Yet man's activity has, with tragic frequency, exploited this richness and lowered the capacity of his environment to sustain life. This was clearly documented as early as 1864 by George Perkins Marsh in his great but neglected book *Man and Nature*.[3] Today when we are being reminded that the survival of our species may be at stake (as though this were a

[3] George Perkins Marsh, *Man and Nature*, 1864. Reprinted by Harvard University Press, 1965, David Lowenthal, ed.

novel discovery), it is proper to recall Marsh's concluding statement: "Every new fact, illustrative of the action and reaction between humanity and the material world around it, is another step toward the determination of the great question, whether man is of nature or above her."

Although ecological knowledge is about where astronomy was in the 1700s and includes a fair amount of intuitive judgment based on experience, there is no longer any serious question that natural communities of plants and animals represent what is called in physics an open steady state. That is to say that they are in working equilibrium within and without, receiving energy from the sun, keeping themselves in repair, and—thanks to the diversity of organisms making them up—recycling the materials they use. In other words, their waste products are not dissipated or locked up beyond recovery, even though there have been times of exuberant activity when a surplus of hydrocarbons was produced in the form of gas, oil, and coal. These reserves of energy-containing materials, consisting as they did mostly of elements abundant in air and water, did not seriously impound the mineral nutrients needed for plant growth. Although the discovery that their energy could be added to that supplied to man by food has done much to improve the lot of mankind, it has also contributed greatly to the present critical unbalance between life and environment.

The remarkable cycle of use and reuse of essential materials that takes place in the absence of human interference is only one phase in the economy of natural processes. The living cover of vegetation tends to protect the land surface from the impact of wind and water which would otherwise carve it up and move its materials about in violent fashion. The resulting stability gives an opportunity for plant and animal communities to become organized in the direction of greater efficiency. One of the expressions—and so far as life is concerned, one of the great benefits—of this continuing relationship between community and environment is the genesis of soil, whereby the raw mineral and organic materials of a particular site are organized in ways that tend to bring the site up to the maximum fertility possible under existing conditions. Thanks largely to Russian and American scientists, we know that this organization is expressed in soil profiles, resultant of climate, parent material, and the activities of plant and animal life made possible by these physical factors. Roughly speaking, the top layer of an undisturbed profile contains organic matter (humus) added by living organisms; the layer below it consists of parent material modified chemically and physically; and the bottom layer consists of parent material little changed. The uppermost layer is at once most fertile and most vulnerable to exposure. It is considerably thicker in natural grasslands, which are the world's granaries, than in forested regions. In the latter, with their higher rainfall, organic acids from the top layer tend to leach out soluble nutrients from the mid-profile; thus where forests have been cleared for agriculture, an infertile surface is quickly exposed by erosion.

Such, from the equator poleward and from the humid seaboards to-

ward the arid continental interiors, has been the constructive and stabilizing role of living communities of plant and animal life in the absence of
our own species.

Here then is a model or standard by which we can measure the extent
to which our land-use practices operate for or against a permanent and
healthy relationship between man and the landscape. Do these practices
insure a stable surface in the face of wind and water? Do they conserve
moisture and fertility? Do they insure the use and reuse of wastes? And do
they reckon with the kind of environment in which they are carried on?

Such are the questions raised in any ecological analysis of land use to
determine the direction, whether constructive or destructive, in which our
operations are moving. To this and the perceptive kind of fiscal accounting
mentioned above must be added a third approach—the scrutiny of our
values and their effect upon the way we treat environment.

Here we encounter the most difficult, yet necessary, kinds of problems.
Matters of custom, sanction, and belief are deep-rooted and hard to
change. Economic and ecological analysis involves technical questions that
can be tackled more or less impersonally and whose answers speak for
themselves. But a whole cultural system goes on trial when we show that
practices that are perfectly legal and profitable to the individual are being
carried on at the expense of society and its future welfare.

Throughout the 1920s men with sufficient capital or credit found that
they could carry on large-scale mechanized wheat farming on the High
Plains regardless of lower prices for wheat that had proved disastrous to
smaller operators farther to the east. In the east wheat had been an essential part of the rotation on family-sized, mixed grain, hay, pasture, and
livestock farms. But the cost per unit of product was too great to permit
the smaller farmer to compete with the "suitcase farmer" who could plow
and plant a big holding in the High Plains in a few autumn days, relax in
Florida or California during the winter, and then return in time for a brief
tour of harvesting and marketing by machine. In order to practice this system, the native cover of the low-growing, turf-forming grama and buffalo
grasses and the legumes mixed with them had to be plowed up. And because rainfall was low and evaporation high, moisture was conserved by
allowing it to accumulate on fallow land for a year, in between crops. At
the same time, as enthusiastic agricultural specialists pointed out, nitrogen
that had been fixed by bacteria on the roots of native legumes could be
fixed by free-soil bacteria during the fallow years.

Thus it was with the blessing of all concerned—banks, chambers of
commerce, local newspapers, implement dealers, and technical advisers—
that wheat production on the semiarid High Plains not only continued but
expanded during the Cloud Nine Decade of the 1920s. Toward its close I
was present when a representative of one of these several promotional
agencies, a man who flew his own plane, reported jubilantly that he had

just returned from the high country and had learned that thousands more acres had been plowed up for fall planting of wheat.

By that time I had spent a good deal of time in western Oklahoma and adjacent states and had a fair understanding of the character and limitations of that region. In contrast to the native shortgrasses of the High Plains which were perennials and soil binders suited to the climate, wheat is not a perennial; the wheat roots do not propagate a new plant each year, and thus a wheat crop offers less protection to the soil than had the native grasses. Dutifully, I tried to explain to this enthusiast that the long period of favorable rainfall would not continue indefinitely, that groups of moist years would be followed by groups of dry years, that trouble was ahead, and that some moderation was in order. Result: a complete brush-off and wasted breath so far as I was concerned.

The autumn of 1929 brought Black Thursday, October 24, with its complete financial collapse but saw no letup in the planting of wheat in the shortgrass country. The following year soil moisture was sufficient to permit germination and harvest of wheat far in excess of the ability of hungry people to pay for it. Even this, plus signs of incipient drought, did not discourage planting in the autumn of 1930. Now, however, there was not enough moisture in the soil; wheat failed to come up, or did so in such feeble amounts that there was little ground cover.

The autumn, winter, and spring of the next year was a ghastly repetition, bringing reports that the bare soil was drifting badly in many places because there was no rooted plant life to hold it. Since there were also heavily vested interests concerned to maintain the economic prestige of "factories in the fields," there were attempts to discount alarming reports from the dry-farming zone farther west. These attempts increased as news became more disturbing, and continued, in fact, until the evidence was so powerful that to deny it was clear folly, bad faith, or both.

Wind seldom rests in the treeless grasslands of the interior—least so in the early months of spring. Once started, it did not take long for trouble to spread far beyond the limits of the center of wind erosion. One day, as we were working in our laboratory a hundred or more miles east of the source of dust, the air took on a weird gray-green tint. Looking out of the window toward the west, we saw the horizontal top of a black, rolling cloud rise from the distant earth and move toward us—much as one might lift the edge of a blanket high, while its lower edge trailed the floor as he moved it in front of him. It was unlike any storm cloud that we had ever seen. Hope that it might be a rain cloud was dispelled as it reached and passed over and through us.

Dust, like wind, is familiar enough in the far Midwest, though it is not always highly visible. I had learned of its insistent presence first in Nebraska, where efforts to hone microtome knives to a perfect edge ended in disappointment until I found that a seemingly clean stone would have enough gritty particles on it to gouge the fine steel. To free the hone from this invisible contamination was as important as to sterilize nutrient agar in culture tubes.

But the dust now coming in was something else again, coating everything with a heavy film, renewed each day. Only later did I learn that its mysterious advent challenged meticulous housewives a thousand miles east of us. On one notable occasion it drifted into a congressional hearing room where Hugh Bennett was pleading for funds for soil conservation. Hugh Hammond Bennett, be it noted, was the first scientist in the Department of Agriculture permitted to warn of the dangers of soil erosion. This he did in a pamphlet entitled *Soil Erosion, A National Menace* (1927). He later became chief of the Soil Conservation Service, set up first as the Soil Erosion Service, following the Dust Bowl disaster. With the help of this spectacular demonstration Bennett got his funds. With greater or less intensity, such dry storms persisted, and so did the drought. When it finally ended in a drenching rain in the late summer of 1935, the guests at a country club dance in an Oklahoma city are reported to have run outdoors, spread themselves on the hoods of their cars and let the welcome rain soak their party clothes.

Meanwhile, in line with my professional studies, I continued to make field trips over Oklahoma and neighboring states. In this work I was fortunate to have the company of one of the great geographers this country has produced, the late Warren Thornthwaite, a specialist on climatology but also a man of broad training and interests, including population. With the advent of drought, crop failures, and dust storms, the contiguous areas of Kansas, Oklahoma, Texas, New Mexico, and Colorado had been dubbed the Dust Bowl. From it came reports of car engines out of commission from dust, jackrabbits blinded by it, cases of dust pneumonia, buried fences, and so forth, as well as Bunyanesque yarns. An incredulous editor sent out reporters; disbelieving their accounts, he sternly commanded the truth. Jobs were getting scarce, so they decided to soften their stories.

However, there was nothing to soften what Thornthwaite and I, later in company with government scientists, saw for ourselves. Entering the region we could spot a distant town by its heap of repossessed farm machines: tractors, combines, and the like. Fences were buried in drift, sometimes to the post-tops, and vacant one-story houses over the windowsills. What had been wheat fields were barren and wind-carved down to the plowsole—the hard layer of mineral soil under the material that had been loosened in tillage and planting.

Perfectly symbolic of the desolation was an abandoned and wrecked homestead with the skeleton of a dead cottonwood. Among its branches was a raven's nest, made of the only material that was available, namely pieces of barbed wire. Fortunately for our reputation, photographs of this spectacle are on file and have been published. As time went on, the number of abandoned houses and the amount of idle machinery increased. So too did traffic bound for California—a sorry parade of jalopies and antiquated pickups, loaded with as much in the way of humanity and house-

Above: On May 21, 1937, Springfield, Colorado, was engulfed by a dust storm, the result of mismanagement of the land. The dust clouds, here shown approaching Springfield, reached the city limits at exactly 4:47 P.M. and plunged the city into darkness for about one-half hour. *Below:* Buried farm machinery after a dust storm in Gregory County, South Dakota. (SCS, Fred C. Case; Soil Conservation Service)

hold odds and ends as they could carry, and frequently more. Breakdowns, so vividly described by John Steinbeck in *The Grapes of Wrath,* were frequent. Once we saw a disconsolate family standing around the ashes of a rickety vehicle that had been loaded with all they owned.

Thornthwaite later made a scholarly study of population shifts in the Great Plains during the 1930s. His paper furnishes a statistical backbone that amply supports the story of mass migration due to the breakdown of an economy that had developed in buoyant disregard of what I am pleased to call natural history. Although the record showed clearly enough that drought had been a recurring and inevitable feature of the semiarid country, its devastating arrival and persistence was looked upon as an Act of God, not a phenomenon to be taken into account beforehand.

One experience during this time should be recorded. In 1933, when troubles were at their height, with abandoned farms and city breadlines, I went by train from Oklahoma to Utah. Passing through western Kansas and eastern Colorado I was struck by the sight of what seemed like heaps of sand beside the railroad. Actually it was wheat, for which, despite a hungry nation, the market had collapsed. And to top this story, when I arrived in Salt Lake City I found no breadlines. There was food and work for all, thanks to the Mormon social system based on cooperation and built with a respect for natural processes. Even had there been a series of crop failures, hunger would have been averted by a system of community warehouses.

Notable also was a development in west Texas, as hard hit as any part of the Dust Bowl. Here a group of farmers, perhaps a bit more secure financially than the migrants elsewhere, determined to see it through. Recognizing that improvement and future security could be had only by organizing, they set up voluntary soil conservation districts, which were later to serve as a model for national and state legislation. One of their first problems was to get a living cover on the soil to tie it down. In developing appropriate practices to that end, it soon became evident that weeds, hardy pioneers that prepare the way for more permanent vegetation, could be useful allies.

Here was a practical application of the principle of biological succession. The first plants to invade an unoccupied area tend to be annuals that can endure extremes of light, moisture, and temperature. Their roots anchor the mineral substrate; their wastes and those of associated animals add organic material. But the conditions they create are not favorable to their offspring, which, for example, cannot survive shading. A sequence of perennials follows, until stable communities in balance with local conditions take over.

As conditions became steadily worse, pressures on government and public agencies increased, coming to a focus on the science departments of the university where I was then stationed—early in 1935. An emissary came around, asking what we were doing in the way of studying the dust,

and so forth. Actually this query was beside the point, for we knew well enough what the dust was, where it came from, and why.

The real need was for a better public understanding of the inexorable relation between soil, vegetation, and a semiarid climate where periods of drought were sure to recur. And to place the matter in perspective it seemed necessary to show that the dust storms were not an isolated bit of bad luck, but an example of the kind of penalty that had throughout history been exacted wherever the "laws" of nature were violated—as they generally have been—in making use of the environment.

Exploitation of the High Plains had converted them into a *tierra desierta*, as the Spanish call wasteland. Unlike the great climatic desert which lay to the southwest, this was man-made—in effect an enlargement of the natural desert. Here was the suggestion for a title, and so the summer of 1935 was given over to preparing a manuscript under the rubric of *Deserts on the March*, aided as I was by the skilled and enthusiastic editor of the University of Oklahoma Press, Joseph Brandt.

Beginning with some of the principles governing a healthy relation between life and the landscape, I tried to show how consistently man had violated them, to his own ultimate damage, through erosion of soil, pollution of water, and even contamination of air by dust. At the end I suggested the need for the right kind of education, the use of taxation to encourage proper land use, and the need for ecological advice at the local community level. In other words, I felt then, as I do now, that when one tries to arouse public concern on an issue, he ought to do the best he can to suggest workable solutions.

I was completely surprised at the reception of this little book when it appeared in November of 1935. I regret that I did not know, when writing it, that 71 years earlier the classic *Man and Nature* (mentioned previously) had been published by a learned and widely traveled Vermonter, George Perkins Marsh. In it he made an overwhelming case for the extent to which human activity had despoiled the generous earth, for he had seen the ruins of ancient empires and had lived in the midst of a ruthless pillage of our own landscape.

Although fully aware that human numbers as well as values and practices were a source of pressure on the environment, I relied on the best authorities then available who reported that population increase was leveling off. This, of course, gave promise of an eventual equilibrium, now more remote than ever thanks to an upsurge in the birth/death ratio—an outcome that not even the experts anticipated. Prophecy is always risky business, even for experts, particularly where human behavior is involved, as it is in determining birthrates. Changing circumstances may again reduce them, as during the depression, perhaps as a consequence of public concern. Unless this happens, all biological experience indicates that eventually human numbers must come into balance with the limitations of environment in some other way, for Nature will have the last word.

At this time, before the revolutionary developments of World War II, it was impossible to foresee the present extent of urbanization and chemical contamination of the atmosphere. Nor could it have been predicted that a strange word such as "ecology" would appear on Wall Street, in political pronouncements, and almost daily in editorial columns.

Although the great drought was broken by rainfall during 1935 and later, the distress of the average farmer continued, coming to a head with a march on the state capital of Oklahoma a year or two later. Faced by an angry assembly, the harassed governor was obviously relieved when a young newspaperman stepped up and whispered in his ear. At once the executive rose and announced the appointment of a committee, including myself, to draft remedial legislation on the subject of erosion control. This was a matter to which the newly formed Soil Conservation Service—as well as others—had given thought, so the committee, which went to work at once, had no lack of advice. What it came up with was a Soil Conservation District Law, permitting landowners and operators in a county or valley to organize, establish regulations, and receive expert advice from representatives of the federal SCS, which, however, they were free to take or leave.

This proposal, finally enacted into law in 1937, helped set a pattern that by 1948 was adopted throughout the nation, despite resistance in a few states.

Although Oklahoma passed the new law early, it was not the first state to do so. Bearing the blessing of Washington, this bill had an unmistakable New Deal aroma, unwelcome to the state legislature and the powers controlling it. The business of getting it past this formidable barrier was an education in practical politics. Fortunately the head of an important utility saw the relationship between sound land use and the permanent economy so essential to the welfare of his industry. But it was, finally, our success in convincing the legislative leadership that the bill was not a doctrinaire whimsy but a sound and solid proposal on its own merits that carried the day.

In retrospect, though erosion by water had been a constant drain upon American agriculture from its beginning, it was the spectacular drama of wind erosion that crystallized concern for better soil management. Before the decade of the 1930s had ended, the air traveler could observe a vast improvement in land use, marked by strip-cropping, contouring, and the use of winter cover crops. Regrettably, however, this was less evident in the best areas than in those less fit for agriculture.

It would be a pleasure to record that in the Great Plains the lesson of inevitable drought and the need for precaution had been learned beyond erasure. For a time there was a visible improvement in restoring idle land to cover and keeping the looser, more vulnerable soils from cultivation. On firmer soils, less likely to be blown, crops were planted in furrows between ridges that would give protection against wind damage.

But another war, greedy for resources as well as lives, intervened and eventually ended. In 1948, returning from Mexico via the Southwest, we stopped for lunch at a place whose festoon of large cars indicated a prosperous and appreciative clientele. As we ate, a voice at a neighboring table boomed out: "I don't care now if she blows away to Hell. I've got my money's worth out of her." Once again a conjunction of moist years and favorable demand had encouraged speculative operations and blurred the lessons of a great drought and depression.

Perhaps this experience exemplifies, as well as any, the fundamental challenge represented by the Dust Bowl. If human beings are to survive, there must be some reasonable harmony among them and between them and the world of nature of which they are inescapably a part. Such harmony can never be achieved so long as the standard of what is right depends solely upon what is possible and profitable to the individual, or the corporation that is invested by law with the privileges of an individual.

Herein lies the key to the revolution that is enveloping us, and that is symbolized by the protest of youth worldwide. The meaning of justice may puzzle us as it did Socrates, but it is beginning to embody the dictum attributed to Taoism: violence toward nature, like violence toward man, is evil. More than this, and thanks to the insight afforded us by science, we are beginning to see that violence toward nature is violence toward the generations of men as yet unborn.

2. THE COLLAPSE OF CLASSIC MAYA CIVILIZATION

Jeremy A. Sabloff

The ancient Maya built a great civilization during the first millennium A.D. in the jungle lowlands of what is now Guatemala, Mexico, Honduras, and British Honduras. After more than five hundred years of growth and development, Classic Maya civilization suddenly collapsed in the ninth century A.D. By studying the nature of the collapse of this civilization, important lessons may be drawn concerning the way a culture is constrained by what the land is like, how many people live on the land, and how they treat it.

The reasons for this collapse have perplexed archaeologists through the years, and many hypotheses have been advanced to explain it. However, none of the hypotheses has been found completely acceptable, and in recent years there has been a growing realization among archaeologists that the explanation is bound to be complex and difficult to discover. Many students of the problem of the Maya collapse, myself included, now feel that the problem is related to general trends in the development of civilization in all of Guatemala and Mexico.

In order to solve the problem of the disappearance of Classic Maya civilization, it is necessary for the archaeologist to generate explanatory hypotheses that take into account the complexity of the problem and that relate it both to the general historical and ecological situation in Guatemala and Mexico and to the specific ecological situation of the jungle lowlands. I shall attempt to show here how the anthropological perspective of the archaeologist offers a means for testing hypotheses about man's place in his ecosystem and the relationship of the ecosystem to the fall of civilizations.

Classic Maya Civilization

The Classic Maya civilization flourished between A.D. 300 and 900, many centuries before the arrival of the Spanish conquistadors in the sixteenth century. The center of the Classic civilization was the Southern Lowlands (see Figure 1), where the civilization evolved

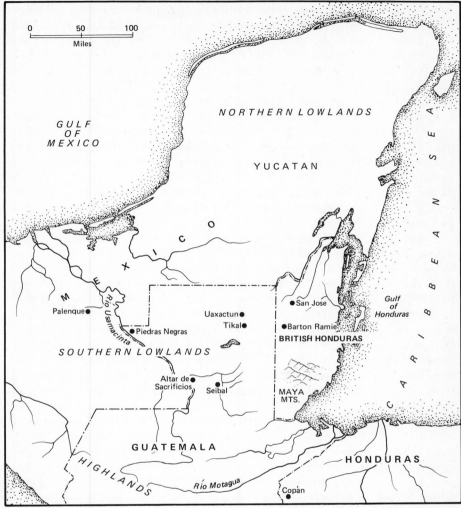

FIGURE 1 Map of the Maya area. Classic Maya civilization developed and then collapsed in the Southern Lowlands.

locally from a simpler cultural base. Although our attention here will be on the Classic civilization of the Southern Lowlands, it should be noted that there also were important Pre-Columbian Maya cultural developments in the Northern Lowlands of Yucatán and in the highlands of Guatemala and Chiapas, Mexico. Maya civilization flourished in both these areas after the fall of the Classic Maya until the Spanish Conquest. In fact, numerous Maya-speaking peoples with a Maya culture can still be found today in the highlands of Guatemala and Chiapas and in the northern part of the Yucatán peninsula.

The Classic civilization is renowned for its many achievements in astronomy, calendrics, hieroglyphic writing, architecture, sculpture, ceramics, and other arts. Maya astronomers made highly accurate solar and lunar observations, which are still marveled at by modern scientists, and they developed a calendrical system consisting of interlocking cycles of varying lengths. They had a numerical system that was based on the number 20 and included the zero. This enabled them to count places arithmetically and to invent their complex calendrical system. They also built large ceremonial centers, with elaborate pyramids, palace structures, and beautifully carved monuments known as stelae. The stelae were usually erected every ten to twenty years, often to commemorate the ascension of a new ruler. In addition, the Maya made excellent polychrome pottery with well-executed realistic painting. In contrast to the cultural situation of the great early civilizations of the Near East, the Classic Maya did all of their carving and building, as well as their agriculture, with stone tools; metallurgy was not part of their technology.

How the Maya governed themselves is a subject of much debate among archaeologists. It is generally accepted that a group of high-ranking individuals lived year-round in or near the ceremonial center and spent their time on a variety of religious and political duties. Some additional people who lived near the center were involved in specialized craft pursuits, although it is debatable whether these were full-time activities. The bulk of the population, however, lived in the sustaining areas surrounding the major centers, grew food for themselves and for the elite residing in these centers, and provided a ready labor pool for the great ceremonial building projects. Recent analyses of Classic Maya burials and associated burial goods in the ceremonial centers and in outlying peasant houses have shown that the gulf separating the elite and the peasants may have widened through the Classic Period and that any class mobility that may have existed in the earlier part of the period was virtually eliminated by the end.[1] The elite may have been ranking members of various large family groups (lineages) and may have been responsible for ceremonies in the lineage temples within the ceremonial center, but this proposition is speculative. It is certain, however, that many high-ranking personages were buried within the temple pyramids in full splendor, surrounded by rich burial goods.

Environment and Subsistence in the Southern Lowlands

The great Classic Maya civilization, whose outline has just been sketched, grew and flourished in a tropical rain-forest environment. What we know about this environment comes mostly from our observations of the same region today. The direct archaeological data are quite meager, because there is virtually no preservation of seeds and plants over long pe-

---

[1] William L. Rathje, "Socio-political Implications of Lowland Maya Burials: Methodology and Tentative Hypotheses," *World Archaeology*, Vol. 1, No. 3, pp. 359–374 (1970).

riods of time in the moist lowland climate. But modern observations of the environment, as well as of agricultural productivity, can be applied to the ancient condition, because there is no evidence for any significant climatic change over the past two millennia in the Southern Lowlands.[2]

The environment of the Southern Lowlands is dominated by a tropical rain forest of mahogany, Spanish cedar, breadnut, sapodilla, and various palm trees. Large savanna grasslands and swamps are interspersed with the forest cover. Temperatures usually range between 65 and 95° F and the rainfall is high, averaging between 60 and 90 inches a year. Underneath the rain forest is a limestone shelf that slopes gently downward to the north. There are some major lakes in the central part of the area and several important rivers such as the Usumacinta, the Pasión, and the Belize. The topography is rugged in some regions of the Southern Lowlands, although the only large rise is the Maya Mountains in British Honduras.

Classic Maya agriculture was apparently based in large part on the cultivation of maize. It is still the major crop in the Southern Lowlands today, and documentary sources from the time of the Spanish Conquest indicate that it was the single most important crop well into the past. The Maya almost certainly grew other crops such as beans and squash. Some skeletal evidence in ancient refuse deposits suggest that to supplement their diet, the Classic Maya hunted deer, tapir, wild pig, and other game, and fished in the freshwater rivers.

The agricultural system used by the Classic Maya to cultivate maize and other crops was almost certainly a shifting system known as slash-and-burn cultivation. This system is used today in the Southern Lowlands, and archaeologists cannot imagine any other system that could have been used in the tropical rain-forest environment of the Classic Maya. In the slash-and-burn system a plot is cleared during the late fall and early winter, and the dead wood is burned during the following dry season. Today, the plots are cleared (that is, slashed) with metal machetes, but in the past they were cleared with stone axes. After the burning, which leaves a rich ash over the soil, the field is planted with several crops mixed together, but with maize predominating. A pointed wooden digging stick is used to prepare holes for planting the seeds. Although a good crop can be harvested for one or two years, the soil soon becomes depleted. First, soil loses nutrient minerals to the growing crops. Second, once the forest cover is removed, the soil loses the rich mass of decayed plant life which accumulates on the jungle floor. In addition, the soil is depleted by the rays of the sun which did not penetrate the rain-forest cover before clearing. The sun's rays kill many useful organisms and dry out the soil. Thus, after one or two plantings a plot must lie fallow for a long period of time, often for six to eight years and sometimes longer.[3] During that time a rich secondary growth returns, and if enough time is allowed, a new forest cover appears

[2] Ursula M. Cowgill, "An Agricultural Study of the Southern Maya Lowlands," *American Anthropologist*, Vol. 64, No. 2, p. 107.

[3] *Ibid.*, p. 108.

as well. When the nutrients in the soil are restored, the plot can be cleared again, and the cycle starts over. Obviously, at any one time, much land has to remain in an unusable state.

If a plot is used for more than two years in a row or if it is not allowed to lie fallow for the necessary length of time, the crop yields drop considerably. As the Maya population increased during the Classic Period, more food was needed. As a result, great pressure was placed on the slash-and-burn system.

There is a popular impression that the slash-and-burn agricultural system is inefficient, does not provide good yields, and cannot support large groups of people. This impression is generally false. The normal yield of slash-and-burn cultivation of maize in the Maya area today is relatively high. It has been estimated that it could support 100 to 200 people per square mile.[4] If the shifting system of plot rotation was followed faithfully by the Classic Maya, as one assumes it was, it almost certainly could have supported their great ceremonial sites throughout *most* of the Classic Period. But as the ceremonial centers became more like urban centers toward the end of the Classic Period, great strains, as we shall see, were placed on the system.

The End of the Classic Period

In the ninth century A.D., Classic Maya civilization collapsed. Virtually all the major activities in the ceremonial centers, such as the construction of temples and palaces and the erection of monuments, ceased by A.D. 900. The great centers were abandoned, and the jungle reclaimed the land where Maya priests and officials had once ruled. Moreover, there appears to have been a great depopulation of much of the Southern Lowlands, although scattered peasants continued to live in the countryside. Some sites, especially those near the lakes in the central area, continued to be occupied, but on a much diminished scale.

There is some disagreement among archaeologists over just how complete the abandonment of the Southern Lowlands was. The Maya remained at a few relatively small sites for a number of years after the disappearance of Classic Maya civilization. Some of the abandoned centers apparently retained their sacred character; occasional offerings were made throughout this period by the scattered rural populace, who were, in some cases, the remnants of the once great Classic Maya. In addition, immediately following the collapse at Tikal, the greatest center in the Southern Lowlands, small groups of people camped out in large palaces in the center, scattered their refuse, broke into tombs, dragged stelae from their original locations and reset them in the central plaza.[5] These scavengers, however, may not have all been Classic Maya peoples; we surmise this because some of their

[4] Ursula M. Cowgill, "An Agricultural Study of the Southern Maya Lowlands," *American Anthropologist*, Vol. 64, No. 2, p. 109.

[5] William R. Coe, "Tikal: Ten Years of Study of a Maya Ruin in the Lowlands of Guatemala," *Expedition*, Vol. 8, No. 1, pp. 5–56 (1965).

pottery was not of local manufacture, and their way of life was not typical of the Classic civilization.

Thus all the activities associated with the ceremonial centers and the political-religious elite—that is, all the hallmarks of Classic Maya civilization—were ended precipitously between A.D. 800 and 900. In addition, much of the peasant population that supported the centers disappeared from the Southern Lowlands.

Reasons for the Collapse of Classic Maya Civilization

Explanations of the collapse of Classic Maya civilization in the Southern Lowlands range from hypotheses that emphasize single catastrophic factors to ones that use complex combinations of factors.[6] Although there is no consensus among archaeologists, there is a growing acceptance of the idea that the hypothesis that will prove to be of greatest utility will be a multifactor one.[7]

To my mind, three of the key factors in the disappearance of Classic Maya civilization were the limitations of the lowland environment, overpopulation, and incursions of "Mexicans" (Toltecs) or Toltec-influenced peoples into the Southern Lowlands. These factors were not independent and they all affected each other systemically.

When the Southern Lowlands are considered in relation to the environment of Highland Mexico, their limitations become quite significant. The Southern Lowland environment is basically uniform: its climate does not vary considerably, its resources are evenly distributed, and crops mature at about the same time throughout the area. But even more important, the lowland environment could not support a more intensive form of agriculture than the slash-and-burn system, given the technology of the Classic Maya. The slash-and-burn system, in turn, affected the patterns of ancient Maya settlement. Because of the need to bring different land under cultivation every two years, a large amount of land is needed to support each person. People cannot all be concentrated in one spot, because some of them would have to walk impossibly long distances every day to get to their fields. Therefore the slash-and-burn system is usually associated with a relatively dispersed settlement pattern—as it was among the ancient Maya throughout most of the Classic Period—and not with a pattern featuring large, concentrated urban centers.

In contrast to the Southern Lowlands, the environment of the Central Mexican Highlands, around modern Mexico City, is relatively diverse. The mountainous topography is quite rugged and contains a number of valleys.

---

[6] General references and suggestions for further readings on specific hypotheses about the collapse are given in Suggestions for Further Readings at the back of the book.

[7] Richard E. W. Adams, "Maya Archaeology 1958–1968, A Review," *Latin American Research Review,* Vol. IV, No. 2, p. 29 (1969).

Above: The temple of the Giant Jaguar, at Tikal, the largest city of the Classic Maya. The limestone temple was built around 700 A.D.; it towers 145 feet about the Great Plaza. *Left:* A stele from Seibal, erected in 850 A.D. This stele was one of the latest monuments carved in the Southern Lowlands. Within 50 years or so after the erection of this stele, virtually all Maya sculpture had stopped being made. (Monkmeyer Press Photo Service; Gair Tourtellot)

These valleys have many diverse microenvironments within them (such as valley floor, lower slopes, and higher hills), and they also vary in climate and resources among themselves. Furthermore, the ready availability of water in the many mountain rivers that bisect the valleys enabled the ancient inhabitants to practice irrigation. The environmental diversity both within and among the valleys also promoted both specialization and trade.

In the Mexican Highlands irrigation and highly developed trading networks were among the factors that stimulated the rise of a secular elite who controlled the irrigation systems and the trade, regulated the markets, and provided safety for the traders. Moreover, the intensive, nonshifting agriculture, which was made possible by the use of irrigation, allowed the formation and growth of large cities, which were the centers of trade for many valleys and the seats of secular and religious authority.

In other words, the environment of the Central Mexican Highlands permitted the growth of the state in the first millennium A.D., whereas that of the Southern Lowlands limited the Classic Maya to what is known as a ranked or stratified society.[8] The lowland environment evidently was rich enough to permit the rise of a great civilization with ceremonial centers and a relatively dispersed population, but it could not normally support the growth of compact, secular, urban centers and a powerful state society. As the states of the Central Mexican Highlands began to expand both economically and politically (militarily) in the first millennium A.D., the Classic Maya were put at a serious economic and political disadvantage. Given these conditions one might expect that the Classic Maya civilization would have been absorbed into the expanding Mexican empire. Instead, the Classic Maya civilization collapsed and the Southern Lowlands were depopulated.

Turning now to the question of overpopulation or maximal population in the Southern Maya Lowlands, this factor in and of itself probably could not and did not cause the disappearance of Classic Maya civilization. But it could have left the civilization vulnerable to collapse on many counts. There is clear archaeological evidence of growth in population during the Classic Period in the Southern Lowlands. There were more centers and there was much additional building in the Late Classic Period. Moreover, recent intensive surveys and excavations of house units surrounding the ceremonial centers are beginning to provide reliable data on population size and growth trends, data which were lacking until the last decade. At Barton Ramie, a site in British Honduras, archaeologists have excavated 25 percent (65 mounds) of all the house units (or house mounds) that have been found there. In the Early Classic Period (A.D. 300–600) 50 of the mounds were occupied, whereas in the Late Classic Period (A.D. 600–900) all 65 were occupied. Furthermore, the house units were much more complex in the later period. These data permit an estimate that the population

[8] Morton H. Fried, *The Evolution of Political Society*. New York: Random House, 1967.

in Early Classic times was approximately 1500 people and increased in Late Classic times to about 2000 people.[9]

Significant increases have been found at other recently excavated sites, too. At Tikal, surveys and excavations have revealed that by the later part of the Late Classic Period the site had probably become an urban center. It has been estimated that 49,000 people occupied this great site. This figure is almost certainly a very large increase over Early Classic times, although no exact figure is available for the earlier period. In terms of population density, there were approximately 1600 people per square mile in the central core area of 25 square miles, and 250 people per square mile in the peripheral 40 square miles of the site.[10] Compare these figures with U. Cowgill's estimates, given earlier, that slash-and-burn agriculture (of maize) could support 100 to 200 people per square mile in the Southern Lowlands.

Thus what occurred at Tikal and other centers was an increased centralization or nucleation of settlement during the Classic Period. That is, not only did the population increase absolutely but it became more concentrated, even though this process was almost certainly nonadaptive for the Classic Maya in the long run. As noted earlier, slash-and-burn agriculture is best suited to a dispersed settlement pattern and cannot, by itself, support urban centers for extensive periods of time.

If large population increases on the order of 33 percent (the figure for Barton Ramie) occurred throughout the Southern Lowlands from Early Classic times—when there already was a substantial population in the area—to Late Classic times, then great demands must have been placed on the soil. It is possible that the ancient Maya misused the land by planting plots for more than two years or by not letting them lie fallow for six years or more. The beginnings of such a trend can be seen in a modern Maya community in Chiapas where there is currently a land squeeze.[11] Land misuse of this kind would have cut yields considerably and would have quickly depleted the soil, so that some areas might have become virtually unusable in a short period of time. Unfortunately, it is difficult to test this hypothesis archaeologically. It has even been suggested that the Classic Maya might have tried intensive nonshifting agriculture which brought about soil erosion and loss of soil fertility very quickly.

Weeds and grasses originating in the neighboring savanna may have interfered with Classic Maya agriculture by invading lands previously used for crops. Today, with metal tools, the modern farmer in the Southern Lowlands is able to cope with grasses and even roll back the savanna,

[9] Gordon R. Willey, William R. Bullard, Jr., John B. Glass, and James C. Gifford, *Prehistoric Maya Settlements in the Belize Valley*, Papers of the Peabody Museum of Archaeology and Ethnology, Harvard University, Vol. 54 (1965).

[10] William A. Haviland, "A New Population Estimate for Tikal, Guatemala," *American Antiquity*, Vol. 34, No. 4, pp. 429–433 (1969).

[11] Richard Price, "Land Use in Maya Community," *Internationales Archiv für Ethnographie*, Vol. LI, pp. 1–19 (1968).

bringing more land (although perhaps less fertile land) into cultivation. Without metal tools, the Classic Maya farmer was generally limited to farming on forest land, and grass incursion could have presented a serious problem for him. The slash-and-burn agriculture of the Classic Maya was not adaptable to the needs of an increasing population and a population living increasingly in nucleated settlements.

The Classic Maya probably could not trade for more foodstuffs, because they probably had few goods available to offer to their mercantile-oriented neighbors in surrounding regions. Some Maya might have migrated to the Northern Lowlands, but any kind of extensive expansion by the Classic Maya was blocked and made impossible by the spread of the Toltecs or Toltec-influenced peoples from the Central Highlands into the Gulf Coast Lowlands, up the Usumacinta and Pasión rivers, into the Guatemalan Highlands, and into Northern Yucatan. Whether the Central Mexican state expanded for economic, or military or other political reasons, or some combination of these, is difficult to say, but by the end of the first millennium A.D., it either occupied or had influence over much of Mexico and Guatemala.

Recent excavations by the Peabody Museum, Harvard University, at the sites of Altar de Sacrificios and Seibal have uncovered artistic, architectural, ceramic, and settlement data which strongly support the idea of a "Mexican" or non-Classic Maya invasion of at least part of the Southern Lowlands. Some of the relevant data include the presence of non-Classic Maya designs and figures on the stelae, figurines, and pottery, the appearance of foreign structures such as a round temple, the shift in the importance of ceremonial complexes at Seibal, the centralization of religious activities, and the influx of large quantities of a foreign pottery known as Fine Orange. Recent neutron activation analyses of the mineral content (trace elements) of this kind of pottery indicate that much of it was manufactured at one site or a series of nearby sites, perhaps in the Gulf Coast Lowlands.[12] It remains to be seen if many other sites, in addition to Altar de Sacrificios and Seibal, were invaded or occupied by non-Classic Maya peoples at the end of the Classic Period. Nevertheless, it is almost certain that the invasion of at least part of the Southern Lowlands was one of the factors that caused the disappearance of Classic Maya civilization.

Although an invading army probably could not, by itself, have caused the abandonment of the major centers and depopulation of the Southern Lowlands, it could have directly wrecked the agricultural system. Or it could have forced the Classic Maya peasants to neglect their crops in order to defend their centers. The loss of a major crop at a time of great stress

---

[12] The analyses have been made by E. V. Sayre, G. Harbottle, L.-H. Chan, and their associates at the Brookhaven National Laboratory. See Edward V. Sayre, Lui-Heung Chan, and Jeremy A. Sabloff, "High Resolution Gamma Ray Spectroscopic Analyses of Fine Orange Pottery with Brief Comments on Their Archaeological Implications," in *Science and Archaeology* (edited by R. Brill), M.I.T. Press (in press).

due to population pressures and, perhaps, an overused soil could have toppled the whole sociopolitical system. The peasantry could also have chosen this time to revolt against an elite that might have made increasing demands on it in the face of threatened invasion. However, this idea of peasant revolt—which has been developed by J. Eric Thompson and has achieved a high level of popularity among some archaeologists—is still hypothetical and has little concrete evidence to support it. The invaders might also have attempted to replace the indigenous Maya elite at some sites, but the disruptions, famines, diseases, unfamiliarity with Maya agricultural practices, and other factors would have doomed their efforts. One can conjecture endlessly; the concrete evidence is too scanty to test at the moment the *exact* reasons for the widespread effect the invasion had on the Southern Lowlands.

Summary

The details of the disappearance of Classic Maya civilization still remain an archaeological mystery, but the basic causes have recently become clear. We can now see that the causes were multiple and systemically reinforcing. One of the root causes was overpopulation and an insufficient food supply, which in turn probably led to misuse or overuse of the soil. Although these factors may not have been sufficient in and of themselves to have directly caused the collapse, they would have created a precarious man-nature balance in the Southern Lowlands. This situation would then have made it possible for the foreign invasion—which definitely affected the western and southern parts of the Southern Lowlands and may have affected the whole area—to cause collapse and depopulation. It could also have helped bring about a peasant revolt. In fact, it is quite likely that overpopulation and foreign invasion, as well as ecological catastrophe and peasant revolt, were all involved in the collapse of Classic Maya civilization.

In brief, the tropical rain-forest environment of the Southern Lowlands had a certain limited potential for supporting people with a stone technology and slash-and-burn agriculture. Such an environment permitted the development of a certain kind of civilization with ceremonial centers and a relatively large overall population. But when this population became much larger and more centralized in Late Classic times, the demands on the soil may have exceeded its potential, and the soil itself may have become at least partially depleted. This dangerous situation would have left the Classic Maya vulnerable to a number of internal and external pressures, any or all of which may have caused the Classic Maya civilization to crumble and disappear.

Several ecological lessons are to be learned from the disappearance of this civilization. Man is part of an ecosystem, and his culture is limited by the environment he inhabits. If he overtaxes the resources of his environment through population increase, nonadaptive settlement patterns, and misuse of the soil, then his whole cultural system is put in jeopardy. A civilization that appears to be prospering may in fact possess a population that has reached a dangerous level, and may have devised agricultural methods to support it, which, though sufficient in the short run, are disastrous in the long run. This precarious situation may remain in balance for a number of years until calamity suddenly strikes. The ancient Maya experienced disaster when they were invaded by peoples who came from a highland environment that offered greater potential than the Southern Lowlands. The collapse of Classic Maya civilization offers us the unfortunate, but illuminating, example of how quickly, and possibly unexpectedly, complete disaster may befall a great and thriving, but ecologically vulnerable, civilization.

II. THE QUEST FOR ENVIRONMENTAL QUALITY: TEN CONTEMPORARY CASE STUDIES

These are partisan essays. The authors of the first two case studies are concerned with the environment where so many live, the city; opposition to what they have tried to accomplish takes the form primarily of inertia and lack of imagination. The next essay, on abortion reform, reveals how laws which are closely tied to community values will evolve as those values evolve. This and the remaining essays (which are concerned more directly with man's interaction with his natural surroundings) all describe situations in which opposition, far from being apathetic, is forceful and well articulated, and many readers will not agree with every author. But no reader will come away from these case studies with the feeling that all will be well in the world if only a few billion dollars are spent on pollution abatement while contemporary values go unchallenged.

A. ONE MAN ALONE

The urban landscape is as soft as flesh. When it erodes,
life erodes; and the scars last longer than dust bowls.
Dr. H. Lyle Stotts is bandaging urban sores and waiting
for the laws and action groups to catch up.

3. WINDOW TO THE CITY: THE EMERGENCY ROOM

H. Lyle Stotts

Tearful, towheaded Robert, still clad in his swimming trunks, was carried into the emergency room by his father. A blood-soaked towel covered his right foot. Robert had stepped on a broken bottle while wading in a stream that flows through the nearby municipal park. When I removed the towel, a 2-inch laceration crossing the ball of his foot was revealed. It would require my services, for I am an emergency room physician in a large community hospital in Bridgeport, Connecticut.

The arrival of this case had interrupted my conversation with Hector S., who had already anxiously waited for more than an hour to see me. Hector S. is Puerto Rican, 25 years of age, and emaciated. He had been brought to the emergency room by a minister who had had much experience with Hector's kind of problem, for many in his parish were so afflicted. Hector lived in the local ghetto and his problem was heroin addiction. He supported his habit by dealing in drugs, allegedly purchased in New York City 60 miles away, and by crooked gambling, his victims being other ghetto residents. He had come to the emergency room to seek legal commitment to the state mental hospital, a request I intended to grant, but with little conviction that it would be an effective instrument of change, for Hector had had three previous commitments, each of which had represented but a short interlude in his destructive life-style.

Even as I considered the extent of Robert's laceration, I took note of the young woman being wheeled into the emergency ward. She sat rigidly upright, and each of her hands tightly grasped the sidearms of the wheelchair. Her demeanor mirrored anxiety, almost panic, and her breathing was rapid and noisy. Between gasps she uttered a single word: "Asthma."

Cases such as these are reflections of the urban environment, and account for a part of the more than 35,000 cases that visit our emergency room each year. Bridgeport Hospital is the largest of three hospitals that serve a population of more than a half million people. The city of Bridgeport, located on the North Shore of Long Island Sound, is an industrial center, the home of

31

such well-known manufacturers as General Electric, Remington, Sikorsky, and Carpenter Steel, and has a population of 175,000. The remainder of the area served by Bridgeport Hospital is composed of smaller communities fitting the description of suburbia. Within the area described can be found most, if not all, of the problems of the urban environment. Many of the human consequences of these problems end up in the hospital emergency room. In this sense, then, the emergency room is one window through which we may view our environment. Indeed, the health of a community might be gauged by the frequency and kinds of problems that filter through the doors of the emergency room; and this will be true whether we use as indices the physical ills, such as bronchitis, emphysema, asthma, and gastroenteritis, or the sociopsychological ills, such as drug addiction and violence. Thus we may, with helpful instruction, take a more extended look through this unique window.

As the simplest example of what I mean, consider the case of Robert and his cut foot. We see hundreds of similar cases each summer, many of which are the result of broken glass, which is increasingly lining our beaches, playgrounds, vacant lots, and parks. The logical sequence of increasing numbers of people, increasing consumption, throwaway bottles, broken glass, increasing exposure, and finally injury is obvious—when we think about it. But our first response, environmentally, with regard to throwaway bottles has been aesthetic—reason enough to ban them I admit, but not the only reason. Injury is now an additional reason when viewed from the perspective of the emergency room. Consider, too, economics. There are costs involved in the care of these injuries—costs, I might add, that are not reflected in the price of the throwaway bottle.

The relationship between broken glass and lacerations is clear. And there are still other features of our environment which we have no difficulty associating with illness or injury. We see increasing numbers of automobile accident victims, coinciding with increasing numbers of automobiles. We have occasionally seen patients who have acquired viral hepatitis and who have a history of having eaten raw clams harvested from the polluted east coast shores. And we have seen machinists who have developed a severe and aggravating skin rash caused by oils used in their work.

The medical literature is laced with studies that relate various air pollutants to asthma, bronchitis, and emphysema. Indeed, even in the absence of these pathologies, 10 percent of the general population experiences increased coughing, sputum production, and shortness of breath when exposed to airborne irritants at the concentrations found in our major cities.° Hydrocarbons, sulfur oxides, carbon monoxide, oxides of nitrogen, and particulate matter are among the contaminants which, in association with specific atmospheric conditions, have been indicted as causes of discomfort and illness. In addition, there is ample evidence that water contamination can cause gastroenteritis, hepatitis, skin disease, and eye irrita-

° See the following essay in this book. [J.H. and R.H.S.]

tion. In view of this evidence, it is hardly surprising that we regularly see illness related to the deteriorating quality of our air and water.

In my profession we are in the habit of relating frequency of problems to cause. When the weather is good, particularly in the summer and on weekends, we see more lacerations. We believe we understand this, for when lawns are mowed, when houses and yards are repaired, when people are on the beaches, when children are outdoors at play, the frequency of lacerations increases. Similarly, during the outbreak of Hong Kong flu several winters ago, we were not surprised to discover that a disproportionately large number of those stricken had come from the ghetto, where we believed crowding was the most acute, and heating facilities were the poorest.

But we surely have not exhausted the list of cause-effect relationships between features of the environment and illness and injury. On an evening not long ago, two patients with emphysema and two patients with asthma appeared at the emergency room within the space of three hours, an incidence that must be considered unusual for us. I saw cases of asthma much less frequently when I had a suburban practice only 15 miles away. This particular evening occurred at the end of a three-day period of cloudless, uniformly warm weather, and I must wonder what the levels of various pollutants were during this period.

What of the cause-and-effect relationships of which we know nothing? Presumably every departure from health has a cause or combination of causes, but it is a confession of truth that we treat many illnesses for which we do not know the origins. In the span of a single week, we often see a hundred or more patients whose complaints include vomiting, diarrhea, or abdominal pain. Most of these patients are not ill enough to justify admission; most are diagnosed as having gastroenteritis, and most are treated symptomatically—and it is appropriate to do so. But that is not to say that we know what may have caused this symptom triad, whether it be bacteria, viruses, or any one of a thousand other possible environmental contaminants. Part of the reason for our failure may lie in the fact that we have not been trained to think about illness from an environmental point of view. But another reason—and one which I wish to expand upon—is that there has been too little coordination between pollution detection agencies and the health services. I shall be specific; still my remarks may have general meaning for other communities.

Making and reporting measurements of specific pollutants in the air in and around Bridgeport are the responsibility of an agency which is funded by the federal government. The reports of the agency's findings are delivered to the upper echelons of our bureaucracy for the avowed purpose of helping to determine standards of pollution, but they are also available to state and local pollution abatement boards to be used to determine violations of standards. However, so far, such reports have not been requested by our local hospitals and physicians nor has the pollution agency requested information with regard to the kinds of illnesses being seen at the hospitals. And yet information of this kind will have to flow freely between the two, as well as into the community, if we are to become more sensitive to environmental health problems.

I suggest that we need to have answers regularly to such questions as the following: What are the specific pollutant levels in air and water? In what part of the city are they the highest? Which factories or sewage do they come from? What illnesses are most prevalent? Where do the people live who are acquiring these illnesses? To answer these questions there will have to be greater cooperation among the hospitals. Each of the three hospitals in Bridgeport is located in a different section of the city. Presumably they perform many services that are common to all; but we may equally presume that each reflects a dimension of the community environment that is unique. Each hospital keeps elaborate records; each has computers capable of performing all kinds of miracles; and each has a staff of educated people. Yet no lines of communication have developed between them and the pollution detection agencies, nor even between one hospital and another. If the information that each has to offer could somehow be joined, it would benefit the city. The questions that have been posed could be answered. Public sensitivity would be heightened. And support could be gathered for constructive change.

Having looked at the effect of the urban environment on physical health, I wish to turn now to other by-products of a more social nature. Consider, first, narcotic addiction. There are a number of parameters by which a community might judge the extent of addiction, the most obvious being the crime rate—theft in particular. But the extent of the activity in the hospital emergency room is an additional index, despite the fact that only a small percentage of the addicted seek or require medical attention.

Overdose, infection, and requests for help to break the habit are reasons why addicts come to us in the emergency room. Some of our impressions of drug addiction are probably accurate representations of the overall picture, but some may deserve additional study to clear up the distortions. It appears to us in the emergency room that narcotic addiction is increasing, particularly since January 1969, for we are now seeing one or more addicts a week. This is all the more alarming when I am reminded that we are seeing only the visible part of the iceberg. We are also seeing as many addicts coming from suburbia as from the inner city, an observation that gains in significance when we consider the inner-city bias of our sample. (Although we see all of the overdoses and all of the cases of hepatitis, addicts who are from the more affluent segment of our society probably make greater use of the private physician than does the ghetto resident, when they have a less serious infection or are just seeking help with their habit.) Clearly, addiction is no longer a problem peculiar to the ghetto. A final impression: we see, of course, addicts who are black, white, and Puerto Rican, but it is of interest that whites are the most numerous, and blacks the least.

Above: New Yorkers seek relief from the summer heat at a Staten Island beach. (DPI)
Below: Residents on this block in New York City demonstrated in 1970 against a curtailment of garbage collection by piling a mound of garbage in the street. (UPI)

Many of the addicts we see are experiencing symptoms of withdrawal when they arrive, and thus there is an added dimension of urgency to their plea. As physicians, we may offer such a patient the dubious benefit of commitment to the state mental hospital where he will, at least, be forcibly removed from his source of supply. Some few will accept this, but many more will not. The addict, either through previous commitment, or by talking to other addicts "on the street," has learned that to be so committed in many situations means to go through withdrawal "cold turkey," that is, without the benefit of ameliorating drugs. We in the emergency room may not, by law, administer narcotics, even the synthetic, methadone, to the patient. This seriously handicaps our efforts to help the addict, because as his symptoms of withdrawal increase, whatever marginal motivation he may have had begins to dissolve. In fact, addicts have told me that to avoid the pain associated with withdrawal is often a greater reason for "shooting up" than to obtain euphoria. Whether or not methadone ought to be used for long-range treatment, it clearly ought to be made available to hospitals as a short-term means of shoring up an addict's evaporating motivation for treatment.

Even if we could grant this temporary respite, we would still be faced with deciding which of several existing programs would best serve the needs of the patient. It would be useful to have a central input center to which all addicts could be sent, and from which they could later be channeled into appropriate programs. Such a center would have the added advantage of being able to evaluate the conflicting claims of the various treatment programs, each of which defends its approach with almost religious zeal. After a long period of foot dragging, this possibility may soon be realized; early in 1969 one of the emergency physicians in our group was appointed by the mayor of Bridgeport to develop a comprehensive drug treatment program.

In addition to heroin addicts, we occasionally see patients who have taken LSD and who are having a "bad trip"; that is, they are uncomfortably agitated or, in some instances, terrified. Most frequently these patients are white and middle class, prototype suburbia. On the other hand, we have not seen, so far as I know, a single case of marijuana toxicity or sequelae from marijuana use. Considering the alleged extent of its use, this may be significant.

The violence of the city is also visible at the emergency room when the victims of shootings, knifings, beatings, muggings, and other violent crimes arrive for treatment. The overwhelming majority of such cases come from the depressed and crowded section of the city. I am reminded every day of the appropriateness of the remarks made by Desmond Morris in his provocative book *The Human Zoo.*

> Under normal conditions, in their natural habitats, wild animals do not mutilate themselves, masturbate, attack their offspring, develop stomach ulcers, become fetishists, suffer from obesity, form homosexual pair-bonds, or commit murder. Among human city dwellers, needless to say,

all of these things occur. Does this, then, reveal a basic difference between the human species and other animals? At first glance it seems to do so. But this is deceptive. Other animals do behave in these ways under certain circumstances, namely when they are confined in the unnatural conditions of captivity. The zoo animal in a cage exhibits all these abnormalities that we know so well from our human companions. Clearly, then, the city is not a concrete jungle, it is a human zoo.[1]

When I see the harvest of violence as it passes through the emergency room, I think about the frustration from which it has arisen. There is escape for the affluent, for they can create a pleasant, though artificial, surrounding which walls out the crowding of the city, or they can migrate to suburbia. But escape is not easily available to the least privileged residents of the inner city. Whether this accurately accounts for the violence we are called upon to treat or not, the fact remains that we see almost every day a wife who has been beaten by her husband, a husband who has been stabbed by his wife, barroom friends who have caused each other violent injury, a battered youth who has been set upon by a street gang, an old man who has been assaulted for his money, or a motorist who has been struck by a rock thrown through his open car window; the setting out of which these problems arise is the depressed section of the inner city. Still, common as these problems are, they do not represent the prevailing style of life of most inner city residents. On the contrary, ghetto dwellers constantly petition the administration for protection. The response, when it comes, is predictable: more police power and increased surveillance in high crime areas.

Such responses, as well as an effort to make the courts work more swiftly, are appropriate and necessary. Still, one must realize that such measures will not begin to reduce whatever anger and frustration are due to overcrowding. To reduce overcrowding we shall have to proceed on a number of fronts, including housing and employment. But we must especially deal with overpopulation.

Religion and lack of education are two factors that interfere with effective birth control in the crowded areas of the city, but they are not the only ones. Consider the case of Rose B. She is 32 years old, pregnant for the twelfth time, and she and her children are on state aid. In fact, Rose has never been without state aid over the past 14 years. She had demonstrated remarkable fertility and remarkable carelessness. Still it is noteworthy that an early attempt to seek medical help was frustrated both by law and by custom. When she was 22 years old, and at the end of her fifth pregnancy, she requested tubal ligation. Tubes here refer to the Fallopian tubes which connect the abdominal cavity with the uterus. The egg, after it is erupted from an ovary, travels down these tubes to be fertilized by a spermatozoon prior to implantation on the wall of the uterus. Ligation, or cutting and tying of the tubes, therefore prevents the joining of egg and sperm. It is a simple surgical procedure.

[1] Desmond Morris, *The Human Zoo*, London: Jonathan Cape, 1969, p. 8.

It was this procedure that Rose requested of not one, but three doctors. All of them refused her request. Legally, her request could not be granted, and it is still against the law in this state as of this writing (June 1970). Yet, changes of custom often precede legal change, and in the matter of tubal ligation, this is surely so. Despite the law, tubal ligation has been performed for at least the past ten years to prevent conception. Physicians within each hospital have formulated guidelines to help decide when this procedure may be performed. The criteria, aside from those related to maternal health, have been based upon the age of the women making the request and the number of children she has had. A common criterion has been five children and 25 years old, four children and 30 years old, and still fewer children with advancing age.

Thus Rose with five children at age 22 would even today have trouble getting her request for tubal ligation granted. Surely, we shall hold Rose accountable for her large, unplanned and, in part, unwanted family, but we must also lay some of the blame on the law and on a custom which has placed greater stress on the patient's age than on the number of children she has had or on her ability to take care of them. Consider also, Ellen W., age 19, unmarried, in her fourth pregnancy, and on state welfare. I asked her if she wanted to avail herself of abortion followed by tubal ligation and her answer was twice affirmative, yet neither is available to her. One may wonder whom the laws or customs are supposed to benefit in this matter; surely not Ellen, her future unwanted children, or society.

In our society, having too many children serves as an impediment to upward economic movement, even when the children are state supported. Thus it might be supposed that there would be increasing support for the small family concept among those on welfare. But this is not the case. Nor do we hear the leadership of minority groups enthusiastically endorsing family size limitation. We ought to be asking ourselves why.

Some illumination may be provided by the following case. Mrs. C. had recently given birth to her third child. Because she was a state welfare recipient, she had been attended throughout her pregnancy and subsequent delivery by the resident house staff of the obstetrical department, rather than by a private physician. She was taking birth control pills at the time I talked to her, and although she admitted that it was in her own interest to do so, nevertheless she resented being singled out to be offered birth control advice. Her feeling in this regard arose from the following situation. Planned Parenthood has made much progress in extending knowledge and help in family planning. In our hospitals, a representative of that organization regularly visits every welfare mother before she leaves the hospital after having had her baby. Representatives of Planned Parenthood may not, however, approach private patients, a prohibition emanating from the ethic of the privacy of the private patient, and the inviolate relationship of doctor and patient. The welfare patient to whom I had been talking had noted that the representative of Planned Parenthood had visited her while passing by the other patients whom she knew to be private. She admitted that this caused her to feel a sense of hostility, born out of defensiveness about her welfare status. Fortunately, her hostility did not cause her to ignore the sound advice she had been given, but when she told me her story, I could not help but wonder if this same hostility causes welfare-recipient leadership to speak in such low tones with regard to family size limitation.

Privacy is, without doubt, a vanishing luxury that needs all the protection it can get; but we should perhaps consider new ways to get the birth control message equally to all, a way that will neither invade privacy nor incur rancor. In our hospital, for example, we have no reservations about allowing the makers of soaps, detergents, throwaway diapers, baby foods, powders, water softeners, and nose wipers to advertise the virtues of their products by giving free samples to every patient who has a baby. Why not, indeed, give Planned Parenthood, and Zero Population Growth, too, the opportunity to distribute propaganda that is more cogent to survival than advertisements for sweetly scented baby powder.

Welfare patients are by no means the only ones with whom I discuss population problems. I am aware that from an ecological standpoint it is the affluent who will do the most to spoil the integrity of the earth, for they have the economic ability to satisfy their appetite to consume. However, in conversations with moderately affluent Americans who might have been silent had I not sought their opinions, I find little inclination to take seriously the reported dangers of overpopulation. When such dangers are suggested to them, the most common reply is, "There is still plenty of land left." They point to the wooded land still remaining in Maine, or Vermont, or some other place that has not yet been urbanized. And when the discussion turns to a consideration of actual family size, I have rarely encountered individuals who have based their decisions upon ecological concepts. Instead, personal economic reasoning reigns supreme. Often I have heard the assertion, "With the cost of things today, we can't properly feed, clothe, or educate more than two (or three, or four) children." It has become clear to me that most people still do not share my point of view that they have a social obligation to limit the size of their families; today the greatest immediate deterrent to fecundity is financial.

Experiences in the emergency room can also help us to examine our institutions, as the following case will show. Juan P., age 2, the youngest of five children of a family on state aid, developed a running nose, a cough, and a fever during the day. His mother took him to a nearby doctor's office where he was examined, given a shot (presumably penicillin), and a prescription. Later that night Juan was still feverish, and began to vomit. This new development frightened his mother. She therefore called the doctor who had seen him earlier. Since he could not be reached, she took him to one of the hospitals, not ours, where he was examined, given another shot, and a new prescription. The next day Juan was still sick. His mother, believing he had not been given the right medicine, brought him to our hospital for still another examination, but she did not think to bring any of the medicines previously prescribed for him. This example of duplication of services, record keeping, and medications is not at all uncommon. Nor is this the full extent of it, for inherent in our method of dispensing health services is a fragmentation and inconsistency of care that offers no advantages to the patient and greatly adds to the cost. In our well-intended effort to achieve for the welfare patient the same freedom of choice that is accorded the private patient, we have managed to fragment his care, allowing him to alternate between a doctor's office and any of the several hospitals. Doctors whose practices border on areas having large numbers of welfare patients are overly busy and simply cannot be available twenty-four hours a day. Thus, it is understandable that the patient turns to the one health source upon which he can most consistently rely—the hospital emergency room.

Considering the many cases typified by Juan, and the growing health needs of the underprivileged, there may be merit in considering the dispensing of such care from a single source—perhaps a community hospital outpatient department that operates 24 hours a day. This would make both medical and economic sense. Adequate health records would result and needless duplication of instruction and medication could be avoided. Efficiency should not, I think, be the overriding goal, though this would, of course, result. Above all, a single health source would provide a better opportunity for education of the kind that a private physician is in the habit of providing for his private patients. Welfare patients frequently need the most health education, yet the design of our present health care system provides them with the least, and even can be self-contradictory when a patient is allowed to yo-yo between the multiple treatment centers.

It has been my intent in this essay, utilizing examples and opinions, to show that a hospital emergency room can be much more than a treatment facility; it is also a window to the community which can tell us much about its people, its institutions, and the environment.

B. INITIATIVE IN GOVERNMENT

Implicit in all three of the following essays is the idea that compromise is essential in government, whether at the local, state, or federal level. All three of these essays are success stories of a kind, but the success is incomplete.

Commissioner Heller and Dr. Ferrand helped the New York City government to reduce the concentrations of sulfur dioxide in the air. However, these improvements may be nullified, or at least further progress may cease, if the demand for energy within the city continues to grow at the present rate.

As a state legislator in Colorado, the Honorable Richard D. Lamm was in a position to push for the liberalization of abortion laws. So were several thousand other legislators around the country, but he was the first to get a bill through. It is hardest to be first; now many other states have followed Colorado, and a few have gone beyond.

To confront the fact that we are running out of certain materials is one way to become conscious that the earth is finite. Realizing that helium, a material with unique properties, was being wasted thoughtlessly, the United States Department of the Interior developed a conservation program. It is now under attack. The story is told by an "outside expert," Dr. Price, who explains how the deficiencies now confronting the program were built into the program at its inception. We share her concern that a better program be devised and that helium conservation be continued.

4. LOW-SULFUR FUELS FOR NEW YORK CITY

Austin Heller and Edward Ferrand

The life of the air resource management strategist is strenuous and sometimes vexing, but never dull. Mankind, fearing ecological disaster, asks with increasing urgency that rapid solutions be found. Unfortunately, past generations have left us insufficient information to make sound decisions. Now, as we strive to improve the air we breathe, we feel this lack very strongly.

The air resource manager has an urgent need for better information about the environment at different locations and at different times—about the relationships between emitters and the pollution they produce—and most of all, about the effects of pollutants on all living and nonliving parts of our world. We know that breathing a contaminated atmosphere can be injurious to health. But we know much less than we should about the effects of known contaminants on human health. Therefore prudence dictates that we adopt the attitude that most, if not all, contaminants may be potentially hazardous.

However, man's resources are limited and often needed elsewhere with even greater urgency. This suggests that a dual-approach strategy should be used. First, we must commit ourselves to policies more farsighted than those of our predecessors. This implies that we promptly set up the machinery to manage the environment on a scientific basis—a long-range plan becomes necessary. The other phase of the strategy is to abate known pollution sources using current technology.

We shall begin by describing the harmful effects of high concentrations of atmospheric sulfur dioxide on materials, plants, animals, and men. At the concentrations found in many urban centers, sulfur dioxide is considered capable of causing temporary or permanent damage to the human respiratory system. We shall then explain how, worldwide, large amounts of sulfur dioxide are emitted into the air during the combustion of fossil fuels. We shall conclude our essay by relating the story of how sulfur dioxide emissions in New York City have been reduced by over 50 percent between 1966 and 1970.

Effects of Sulfur Dioxide

A useful way to describe the amount of sulfur dioxide in the air is to give the concentration by volume of the gas in the ambient air. Measurements customarily determine the average concentrations during an hour-long or day-long period at fixed stations, and the results are usually stated in "parts per million" (ppm), where 1 ppm means there is one volume of sulfur dioxide in every million volumes of air. For purposes of orientation, the average hourly concentration in upper Manhattan in 1969 was 0.11 ppm (down from 0.21 ppm in 1964), and the peak hourly concentration in 1969 was 0.8 ppm (down from over 2.0 ppm in 1964).

Sulfur dioxide in the ambient air affects materials, vegetation, animals, and man. It prolongs the drying time of painted surfaces and makes them less durable; it increases the corrosion rate of metals; it attacks building materials such as limestone, slate, marble and mortar, increasing the rate of deterioration of sculpture and monuments. Fabrics that are sensitive to acidic substances (for example, cotton, rayon, and nylon) will lose their tensile strength when exposed to sulfur dioxide and its products. Paper and leather deteriorate much more rapidly in air polluted with sulfur dioxide, thereby creating serious problems in the storage of valuable documents.

Sulfur dioxide is converted to sulfuric acid and sulfate particulates at rates which increase with the relative humidity. These particulates form suspensions, or aerosols,° and usually constitute from 5 to 20 percent of the total suspended particulates in the urban atmosphere. A large percentage of the sulfate particulates are between 10^{-4} and 10^{-5} centimeters in diameter, at which size they contribute effectively to reducing atmospheric visibility.

There is no question that plant damage has occurred because of sulfur dioxide in the air.[1] In acute exposures this gas is absorbed through the breathing pores in the plant and accumulates as sulfates at the tips or edges of leaves or needles. A bleaching or a reddish-brown color then appears at these areas. Long-term chronic exposure to low levels of sulfur dioxide also causes gradual fading of the green color. The time interval over which the dose is applied is important: a certain quantity of sulfur dioxide absorbed by the plant over an extended period can cause much less damage than the same amount taken in during a shorter time interval. Some plants are particularly easily damaged: Alfalfa and barley are among the most sensitive crops; ponderosa pine can be injured by concentrations as low as 0.25 ppm; and apple and pear trees, the most susceptible of the fruit trees, have shown injuries after an exposure of only 6 hours at 0.5

° An aerosol is a suspension of finely divided liquids or solids in a gas. [J.H. and R.H.S.]

[1] The data in this paragraph are drawn from "Air Quality Criteria for Sulfur Oxides," The National Air Pollution Control Administration, U.S. Department of Health, Education, and Welfare, January 1969.

ppm. Annual mean concentrations as low as 0.03 ppm are believed to cause chronic symptoms in certain plants.

Animal studies have shown that at high concentrations more than 95 percent of the sulfur dioxide is absorbed in the upper (nasal) part of the respiratory system. This high efficiency may result because the irritant gas has caused excess mucus, which in turn absorbs more of the gas; at lower concentrations (0.1 ppm), when there is less irritation and therefore less mucus is formed, a much lower percentage (40 to 50 percent) is absorbed in the nasal passages.

Studies using sulfur dioxide containing radioactive sulfur ($S^{35}O_2$) have shown that traces of sulfur still remain in the trachea and lungs a week after exposure.

Inasmuch as sulfur dioxide is gradually converted to sulfuric acid and sulfate aerosols, it is important to study the effects of inhalation of such materials. Tests on animals show that particulates greater than a few microns in diameter (one micron $= 10^{-6}$ meters) are prevented from reaching the alveoli in the lungs by protective mechanisms in the upper respiratory system. There is also evidence of synergism * between particulates and sulfur dioxide gas—a result which is particularly significant because some of the same combustion processes which produce sulfur dioxide also produce large quantities of particulates.

Identification of the specific harmful effects of atmospheric sulfur oxides on man is extremely difficult because there are so many different contaminants, known and unknown, in the air we breathe in cities. Moreover, the susceptibility of human receptors can vary greatly, depending upon general state of health, age, ethnic background, hygiene habits, and other factors. Attempts to deduce the effects of environmental exposures by extrapolation from laboratory studies on human subjects have not been very successful.

Acute episodes, that is, intervals of short duration during which air pollutant levels were inordinately high, have produced the most dramatic evidences of health effects. In these instances, fatalities occurred predominantly among the susceptible, that is, the elderly and those with preexisting cardiac or pulmonary disease. Table 1 gives some examples of acute episodes during which pollutant buildups resulting from extended periods of poor ventilation were reported to produce adverse health effects.

A now almost historic event, commonly referred to as the "Thanksgiving episode," occurred in New York City in 1966. Because of its impact on the media and the general public, it can be said to mark the point in time when very large segments of the general population became aware that the condition of the city's air could lead to a major tragedy. Unfortunately, the

---

° A synergism is an effect that is due to more than one cause and that is larger than the sum of the effects that the causes acting separately would produce. Accordingly, a synergistic effect is likely to be greatly reduced even when only one of several contributing causes is removed. The attack on sulfur dioxide in urban air is motivated in large part by these considerations. [J.H. and R.H.S.]

TABLE 1. Examples of Acute Episodes of Air Pollution

Date	Location	Levels of SO_2	Effects
Dec. 1–Dec. 5, 1930	Meuse Valley, Belgium	Max. 9 ppm (estimated)	63 died. Several hundred severe respiratory cases
Oct. 1948	Donora, Pa.	Greater than 0.4 ppm daily average (estimated)	20 died. 43% affected to some degree. 10% severely affected
Dec. 1952	London	1.34 ppm daily average (4500 micrograms per cubic meter for particulates)	4000 excess deaths
Dec. 1956	London	0.4 ppm daily average (1200 micrograms per cubic meter for particulates)	400 excess deaths
Nov. 1953	New York City	0.86 ppm daily average	200 excess deaths
Nov. 1966	New York City	0.51 ppm daily average	168 excess deaths

SOURCE: *Air Quality Criteria for Sulfur Oxides,* The National Air Pollution Control Administration, U.S. Department of Health, Education, and Welfare, January 1969.

kind of monitoring capability now considered routine was not then available. Nevertheless, over a period of four or five days the unhealthiness of the air was evident and frightening enough to everyone to create a new political force in New York City. Such motivation of political forces is necessary if a strong air quality improvement and conservation program is to be instituted and maintained.

An earlier air quality crisis occurred in 1953 (Table 1). Records of pollutant levels and excess deaths indicate that the 1953 incident was of equal severity to the Thanksgiving episode. Yet no long-term program of any significance developed out of this earlier disaster. The fact that change did occur after the 1966 incident is in part a result of the intensity of worldwide concern with all facets of the environment. The public concern and political activism, which are still gathering momentum, were a potent force for change in New York City because the City had the nucleus of a staff capable of implementing the technological and the political changes that were necessary.

The National Air Pollution Control Administration has reviewed the effects of sulfur oxides on health. Table 2 has been compiled from their report. The values reported therein are representative of the levels of atmospheric contamination now being judged unacceptable.

TABLE 2. Effects of Sulfur Oxides on Health

SO₂ (ppm)	Particulates (micrograms per cubic meter)	Averaged Period	Effects That May Occur
0.25	750	24 hours	Increased mortality
0.21	300	24 hours	Symptoms increase in patients with chronic lung diseases
0.046	100	One year	School children suffer increased severity and frequency of respiratory disease
0.040	160	One year	Increased mortality from bronchitis and lung cancer

SOURCE: *Air Quality Criteria for Sulfur Oxides,* The National Air Pollution Control Administration, U.S. Department of Health, Education, and Welfare, January 1969.

The Sources of Atmospheric Sulfur Oxides

Sea spray introduces sulfur into the atmosphere in the form of sulfate aerosols, and a much smaller amount of sulfur in oxidized form is introduced into the atmosphere by volcanic activity. But most of the sulfur that natural processes introduce into the atmosphere is in the form of hydrogen sulfide gas ("rotten egg" gas), generated in swamps and shallow lakes on the land and in the ocean by the action of reducing bacteria on organic matter. In the atmosphere the hydrogen sulfide gas is rapidly oxidized to sulfur dioxide, and then is further oxidized to sulfur trioxide and sulfates.

Man became an active partner of nature in the discharge of sulfur compounds into the atmosphere when he started to burn coal and roast metallic sulfide ores. Within the last two decades, the combustion of petroleum products has increasingly added to the sulfur dioxide concentrations. An estimate of total sulfur emissions from natural and man-made sources is found in Table 3.

Average concentrations of sulfur compounds in the atmosphere are approximately constant, because sulfur is precipitated from the atmosphere as sulfate in solution in rainwater, and is absorbed directly in gaseous form by the ocean and by plant vegetation. Most of the sulfur dioxide in the atmosphere is converted to sulfate before precipitating; sulfur in the form of sulfur dioxide has an estimated residence time in the atmosphere of 4 to 5 days.[2] (The concept of residence time is defined and a global average con-

[2] C. E. Junge, *Air Chemistry and Radioactivity,* New York: Academic Press, 1963, p. 61. See also E. Robinson and R. C. Robbins, *Sources, Abundance, and Fate of Gaseous Atmospheric Pollutants,* American Petroleum Institute, 1968 (Supplement).

TABLE 3. Annual World-Wide Emissions of Sulfur (tons)

Source	Quantity of Sulfur Emitted [a] (millions of tons)
Pollutant sources of sulfur dioxide (man-made)	73
Hydrogen sulfide from biological processes on the continents	68
Hydrogen sulfide from biological processes in the oceans [b]	30
Sea spray (mostly sulfate aerosols)	44
Total	215

SOURCE: Elmer Robinson and Robert C. Robbins, "Gaseous Sulfur Pollutants from Urban and Natural Sources," *Journal of the Air Pollution Control Association*, Vol. 20, p. 233 (April 1970).

[a] To obtain the quantity of hydrogen sulfide emitted, multiply by $34/32 = 1.06$. To obtain the quantity of sulfur dioxide emitted, multiply by $64/32 = 2$. To obtain the quantity of sulfate ion emitted, multiply by $96/32 = 3$.

[b] There is a larger uncertainty in this estimate than in the others.

centration for sulfur dioxide in the atmosphere is computed in Comment 1: *Average sulfur dioxide concentrations in the atmosphere*.)

Comment 1: Average sulfur dioxide concentrations in the atmosphere

The *residence time* of a substance which has reached an equilibrium concentration in a storage area like the ocean or the atmosphere is the amount of the substance in the storage area divided by the rate at which the substance enters (or leaves) the storage area. Let us apply this concept to estimate the amount of SO_2 stored in the atmosphere.

According to Table 3, each year man-made sources emit 73×10^6 tons of sulfur into the atmosphere in the form of sulfur dioxide. In addition, every year natural sources emit 98×10^6 tons of sulfur in the form of hydrogen sulfide gas, nearly all of which becomes sulfur dioxide. Thus approximately 170×10^6 tons of sulfur become sulfur dioxide in the atmosphere every year. Since the sulfur dioxide molecule weighs almost exactly twice as much as the sulfur atom, this corresponds to 340×10^6 tons of sulfur dioxide per year emitted into the atmosphere (and also leaving the atmosphere—to the extent that overall equilibrium exists).

Taking the residence time to be 5 days, we calculate that $(340 \times 10^6) \times (5/365)$ tons, or 4.7 million tons of sulfur dioxide are present in the atmosphere at any one time. The mass of the atmosphere is 5.14×10^{21} grams, or 5.7×10^{15} tons. (See the data in Appendix 4 in this book.) Thus the concentration of sulfur dioxide, averaged over the whole globe, is about 0.8 part per *billion* by weight, or—since the SO_2 molecule is a little more than twice as heavy as an average air molecule but occupies the same effective volume—about 0.4 part per billion by volume. Thus, when a city's air has a sulfur dioxide concentration of 0.4 part per *million,* this represents a thousandfold concentration of the sulfur dioxide in the global atmosphere. The concentration falls rapidly beyond the region of sulfur dioxide emission because winds mix the polluted air with unpolluted air.

J.H. and R.H.S.

Sources of Air Pollution

Table 4 shows the relative importance in 1966 nationally of transportation, industrial production, electric power production, space heating, and refuse disposal as sources of sulfur dioxide and other atmospheric pollutants. Data of this kind must be used with great care to avoid misleading and irresponsible conclusions that introduce unnecessary complications into the efforts of control agencies. Among the important questions which must be dealt with in any analysis of the significance of these numbers are: What is the effect of these emissions upon air quality and what human exposures result? What health effects might such exposures produce?

The significance of an emission rate is dependent upon the area over which emitted, the height at which emitted, and the local topography and meteorology. It is also dependent on population density and health effects. Thus exposure to 5 ppm of carbon monoxide for 8 hours in the atmosphere is of little significance compared to an exposure of only 1 hour to 0.5 ppm of sulfur dioxide.

Table 5 compares a detailed summary of annual emission rates in New

TABLE 4. Emission of Five Major Air Pollutants in the United States by Source of Pollution (1966)
(Data in millions of tons per year)

Source	Sulfur Dioxide	Particulates	Oxides of Nitrogen	Hydrocarbons	Carbon Monoxide
Automobiles	1	1	6	12	66
Major industries [a]	9	6	2	4	2
Electric power generation	12	3	3	1	1
Space heating (indoor heating)	3	1	1	1	2
Refuse disposal	1	1	1	1	1
Total	26	12	13	19	72

Source: "The Sources of Air Pollution and their Control," Public Health Service Publication No. 1548, Washington, D.C., 1966.
[a] Includes only the six major industrial polluters.

TABLE 5. Emission of Five Major Air Pollutants in New York City by Source of Pollution (1966 and 1970)
(Data in thousands of tons per year)

Source	Sulfur Dioxide	Partic- ulates	Oxides of Nitrogen	Hydro- carbons	Carbon Monoxide
Transportation	20 (**20**)	10 (**10**)	51 (**50**)	189 (**160**)	1564 (**1370**)
Manufacturing	11 (**9**)	6 (**5**)	3 (**3**)	<1 (<**1**)	<1 (<**1**)
Electric power generation	323 (**150**)	13 (**6**)	102 (**110**)	2 (**3**)	1 (**1**)
Space heating	530 (**210**)	29 (**22**)	127 (**130**)	7 (**7**)	20 (**20**)
Refuse disposal	3 (**3**)	32 (**26**)	3 (**3**)	24 (**18**)	46 (**35**)
Evaporation of fuels and solvents	—	—	—	116 (**115**)	—
Total	890 (**390**)	90 (**70**)	290 (**300**)	340 (**300**)	1650 (**1400**)

SOURCE: New York City Department of Air Resources: Emissions Inventory Survey.
Numbers in boldface in parentheses refer to 1970.

York City in 1966 with our estimates for 1970. Comparison will show where progress has been made (primarily in reducing sulfur dioxide emissions) and where it is lacking (most noticeably in the continuation of major pollution from automobiles).

Sulfur and Oil

As is seen in Tables 4 and 5, power production and space heating account for almost all of the sulfur dioxide emitted into the air. The sulfur dioxide in both cases is produced when fossil fuels (coal and oil) are burned and their energy is converted into electrical energy and heat. The oil that is burned in large installations such as factories and power plants is quite different from the crude oil that comes out of the ground at an oil well. Both crude oil and the oil that has been processed at a refinery contain an enormous variety of organic compounds; most of them are hydrocarbons (containing only hydrogen and carbon), but some of them also contain sulfur. The compounds with the lowest molecular weights tend to have the lowest melting points and are liquids at room temperature. (Organic compounds of even lower molecular weights, like methane and ethane, are gases at room temperature. Often found in the same wells, this gas mixture is called natural gas.) The high molecular weight organic compounds form gummy pastes (asphalt and tar are made from them), and can only be liquefied by heating. These compounds are also the ones containing most of the sulfur.

Although there is available energy locked up in all of these molecules (energy that is released when these molecules are oxidized), for many uses, particularly for internal combustion engines, the heavier portions of the crude oil are nearly useless. The heaviest portion of refined oil, called *residual fuel oil*, is consequently relatively inexpensive, in terms of energy content per dollar. Residual fuel oil has become competitive with coal in certain parts of the United States (especially the Northeast) as an energy source, especially in those installations that are large enough that the cost of special heaters for tanks and pipelines is not a major fraction of the total cost. The oil must be heated in order to lower its viscosity and permit it to flow rapidly through pipes.

In 1966 the United States burned more than 650 million barrels of residual oil, of which about 475 million barrels were imported, with Venezuela the most important supplier. More than a third of the total was used along the eastern seaboard from Virginia to Maine. New York City consumed about 10 percent of the total.

Two important characteristics of either crude or residual fuel oil are its *pour point* (the temperature at which it begins to be able to flow according to a well-defined test) and the percent of sulfur by weight. In crude oils, sulfur content varies from 0.2 percent in some Algerian oil fields to 5 percent and higher in certain fields in Mexico. Crude oil from fields in Tia Juana, Venezuela, has a 1.5 percent sulfur content and a pour point of $-40°$ F; the residual oil produced from this crude oil has a sulfur content of 2 percent and a pour point of $+20°$ F.

All the sulfur compounds in fuel oil (among them mercaptans, thioethers, disulfides, and thiophenes) are converted to sulfur oxides during combustion. Less than 2 percent of the sulfur is converted to sulfur trioxide, a fact for which we can be grateful, because sulfur trioxide readily forms corrosive sulfuric acid aerosols.

Of the 26 million tons of sulfur dioxide emitted into the air in 1966 (see Table 4), roughly 20 percent came from oil refining and the smelting of nonferrous metals, 60 percent from the combustion of coal, and 20 percent from the combustion of fuel oil. (See Comment 2: *Sulfur dioxide production from the combustion of fuel oil.*) Hence the control of the sulfur content of coal and fuel oil is a major environmental problem nationally. There are two approaches to the control of sulfur oxide emission: one can remove the sulfur either from the fuel before it is burned or from the gases formed in the combustion process before they are released. Were it practical to remove sulfur dioxide from effluent gases, such techniques could be applied to both coal and oil combustion. But the removal of sulfur dioxide from effluent gases has thus far not been economically feasible under urban conditions, which typically involve power plants that are already in operation and hence are difficult to alter.

Fuel desulfurization is the method that is likely to have the greatest impact on urban sulfur oxide problems. The removal of the sulfur from the coal or oil at a central location would be expected to be more economical than would removal from effluent gases at each of the thousands of coal- or oil-burning facilities in a large city. Techniques for removing sulfur compounds from coal and oil are being actively developed by private industry, primarily because government regulations have imposed schedules with definite target dates. Industry is also responding to an increasingly impatient public.

Large quantities of residual oil will have to be desulfurized to meet the low-sulfur fuel requirements of urban areas. Fortunately, the removal of sulfur compounds from petroleum products, although becoming more difficult and costly as lower sulfur concentrations have to be attained, is technologically feasible.

Comment 2: Sulfur dioxide production from the combustion of fuel oil

Let us verify the figure quoted in the text, to the effect that about 20 percent of the sulfur dioxide emitted nationally each year comes from consumption of fuel oil. We assume that almost all of the sulfur dioxide production due to oil is due to *residual* oil (since distillate oil contains only about 0.3 percent sulfur by weight), and we make use of the fact (see text) that in 1966, 6.5×10^8 barrels of residual oil were consumed in the United States. Choosing an average sulfur content for residual oil of 2.5 percent by weight, and using a value of 330 pounds for the weight of a barrel of residual fuel oil, we find that 2.8 million tons of sulfur were produced in 1966 from oil consumption. (A barrel of oil, a unit of *volume* equivalent to 42 gallons, has a weight that depends on density. The densities of oils range between 0.80 and 1.00 gram per cubic centimeter; Venezuelan crude has a density of 0.85 gram per cubic centimeter and the corresponding residual has a density of 0.95 gram per cubic centimeter.) The 2.8 million tons of sulfur corresponds to 5.6 million tons of sulfur dioxide, because the sulfur dioxide molecule, SO_2 (containing one sulfur atom and two oxygen atoms), is almost exactly twice as heavy as the sulfur atom. (The atomic weight of sulfur is 32 and of oxygen is 16.) But in Table 4 we see that the total United States production of sulfur dioxide in 1966 was 26 million tons. The contribution from oil combustion then works out to 5.6/26, or roughly 20 percent of the total sulfur dioxide put into the atmosphere.

J.H. and R.H.S.

New York City's Sulfur Dioxide Problem

The sulfur content of the residual fuel oil used in the New York–New Jersey Metropolitan area before 1964 averaged around 3 percent. In 1964 amendments to the administrative code of New York City imposed a requirement on heavy fuel oil suppliers which limited the maximum sulfur to 2.8 percent. Subsequently New York City's Local Law 14, which became effective on May 20, 1966, called for a stepwise decrease in sulfur content. From January 20, 1967, to May 20, 1969, the maximum allowable sulfur content in coal and residual oil was set at 2.2 percent, and from May 20, 1969, to May 20, 1971, the maximum was set at 2.0 percent. Thereafter

Above: New York City, view to the south from the Empire State Building, on November 24, 1966, during the Thanksgiving episode. Ordinarily, the temperature of the air at low altitudes decreases with increasing altitude. Because warm air is less dense than cool air, the pollutants emitted at ground level into the warm air are carried upward where they mix with cooler air that is relatively free of pollutants, and this mixing prevents the buildup of high concentrations of pollutants in the air at ground level near the sources of pollution. During an "air inversion" such as occurred when this photo was taken, warm air sits upon cooler air, and the cooler air remains in place, absorbing pollutants for as long as several days, so that high concentrations of pollutants result. *Left:* Much of the sulfur piled in the foreground would have been emitted into the air above power plants had it not been removed from oil in this Standard Oil Company refinery. The low-sulfur oil that is produced here will be bound for markets in cities such as New York with low-sulfur-fuel regulations. (Wide World Photos; Standard Oil Company of New Jersey)

only fuel containing no more than 1.0 percent was to be permitted. These restrictions were based in part on estimates of how quickly low-sulfur fuel oil could be made available with the correct physical properties in the tremendous quantities consumed by New York City.

In a manner that must be deemed atypical of the implementation of laws based on good intentions, the schedule of Local Law 14 was not only adhered to but very much accelerated. By 1968 the City's utility, Consolidated Edison, which in 1966 had been responsible for over 35 percent of the City's sulfur dioxide emissions, was using fossil fuels (coal and oil) containing less than 1 percent sulfur. Municipally owned facilities followed soon thereafter, and by October, 1969, all sectors of New York City had decreased their rate of emission of sulfur dioxide, compared to 1966, by more than one-half.

How did this dramatic decrease come about in spite of the fact that information, especially from oil companies, had indicated that low-sulfur fuel could not be made available in the necessary quantities for some time to come? Needless to say, the capabilities of industry which can be brought out when government officials apply the right amount of pressure at strategic times and places are sometimes much greater than those that appear on the surface.

Following the passage of Local Law 14, New York City embarked on a program of sampling fuel oil depots to insure compliance with its law. At the same time it began to investigate various strategies that would serve to reduce sulfur dioxide emissions.

One strategy has proven feasible and rewarding. Large buildings that used residual oil for space heating were encouraged to convert to temperature-controlled interruptible gas service. According to the scheme, equipment is installed which burns natural gas unless the outside temperature falls below a preset value such as 20° F, whereupon a low-sulfur fuel oil (a distillate containing about 0.3 percent of sulfur) is substituted. It was estimated that there could be a decrease of more than 50,000 tons in the sulfur dioxide emitted into the atmosphere each year (about a 6 percent reduction in the total emissions) if City and Housing Authority installations adopted this plan. Although fuel costs are higher, there are compensations, such as lower maintenance costs, that make such heating economical. An increasing number of older facilities have voluntarily converted to interruptible gas, and many new installations also use this method.

New York City also investigated possible ways in which the schedule for the introduction of low-sulfur fuels could be accelerated. During this search it developed that Consolidated Edison might be able to get a large amount of residual oil containing a maximum of 1 percent sulfur. The utility, having decided that the public interest and its corporate image would be well served by changing to a lower sulfur fuel, made an intensive market search. By the end of 1968 it had converted all its facilities to fuels that contained a maximum of 1 percent sulfur—more than 2 years before the original legal requirement.

The New York City Department of Air Resources anticipated that most of the 1 percent sulfur residual oil that would be supplied to New York City would be a blend of Caribbean and African oils. Previous experience in New York City had been limited primarily to fuel oils made from Venezuelan crudes, which though high in sulfur content, have a pour point as low as 30° to 35° F, flowing well enough to permit handling by relatively simple pumps and suction heaters. African crudes, on the other hand, although they contain small concentrations of sulfur, produce residual oils that have pour points as high as 100° F. An installation that had only a suction heater would be in serious difficulty if this oil were to solidify inside the tanks and pipelines. Any blend of Caribbean and African oil would be expected to have intermediate properties, ones which could not be predicted in detail. Thus uncertainties in handling 1 percent sulfur fuel oil appeared to confront the small users.

Accordingly, the New York City Department of Air Resources decided that operating experience with 1 percent sulfur residual oil of intermediate pour point was necessary. It instituted a program with the New York City Housing Authority early in 1968 to test 1 percent sulfur oil having a pour point of 50° F. This test showed that only minor burner and temperature control adjustments were necessary to use this oil. With this experience in hand, the program shifted from testing to extending the use of 1 percent sulfur fuel in New York City. To set an example, municipally owned installations were required to burn 1 percent sulfur fuel during the 1968–1969 heating season. No new law or regulation was required; all that was needed was to include this restriction in the specifications for oil purchased by New York City.

The reluctance of certain companies to plunge ahead to develop processing capability for low-sulfur oil disappeared when they realized that government agencies were earnest in their resolve to remove sulfur dioxide from the urban atmosphere. Moreover, as soon as one company indicated willingness to make a firm commitment to supply low-sulfur fuel, others had to follow. The State Department and the Department of the Interior have also joined the battle, partly because of a strong interest in seeing that the economies of Venezuela and other Latin American countries are not seriously upset. As a result, the Department of the Interior has made serious efforts to make changes in import allocations that will lead to easing the supply problem for low-sulfur residual oil.

As things stand now, one of New York City's most serious air pollution problems—sulfur dioxide—is on its way to solution. The amended version of Local Law 14, which has been in effect since September 1969, restricts the sulfur content of all fuels to a maximum of 1 percent.

Emissions into the atmosphere have been cut by more than half since 1966. (See Figure 1.) Sharp reductions in both peak values and average values of sulfur dioxide concentrations have also been observed. In 1964 peak concentrations were well above 2 ppm, and the average value was 0.21 ppm at the Manhattan station. In 1969, although the 1 percent sulfur

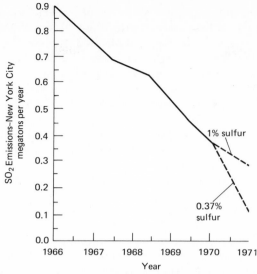

FIGURE 1 Graph of the amount of sulfur dioxide (in millions of tons) emitted from sites within New York City in recent years, according to estimates by the New York City Department of Air Resources.

law did not apply city-wide until October, the maximum hourly average at the station was 0.8 ppm, and the annual average was 0.11 ppm.

The laws to date have been scheduled to make use of the maximum technological capability available at the moment and to insure that rapid development of improved technology is pursued by industry. New York City agencies analyzed every tank at depots in the City immediately after the 1 percent law became effective. Moreover, the oil was followed into the area of the user. A random sampling of installations throughout the city demonstrated almost perfect compliance.

New York City and the surrounding municipalities have reacted in a positive, constructive manner to the problem of excessive sulfur dioxide in the air being inhaled by the public. Pollution control strategies have been successful, because industry has been given both the incentive and the lead time needed to mitigate the effects of one of the worst culprits in the urban environment. Had we been less effective, the 1966 Thanksgiving episode would have very likely recurred, conceivably with even more serious epidemiological effects.

What is of paramount importance is that we have reversed the trend of despoliation of the environment. We did so by the blending of low- and high-sulfur fuel oils to produce a marketable product, which was an interim measure, and by persuading the very large oil refiners to install desulfurization processes, which should have long-term significance.

But we have not been able to put air resource management on automatic pilot. The evidence—the sulfur dioxide levels throughout New York City—indicates that even greater efforts than had been calculated will be required to meet the increasingly stringent criteria for air quality. The greatest improvements have occurred in Manhattan and the immediate surroundings, where most of the residual oil used in New York City is burned. Large areas of the other boroughs use number 2 distillate oil almost exclusively, and therefore their ambient levels were affected by the sulfur regulations only insofar as fewer pollutants were carried there from Manhattan by the wind. Moreover, the winter of 1969–1970 was harsher than usual. An increase in fuel oil consumption for space heating resulted, with the consequence that sections of New York City showed less of an improvement in air quality than would have been expected by the decrease in the sulfur content of residual fuels.

The battle continues on various fronts. Pressure must be maintained on oil suppliers to take the steps necessary to supply desulfurized residual oils in needed quantities. The sulfur content of distillate oils may also have to be reduced and carefully controlled.

At this time, standards governing permissible sulfur dioxide concentrations are being set throughout the United States, in accordance with federal law. The New York metropolitan area has been placed in an air quality region which encompasses, in addition to New York City, three counties in New York State, nine counties in New Jersey, and part of a county in Connecticut. Through the New York State government, standards have been drawn up for that part of the air quality region lying within New York State and have been submitted to Washington, where they now must be coordinated with standards drawn up elsewhere. The standards submitted for approval are summarized in Table 6.

The most challenging of these standards is the one requiring that the annual average of hourly samples at every sampling station be less than 0.03 ppm. To reach this standard, it may be necessary to set up special

TABLE 6. Proposed Standards for Sulfur Dioxide in the New York Metropolitan Region

---

1. During each year at least 99 percent of all the samples that register the average sulfur dioxide concentration during a one-hour period must show a concentration that does not exceed 0.25 ppm.
2. During each year 100 percent of all the samples that register the average sulfur dioxide concentration during a one-hour period must show a concentration that does not exceed 0.50 ppm.
3. During each year at least 99 percent of all the samples that register the average sulfur dioxide concentration during a 24-hour period must show a concentration that does not exceed 0.10 ppm.
4. During each year 100 percent of all the samples that register the average sulfur dioxide concentration during a 24-hour period must show a concentration that does not exceed 0.14 ppm.
5. The arithmetic average of all the hourly concentrations during the year must not exceed 0.03 ppm.

Note: These standards would apply at every sampling station in the region.

low-emission zones in which the sulfur content of fuels burned will be especially low. The determination of such zones requires a more detailed emissions inventory than is now available. Therefore a detailed reporting system of oil deliveries is being given serious consideration. With this procedure in effect, we shall have information about monthly consumption of oil within grid areas one-half mile on each side. To reach a 0.03 ppm annual average, in addition, power plants will have to reduce further the sulfur content of their fuels and, if possible, eliminate some of their power and steam generating plants within New York City.

The summer of 1970 brought an important development in the struggle for clean air in New York City, when the City recognized the need to allow public discussion where policy decisions have large implications for the environment. Thus when Consolidated Edison requested permission to expand an in-City power plant, a public dialogue was allowed to take place within the City government between the advocates of the need for additional power and the Environmental Protection Administration on behalf of clean air. A decision was reached to allow one-half the requested expansion, with stringent controls imposed on allowable fuels and emissions. The issues that were weighed were the need for reserve electric power, especially during the summer peak loads caused by air-conditioning demands, and the backward step such an expansion represents for the environment.

5. ABORTION: A CASE STUDY IN LEGISLATIVE REFORM

Richard D. Lamm

Abortion law reform is still a current controversy, yet in many ways it is history. Colorado, in 1967, was the first state in the United States to reform its abortion law and enlarge the permissible grounds for legal abortion. Yet in the few short years intervening, the debate has moved from reform to repeal. Yesterday's controversies are today's settled wisdom. Politicians who bitterly fought writing into law *any* extensions of the permissible grounds for abortion, today—in maneuvers to prevent complete repeal bills from passing—offer the identical bills they previously argued against.

Torrents of debate have taken place since Colorado's first halting steps at abortion law reform, and Colorado's story must be given the kindness of historical perspective by the reader. The public is apparently ready for abortion law reform or repeal today, but in 1967 either the public was not ready, or at least many legislators *thought* the public was not ready. Colorado's experience is thus instructive not as a present goal or a model of current debate but as a case study in legislative change.

There is a wide spectrum of reasons why people were interested in abortion reform in the 1960s. Some of these people were civil libertarians, some feminists, some former victims of that public health epidemic called illegal abortion, others determined to end that epidemic. The president's Crime Commission has estimated that about one million illegal abortions are performed each year in the United States, and a study by Kinsey concluded that one out of five pregnancies in the United States is terminated by illegal abortion. Few argue with the statement that illegal abortion presents a major public health problem.

At the start of Colorado's 46th General Assembly in January 1967, a small *ad hoc* group of people met to discuss the possibility of changing Colorado's abortion law. It included doctors, feminists, planned parenthood veterans, public health specialists, and some, like myself, who were interested in the demographic aspects of abortion. The prevailing opinion at the meeting was one

58

of pessimism, because it was our feeling that an attempt to change the law at that point in time might hurt more than help the eventual passage of a liberalizing law. There is nothing that so scares legislators collectively as seeing a measure soundly defeated in the legislative process, and its proponents defeated at the polls. Timing is extremely important in the passing of successful legislation, and an ill-advised, premature attempt often has the effect of substantially delaying eventual passage of a law.

On the other hand, there is a certain political taboo on controversial subjects which prevents legislators from raising them. To err may be human, but in politics to err once may be politically fatal. One may be the hardest working, most conscientious legislator in the state; if he becomes associated with one unpopular stand, the voters most often remember the moment of political heresy rather than the years of hard work. Fortunately, in retrospect, I did not in my naiveté realize this in 1967. Only the preceding November I had been elected for the first time as a Democratic legislator from Denver. I had not yet seen the realities of the system, not yet become enamored of being addressed as "Honorable," not yet made the political alliances that urge caution. I thus stumbled across, rather than perceived, a key fact of contemporary political life: that voters realize that these are fast-moving, volatile times and look for leadership and a sense of direction even to the point of voting for someone with whom they do not agree. Voters may not welcome change, but they do recognize it as a central factor of twentieth-century life, and an honestly articulated conviction often will convert rather than offend. I had long held the strong opinion that in these days, when man threatens to overpopulate this finite earth, it was bad social, religious, and political policy to force motherhood on dissenting women. In 1965 these views would have signified defeat; in 1967 they converted rather than offended.

Some in the *ad hoc* group that attended the January 1967 meeting argued that the public was indeed ready for abortion reform, and we agreed to approach a few legislators for their reactions. A steering committee composed of women, doctors, lawyers, and legislators was formed to talk with selected legislators—and the more legislators we approached, the more a cautious optimism began to grow. It soon became apparent that in the last few years an immense change had taken place in public attitudes toward reproduction and its control. Two years earlier, Colorado had succeeded in passing, on a second attempt, a law that provided birth control information to indigent women. On the first attempt at that legislation the bill had never left committee, but upon its second introduction it had passed by a fairly substantial margin. Public acceptance of that law had gone far to pave the way for abortion reform in the minds of legislators. The birth control law had been passed, put into operation, and forgotten. Most legislators were aware that in that case the storms of adverse reaction passed without raining retribution.

Certain issues, like taxation and labor law, lend themselves to partisan appeal; others are by nature nonpartisan. Abortion law reform is by its nature nonpartisan, and our *ad hoc* committee worked hard to keep it that way. A well-respected Colorado Republican legislator, Representative Carl Gustafson, influenced by a recent policy change in his Lutheran church, manifested a great interest in the proposed abortion reform and agreed to join me as prime cosponsor. The background information we found most useful concerned past attitudes toward abortion and the public health aspects of abortion. Highlights of what we learned may interest the reader.

Attitudes toward Abortion in the Past

All societies have faced the problem of unwanted pregnancy. A recipe for inducing an abortion has been attributed to the Chinese emperor, Shen Nung, who reigned in the twenty-seventh century B.C. Egyptian papyri contain information both on birth control (cobwebs) and abortion.

The views on abortion have varied widely both between cultures and within the same cultures. The Hippocratic oath, which states, "I will not give to woman an abortive remedy," did not reflect contemporary Greek attitudes but instead was derived from the Pythagoreans, a minority within the Greek culture. Plato suggested abortion as a method of maintaining the level of population in his ideal state, and Aristotle felt that abortion "before she felt life" was the solution when a woman "had the prescribed number of children." Sonarus (A.D. 98–138), an early Greek expert on obstetrics and a biographer of Hippocrates, wrote, "The fruit of conception is not to be destroyed at will because of adultery or of care of beauty, but it is to be destroyed to avert danger appending to birth." Prisclanius, a fifth-century Latin grammarian, likened abortion to "removing dry twigs to save a living tree, or jettisoning cargo to save a storm-threatened ship."

Some early Christians condemned abortion on the same grounds as they condemned infanticide, basing their objections both on the commandment "Thou shalt not kill" and on the tenet that it is wrong to permit dying without baptism in original sin. Others, like St. Augustine, distinguished between a "living" and "not yet living" fetus, and two famous medieval theologians, Gratian and St. Thomas Aquinas, argued that "life" really begins 40 days after conception for a male fetus and 80 days after for a female fetus. In 1588 Pope Sixtus V in his bull *Effraenatum* defined all abortion as homicide, abolishing the 40- and 80-day distinction, but Pope Gregory XIV in 1591 applied excommunication for abortion of a fetus only after "quickening," which he defined to occur 116 days after conception. The debate continued back and forth until 1869, when Pope Pius IX decided that participation in abortion made one subject to excommunication. This was reaffirmed by Pope Benedict XV in 1917 and by Pope Pius XI in his 1930 encyclical *Casti Connubi,* and is the present Catholic position.

What little Protestant dialogue exists generally regards the fetus not as *homo* (a living person) but rather as a "potential person." Islam considers that life begins 180 days after conception, and Judaism generally says that life begins at a child's first breath. However, none of the above condones abortion on request, without qualification.

There was no English common law crime of abortion, and Blackstone's eighteenth century commentary stated: "Life begins in contemplation of law as soon as the infant is able to stir in the mother's womb." The first law against abortion in England was not passed until 1803, and in the United States, Connecticut passed the first abortion law in 1821. Yet abortion before quickening was not made a crime in the United States until the 1860s.

Abortion and Public Health

There is evidence that the original antiabortion legislation was motivated by considerations of maternal health. During the early 1800s abortion, even in hospitals, was much more dangerous than bearing a child, and the law sought to protect a woman from the risks of abortion by compelling her to complete her pregnancy. Today modern medicine gives us a vastly different picture. In the United States today, according to figures by Dr. Christopher Tietze of the Population Council, it is one-sixth to one-tenth as dangerous to have a hospital abortion as to complete pregnancy.[1] Hungary reports 0.6 deaths per 100,000 operations, and Czechoslovakia in 1964 had 140,000 abortions with no deaths. The fact that a hospital abortion in early pregnancy is now safer than a tonsillectomy obviously ought to be reflected in the formulation of public policy.

There are three ways of inducing abortions in widespread use today. The most common method, "dilation and curettage" (D and C), consists of stretching open the cervix with dilators and removing the products of conception by scraping the uterus. A second method is a new suction method developed in the 1950s in China and brought to the United States after extensive use in Japan and Eastern Europe: a small tube is inserted in the uterus to remove the products of conception by suction. In the third method, used normally after pregnancies beyond three months, a salt or sugar solution is injected into the uterus, which induces labor and the expulsion of the fetus. All these methods present minimal risk when performed by a doctor in sanitary surroundings.

Illegal abortion, by contrast, is accompanied by great risks of death. All estimates of the number of deaths resulting from illegal abortion are imprecise; there are probably between 1000 and 10,000 deaths per year.[2] The Los Angeles County General Hospital reports an average of ten cases a day of women suffering from botched abortions. Illegal abortion is the

[1] C. Tietze, "Mortality with Conception and Induced Abortion," *Studies in Family Planning*, No. 45 (September 1969).

[2] M. S. Calderone (ed.), *Abortion in the United States: A Conference Sponsored by the Planned Parenthood Federation of America*, New York: Hoeber-Harper, 1958, pp. vii, 224.

single largest cause of maternal death in California, and in New York City almost half of all childbearing-associated deaths are related to illegal abortion.[3] The reduction of deaths from illegal abortions presents the greatest single opportunity for reducing the overall maternal mortality rate.

Previous to 1967 the rate at which legal abortions were performed actually went down. Twenty-five years ago there were approximately 30,000 hospital abortions annually, but in 1966 there were only 8000. The reduction was the result, in part, of religious pressures on hospitals and, in part, of improved medical care for pregnant women. Moreover, in 1960 the rate of abortions for private patients was almost four times greater than the rate for ward patients.[4]

Statistics show that there are large disparities between the availability of hospital abortions for the poor and for the wealthy and also for whites and nonwhites. A survey of 60 United States hospitals in 1963 showed that private hospitals had a ratio of therapeutic abortions to live births of 1/256, whereas municipal hospitals had a ratio of 1/10,000. The therapeutic abortion rate for white patients was 2.9 per 1000 live births, whereas that of nonwhite patients was 1 per 13,000 live births.[5]

In Sloane Hospital in New York City private patients received 1 abortion per 55 deliveries, whereas ward patients received 1 abortion for every 224 deliveries.[6] Other studies confirm the obvious: the affluent woman is much more likely to get a hospital abortion than her indigent counterpart, and whites are more successful at obtaining hospital abortion than non-whites.

Creating a Favorable Climate for the Legislation

At the start of our legislative reform effort we found that five states and the District of Columbia recognized some grounds for abortion other than "to preserve the life of the mother." Alabama, Oregon, and Washington, D.C., allowed abortion to "preserve life or health of mother," and Colorado and New Mexico to save the life of the mother or "to prevent serious or permanent bodily injury." Maryland's statute mentioned "safety" of the mother, but this was not defined. Regardless of wording, there were almost no differences between practices in the few states having the more liberal language and practices in the rest of the states.

There had been previous therapeutic abortion laws introduced in various legislatures, but no significant change had ever been effected. Assemblyman Knox, in California, had introduced a therapeutic abortion bill in

[3] Dr. Alan Gutmacher, in a speech given at the University of North Carolina, April 10, 1968, cited in Jaroslav F. Hulka (ed.), *Therapeutic Abortion, A Chapel Hill Symposium.* Chapel Hill, N.C.: Carolina Population Center of the University of North Carolina, 1968, p. 31.

[4] L. Lader, *Abortion.* Indianapolis: Bobbs-Merrill, 1966, p. 29.

[5] F. Lyon, "Abortion Laws," *Minnesota Medicine*, p. 20 (January 1967).

[6] *Ibid.*

1961, and Assemblyman Percy Sutton introduced a similar modification in the New York Assembly in 1965. Neither ever reached a vote in committee. Senator Tony Bielensen introduced a reform bill in the California legislature in both 1963 and 1965, and on the latter attempt it was voted out of one committee but was reassigned to the "finance" committee and buried. The efforts of Assemblyman Sutton in New York were taken over by Assemblyman Al Blumenthal, who introduced his reform measure in 1967. The first state legislature to pass any change was Indiana early in 1967, when it added the grounds of "rape" to a law that had previously allowed abortions only to preserve the life of the mother, but this bill was vetoed by the governor. Harbingers of change were everywhere, yet no state had yet managed to enact even the most mild reform.

The law we proposed was patterned after the model penal code proposed by the American Law Institute in 1962, which by 1967 had still not been adopted by a single state. It permitted termination of pregnancy for two classes of circumstances. (The legal word for such "circumstances" is "indications.") First, it permitted abortion if three licensed physicians certified that continuation of pregnancy was likely to result in either (1) the death of a woman, or (2) the serious impairment of the physical or mental health of the woman, or (3) the birth of a child with grave and permanent physical deformity or mental retardation. Second, it permitted abortion if three licensed physicians certified that the pregnancy was the result of rape (including statutory rape) or incest.

Representative Gustafson and I approached our fellow legislators one at a time, starting with those we suspected from religious belief and from past actions would be the most sympathetic. Our efforts were immeasurably aided by the actions of a volunteer lobbyist, Mrs. Ruth Steel, who two years previously had helped push through the birth control legislation. Mrs. Steel knew most of the legislators personally, and much of our success at obtaining cosponsors was the result of her knowledge of which arguments would be the most persuasive with each particular legislator. The results were startling, and we soon had 53 cosponsors, more than half of the House of Representatives and slightly less than half of the Senate. The large number of cosponsors was a positive sign, but cosponsors have a disturbing history of changing their minds, so we were still a long way from passage.

We realized that we would need the help of people sympathetic to social change if we were to overcome the long-standing taboo on abortion. We felt that physicians and clergymen would be especially helpful, and so we started *ad hoc* medical and clerical committees. We asked the committee members to gather names of fellow doctors and clergymen who would be willing to support or testify in favor of abortion law reform.

The steering committee decided to approach the news media to seek their editorial support. The news media, I think, are the single most important factor in molding community opinion with regard to legislation. It was deemed imperative by our steering committee to negate opposition on the part of the press, at the very least, and, if possible, to seek specific endorsement of the bill from them. A group of us, including a doctor and a minister, called upon all the major news media in the Denver area and explained our bill to them. This interdisciplinary effort proved immensely successful, and won us enthusiastic support in the form of three editorials by the state's most influential newspaper, the Denver *Post*, during the legislative session.

With the cosponsors added to the bill and the committee of doctors and clergymen rapidly growing, it but remained to ensure sympathetic committee assignments in the legislature. The leadership of either the House or the Senate can easily defeat legislation by assigning a bill to a legislative committee where it knows the bill will be buried. Knowledge of the committee system is essential to find a way through the legislative labyrinth. Accordingly, we approached the leadership of both the House and the Senate and requested that the bill be assigned to the House Health and Welfare Committee, which we already knew had a sympathetic chairman and a majority favorable to this legislation. After receiving assurances from the leadership that the bill would get the necessary committee assignment, we introduced the bill.

The Legislative Hearing

The legislative process is most visible during public hearings on proposed legislation. A public hearing can serve a variety of goals: it can build public opinion, isolate opposition, pressure legislators, and serve as a public forum. In the case of abortion reform it was judged that a public hearing was not only advisable, it was indispensable.

The art of passing legislation requires first a determination of what is attainable and, second, the building of a groundswell of support. We determined that a public hearing on a controversial subject like abortion would at a minimum show that the subject could openly be talked about and that "respectable" people believed in liberalizing the existing laws. Once "respectable" people publicly support a "controversial" subject, it both molds public opinion and gives the timid the courage to voice their support. We asked the most conservative and responsible people at our disposal to testify at the hearings; these were ministers, doctors, and lawyers, none of whom had previously been involved in controversial legislation of any kind.

The Colorado legislative hearings on the proposed abortion legislation were heavily attended and widely followed by the press. They were often lively. Children with signs on their backs proclaiming "I'm glad I was born" were brought to the hearings by their mothers. One physician representing the Catholic Physicians Guild, to emphasize his point, placed a bottled fetus on the table in front of the committee chairman. But most of those opposed to the legislation presented reasoned and calm criticism of the legislation. Most, but by no means all, of those testifying against the proposal were Catholics.

On behalf of the legislation we argued that the existing laws were not supported by community opinion. A law, we said, should express some degree of community consensus, and the old abortion laws do not. A large proportion of the community feels that whatever "rights" a fetus has, they should not override the "rights" of women in cases when the two interests conflict.

Two arguments were played down at the hearings. Many of the proponents of the legislation felt that one of the main underlying issues was the population explosion. We believed that many of the religious and moral arguments against abortion were formed at a time when mankind needed more population to fill up an empty earth. But we were reluctant to introduce this issue into the debate for fear that it would expose us too easily to the charge that in our concern for overpopulation we were "getting rid of people." Therefore we kept references to population problems to a minimum and stressed the civil liberties arguments supporting our legislation.

The opponents had a similar problem. Many of them obviously felt that "once a woman has had her fun, she should pay for it." They were more concerned with enforcing their moral code than in protecting the rights of the fetus. But I had ready some of the actual testimony of those opposing venereal disease control legislation on the same grounds in the 1930s, and after a couple of my caustic rebuttals the point was never again raised in testimony.

The main controversy was over the nature of the fetus. Opponents always talked in terms of the "unborn child," while proponents talked in terms of "zygotes," "embryos," and "fetuses." At the last day of debate in the House one opponent quoted the following monologue, which he said appeared in the *Reader's Digest:*

> *October 5:* Today my life began. My parents don't know it yet. I am smaller than the seed of an apple but already I am I. And as unformed as I am right now, I'm going to be a girl. I shall have blond hair and azure eyes and I know I'll love flowers.
> *December 28:* Mother, why did you let them stop my life? We would have had such a lovely time together.

The exact nature of the fetus, we argued, is a matter of spiritual supposition. Life is a continuum, there being life in the sperm and ovum even before fertilization. One didn't lament the "death" monthly of each ovum and the millions of sperm. "Unborn child" begged the question; one might call each sperm an "unconceived child." The proponents and opponents did reach some areas of agreement: that contraception was legal and preferable to abortion; that spermatozoa and ova have no legal status; that many pregnancies are terminated naturally (spontaneously aborted, miscarried, or stillborn), and in these cases the fetus is appropriately viewed as dead human tissue and disposed of without legal interference. The real legislative argument was what rights to give to the developing fetus and what rights to give to the mother, where the mother and fetus had opposing interests.

Given the uncertainty of status, the opponents argued, we must give the fetus the "benefit of the doubt" and deal with it as a human being entitled to full rights including "life, liberty and pursuit of happiness."

The proponents countered by pointing out that the mere fact that we have a crime of abortion shows that the law has always considered a fetus less than a living person, for in every abortion a fetus dies and if it was considered a living person the crime would be first degree murder. The law has not called it murder, it has called it abortion, with penalties historically far less than for murder.

A total of 42 people testified at the House hearings, 23 in favor of the proposal and 19 against. The strongest theme for change, sounded again and again by proponents, was that in a pluralistic society we ought not to write a particular (Catholic) religious doctrine into the law. The chairman called for a vote, and the bill was voted out of committee with only one dissenting vote.

Managing the Bill on the House Floor

Legislators are very sensitive to the intensity of the feelings of their constituents about any issue, for few legislators have enough margin for reelection that they can afford to offend a sizable percentage of their constituency without offsetting benefit. In Colorado and other states, gun control legislation is a case in point: although polls show a majority favoring some restriction on the sale of firearms, the minority opposition is not only more vocal but also votes their convictions. If 80 percent of the people vaguely agree and 20 percent strongly disagree, a politician knows that the benefit is not worth the risk, for the dissenters vote the specific issue, and few politicians have a 20 percent reelection spread.

Often one can help a legislator avoid or mitigate criticism by allowing him to carry tightening amendments. One legislator, who was worried that supporting abortion reform would cost him his seat, was given the job of introducing two restrictive amendments: one that changed the age on the statutory rape provision, and one that added a stipulation that the husband must consent to the abortion. These amendments were well publicized, and allowed him to point to them as his efforts to "tighten" the bill. This should not be cause for cynicism, for most legislation is the result of such ploys and compromises. One may long for a profile of courage and seek out a courageous statesman, but legislation is only passed when one persuades a majority of conventional politicians.

In order to work effectively, one of the first hurdles we had to overcome was to convince the growing number of citizens who wanted outright repeal that our bill would be a significant first step. Legislation is too often lost after being introduced because proponents have not themselves reached a consensus and cannot form the united team necessary to give legislation the needed momentum and unified direction. We won over those who would have wanted a stronger bill by persuading them that both polls and our practical experience gave us the strong impression that the public was a long way from accepting an absence of all restriction whatsoever upon a doctor's authority to approve an abortion. We sensed that the opinion was widely held that the society through its laws has some special responsibility for dictating circumstances under which an abortion may be performed. When the question is put as merely a weighing of two rights—between the health and welfare of the mother and the potential human personality of the fetus—the public seems comfortable and satisfied. If one asks, however, whether a woman ought to have the unfettered right to control her reproduction, if necessary by abortion, public opinion in 1967 answered a resounding No. To attempt legislation on such grounds in 1967, we believed, would have been to exceed the possible and to ensure defeat.

The last stage of the legislative process was to anticipate what would be the principal amendments offered by the opposition on the House floor, and then to assign individual legislators the job of arguing against these amendments. Thus it was arranged that the anticipated attempt to have a residency requirement inserted in the bill would be opposed by a legislator who had heretofore not been particularly associated with this bill. We took similar precautions with the other major weakening amendments: one of them would have required that the decision on when an abortion is allowed be made by a court of law after receiving medical testimony, instead of by an independent medical panel. This, of course, would have been unworkable, for the court dockets in Colorado are lengthy enough that the hearing would have taken place after the birth. Another amendment would have required that an attorney be appointed for the fetus. These amendments were defeated, one by one, with the result that the bill was voted out of the House of Representatives and went to the Senate in substantially the same form as introduced and passed by the House committee. The legislative history of the bill in the Senate was substantially the same as in the House. The bill passed—and, after the House concurred in a minor amendment to provide that no hospital could be held liable for not performing abortions, the bill was sent to the Governor for his signature. All who had written the sponsors supporting the legislation received a letter requesting them to write the governor. The governor early in the session had indicated that he favored some reform of the abortion laws. The opponents staged a campaign for a veto, including picketing the Governor's mansion, but the Governor signed the bill and it became law.

Reflections

Colorado took a bad law and made it better, yet we still have a bad law. We estimated in 1967 that our law would cover approximately 5 percent of the women seeking abortions. There is a wide spectrum of reasons why women want abortions, only a few of which are covered under mental and physical health, fetal deformity, rape, and incest. Most unwanted pregnancies are more correctly classified in the euphemistic medical category, "pregnancies of convenience."[7] A woman does not have to be mentally or physically unstable not to want to add another child to her family; on the contrary, women are rapidly forging a new ethic of parenthood, which requires reasons *for* pregnancies rather than reasons against pregnancies.

Colorado's limited reform, however inadequate, was both short-term improvement and long-term symbol. It immediately became somewhat easier in Colorado for some women to terminate their pregnancies. Before 1967 there had been an average of 30 to 50 hospital abortions in Colorado per year; approximately 80 percent of the women were married. In the first year after passage the number of abortions rose to 142, in the second year to 476, in the third to 946, and the yearly doubling shows no signs of stopping. The percentage of women married has dropped to 50 percent. The availability of therapeutic abortions has increased at public as well as private hospitals.

The greatest effect of the law, however, is that it inspired a number of other, more liberal measures. There is a pollination of ideas among the state legislatures, which causes many states to investigate new legislation passed elsewhere. In the year after Colorado's law was enacted, all but 13 states had repeal or reform legislation introduced in their legislatures. Some passed, some were defeated.[8] But the dialogue which leads to change is now nationwide. Illegal abortion is still epidemic, but the process of cure has irrevocably commenced. Social change is seldom won on a single battlefield, but in numerous nationwide skirmishes.

Colorado picked the legislative route to enlarge the "rights" of women, but it was not the only alternative available. The judiciary is also an institution of social change, and, as was seen in *Brown v. Board of Education* (school desegregation, 1954) and *Griswold v. Connecticut* (birth control, 1965), it is often the courts that act to solve controversial problems. Thus the advocate of change must explore both legislative and judicial alternatives and decide the best one for the issue. In Colorado in 1967 we realized that the courts were a definite possibility. The reasoning of the U.S. Supreme Court in *Griswold* seemed to be broad enough to declare unconstitutional the laws against abortion, for laws against abortion, like laws against birth control, are largely unenforced, force the birth of unwanted children, result in discrimination against lower economic groups, and attempt to impose a particular religious doctrine by statute. In the end, we

---

[7] Over 70 percent of the applications for abortions in the last three years have used the "mental health" indication. Dr. H. G. Whittington, Director of Psychiatry at one of Denver's leading hospitals, has called the mental health indication a "cruel legalistic hoax," and has advocated the complete repeal of all legislative restrictions on abortion.

[8] As of June 1970 abortion laws have been reformed in twelve states (Arkansas, California, Colorado, Delaware, Georgia, Kansas, Maryland, New Mexico, North Carolina, Oregon, South Carolina, and Virginia), and abortion laws have been repealed in three states (Alaska, Hawaii, and New York).

decided against working through the courts because the route seemed long and expensive.

Citizens of other states are looking to the courts for change, and the courts are responding by striking down the challenged laws. For example, the California Supreme Court in 1968, in declaring California's abortion law unconstitutional, states:

> The fundamental right of the woman to choose whether to bear children follows from the [U.S.] Supreme Court and this court's repeated acknowledgement of a "right of privacy" or "liberty" in matters related to marriage, family and sex. . . . There are major and decisive areas where the embryo and fetus are not treated as equivalent to the born child. . . . The intentional destruction of the embryo or fetus is never treated as murder but rather as the lesser offense of abortion.

This language may be broad enough to declare unconstitutional all the legal restrictions on abortion which are written into statute. Complete repeal of all such restrictions is the announced intention of a growing number of people and groups. The governing board of the American Public Health Association has stated that "restrictive laws should be repealed so that pregnant women may have abortions performed by qualified practitioners of medicine and osteopathy" and the Citizens Advisory Council on the Status of Women has recommended that "laws making abortion a criminal offense be repealed." The states of New York, Alaska, and Hawaii early in 1970 passed legislation removing all restrictions on abortions, and other states are in the same process.

The time when mankind needed more population to fill up an empty land is over; the world in 1970 is in serious danger of strangling itself with too many people. Machines, not muscles, produce our crops and fight our wars. Thus earlier arguments are now completely unjustified; instead, there is a widespread feeling that unwanted children are community burdens. People are increasingly demanding control over their own reproductive activities, and they are either going to change the laws in the legislature, as we did, or attack them through the courts, or ignore them, as do the hundreds of thousands of women who each year get illegal abortions.

6. THE HELIUM CONSERVATION PROGRAM OF THE DEPARTMENT OF THE INTERIOR

Charlotte Alber Price

For almost 30 years people believed that helium existed on the sun but not on the earth. Identified by a curious yellow line in the spectrum of sunlight in 1868, helium was not found on earth until 1895. Today, helium has uses that range from the frivolous to the indispensable. The same material that makes children's balloons sail into and beyond the trees is also essential for attaining temperatures close to absolute zero. For helium is both the ideal inert lifting gas and the only material ever discovered which remains a liquid down to absolute zero.

Helium is a minor constituent of the air; it is present at a concentration of 5 parts per million by volume (ppm) and is very expensive to recover. By a set of geological accidents, however, helium also turns out to lie trapped in underground gas formations in concentrations as high as 2 percent, and most of the available helium on earth is found as a minor constituent of the natural gas from the fields of a limited area of Kansas, Oklahoma, and Texas. This gas is available at about one-fiftieth of the cost of obtaining it from the air.

The helium got into the underground gas formations in the first place by the radioactive decay of uranium, thorium, and their decay products. (In these decays the nuclei of helium atoms are emitted.) Uranium and thorium are present in rocks like granite to the extent of a few parts per million, and the radioactive isotopes in question have half-lives of billions of years.* Thus one atom at a time, over the entire history of the earth, helium has been collecting underground. Before this century has ended, we shall probably have released most of that trapped helium to the atmosphere, except to the extent that it is recovered and stored for future use.

Until recently most of the helium in natural gas was never separated from the rest of the natural gas. Instead the helium-bearing gas was piped to consumers and

* A discussion of the terminology of radioactivity is found in the essay on Radiation. [J.H. and R.H.S.]

burned for its energy content. In the process the helium (which, because it is inert, does not burn) was "gone with the wind." °

This essay is about an effort of the United States Government to do something about the waste of helium. Spurred on by a widespread practical concern in the scientific community, the Bureau of Mines of the Department of the Interior greatly expanded its traditional role as the manager of the nation's helium resources by developing a Helium Conservation Program, which was enacted into law in 1960. After working well for nearly a decade, the program got into financial difficulty and became the subject of attack from several quarters. After discussing briefly some of the uses of helium and its sources, I shall describe how the program was established and why it is in difficulty.

The Helium Conservation Program is strongly supported by the private helium industry (which may be receiving an indirect subsidy because of pricing prescribed by the legislation), the scientific community (particularly those members whose work involves low temperatures only obtainable with helium, and geologists who are aware of its scarcity), and the author, who is an economist, but who apparently has a different preference function between the present and the future than many of her colleagues.

The Commodity and Its Sources

Helium occurs in nature in mixtures with other substances, but not in chemical combinations. Almost all of it is helium-4, the isotope whose nucleus is the famed "alpha particle" of radioactive decay.† Helium is the second most abundant element in the sun, after hydrogen, and the second most abundant in the whole universe, again after hydrogen. On earth it is present in trace quantities in the atmosphere, like the other inert gases (sometimes called noble gases).[1] Helium is continually escaping from the atmosphere into space, because it is so light, but the supply in the atmosphere is being replenished by alpha particles which originate from the radioactive decay of uranium and thorium in rocks near the earth's surface.

Because some rocks rich in uranium and thorium happen to have un-

---

° We waste many resources; this story is just more clear-cut than most. A branch of science called thermodynamics provides a universal context for discussing waste. We introduce the reader to thermodynamics in a Postscript following this essay. [J.H. and R.H.S.]

† A second naturally occurring isotope, helium-3, exists, although it is one million times more rare on earth. It is a remarkable substance in its own right. It liquefies at an even lower temperature than helium-4 ($3.2°$ versus $4.2°$ above absolute zero on the absolute temperature scale at atmospheric pressure). [J.H. and R.H.S.]

[1] The proportion of the inert gases in air at about sea level is generally given as argon 0.8–1.0% ; neon 18 ppm (parts per million by volume); helium 5 ppm; krypton 1.1 ppm; xenon 0.09 ppm. The amount of helium in the atmosphere is far greater than in the exploitable underground formations, but it is about a thousand times more concentrated in the underground formations.

derlain airtight "domes," helium is also found in gas formations under the ground. Most of the time the gases trapped with helium are the combustible gases methane and ethane, which are produced when organic matter is buried. This combustible "natural gas" has been greatly exploited for its energy content in the last 30 years, and it is favored over coal and liquid petroleum for many purposes because it is "clean," that is, because it burns without leaving behind ash or giving off sulfur oxides. When helium is contained in natural gas, it serves only to dilute the energy content of the gas, but since concentrations almost never exceed 2 percent by volume, this is never a nuisance, and instead the "nuisance" has always been to recover helium from the natural gas. At what concentrations it "pays" to extract helium from natural gas will depend on the available technology and on demand. In 1960 Congress defined "helium-bearing natural gas" as natural gas containing at least 0.3 percent helium by volume.

Helium is also found in some noncombustible gas deposits. One deposit in Arizona has helium at a concentration of over 8 percent, associated with nitrogen. Other noncombustible deposits are in Wyoming, Utah, and New Mexico and in Saskatchewan, Canada. Helium recovered from noncombustible gas is likely to be more expensive than helium recovered from combustible gases, because there is no natural gas by-product. Indeed, it is hard to see the justification, from a resource policy standpoint, of exploiting these sources of helium at all until the natural gas fields (which are being exploited anyway for their combustible product) are exhausted.

Very little attention has been paid in the United States to the elimination of wasted helium by recovery at the point of use. Practices to assure the recovery and recirculation of helium which are standard in laboratories and industrial plants in several other countries are rarely found here, or even encouraged.

Uses of Liquid and Gaseous Helium

The properties of materials at very low temperatures can, in most instances, only be explored by immersing the materials in a very low temperature liquid. Helium at atmospheric pressure is a gas down to 4.2° Kelvin (degrees Kelvin are centigrade degrees above absolute zero) ° and is a liquid from 4.2° Kelvin all the way down to absolute zero. Every other substance becomes a solid before absolute zero is reached. At atmospheric pressure the coldest liquid other than helium is hydrogen, which solidifies at 14° Kelvin. Thus, to study or exploit the properties of a substance in the vicinity of absolute zero, the substance is immersed in liquid helium.

Out of studies of these low temperature properties (often called *cryogenic* properties) has come the discovery of materials that at very low temperatures have no electrical resistance (a phenomenon known as superconductivity). Superconducting helium-cooled cables may someday become

---

° Relations connecting the Kelvin, Centigrade, and Fahrenheit scales are found in Appendix 3. [J.H. and R.H.S.]

an important method of transmitting electricity over long distances. Super-conducting wires, carrying large currents at lower power loads than ordinary wires, have been wound into coils to create the most powerful magnets ever known. These magnets, already useful in high energy physics and plasma physics research, are on the verge of being useful for commercial applications as well. In the long-term future, one application will almost surely be controlled nuclear fusion, which promises to permit energy to be produced with less air pollution than in fossil fuel plants and with less radioactive waste than in fission reactors.

Although commercial uses of liquid helium are growing rapidly, and the volume consumed grew tenfold between 1962 and 1967, it is still true today that most helium use is a consequence of its desirable (and often unique) properties as a gas. Until World War II helium demand was tied to the fortunes of lighter-than-air ships and thus to congressional appropriations for dirigibles and blimps.[2] Helium, because of this application, is classed as a "munition," and a special license is required to export it.

The use of helium has grown dramatically in the last 20 years (see Figure 1). The ability of helium to diffuse rapidly through microscopic openings, its inertness, and its easy distinguishability, make it ideally suited for use in leak detection. Helium-oxygen mixtures are often prescribed for asthma patients, because the helium, again because of its rapid diffusion, carries the oxygen deeper into lung passages. Helium gas is also used in a

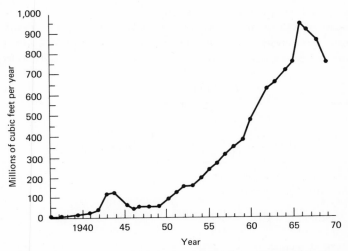

FIGURE 1 Annual shipments of helium for consumption from all plants in the United States, 1937–1969. (U.S. Bureau of Mines)

[2] Neon, the next lightest inert gas, is also lighter than air, but neon is two-thirds as dense as air, whereas helium is one-seventh as dense as air, when these gases are compared at the same temperatures and pressures. Since it is the difference between the weight of the gas and the weight of the air which represents the lifting capacity, neon is not useful for lifting blimps or weather balloons.

mixture with oxygen as a substitute for ordinary air in the breathing mixtures used by deep-sea divers—nitrogen at high pressures is a deadly intoxicant.

One of the most remarkable of the new uses of helium is in the space program. As a result of its inertness, helium, as a gas under pressure, has turned out to be invaluable for moving two highly reactive gases from separate chambers to a central reaction region without interacting with either of them. Thus on board the Apollo flights, liquid oxygen, liquid hydrogen, and liquid helium are found. Oxygen and hydrogen combine (to make water) in the rocket engine, and helium is the agent that brings them together in a controlled fashion. Helium is preferred over any of the other inert gases because of its light weight.

Of the numerous other uses of helium, I shall mention just one more—welding. (A breakdown of helium use by application is seen in Table 1.) Because helium does not combine with other substances, even at high temperatures, it turns out to be desirable to bathe a welding site with helium during the welding process, to prevent corrosion of the surfaces as they are being joined. For this purpose, lightness is not always a particular virtue, and argon is sometimes substituted for helium in welding applications.

The Development of a Conservation Program

From 1937 to 1961 helium production was a government monopoly. The Bureau of Mines built and operated "helium activity" plants, extracting helium from both combustible and noncombustible sources. Helium conservation had taken the form of storing excess production from plants

TABLE 1. Uses of Helium, 1967

Reported Use	Volume (million cubic feet)	Percent of Total
Pressurizing and purging	370	40.8
Controlled atmospheres	105	11.6
Research	101	11.1
Welding	97	10.7
Lifting gas	69	7.6
Leak detection	64	7.1
Cryogenics	51	5.6
Chromatography	22	2.4
Heat transfer	14	1.5
Other	14	1.6
Total	907	100.0

SOURCE: U.S. Bureau of Mines, Helium Activity.

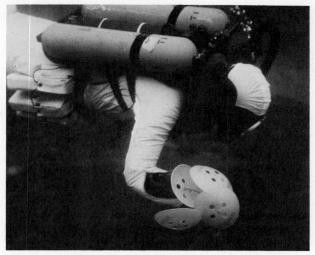

An "aquanaut" associated with the U.S. Navy's Sea Lab II measures the flow of ocean current in Scripps Canyon, 200 feet below the surface off La Jolla, California. The two tanks of gas on his back hold helium and oxygen; replacing nitrogen with helium, which is several times less soluble in body tissue, permits divers to operate in the depth range from 150 to 600 feet without experiencing a disabling intoxication (nitrogen narcosis). (U.S. Navy)

recovering helium from pipeline gas, but recovery and storage were on a small scale.

In the late 1950s the critical shortage of helium—before the newest of the Bureau of Mines plants (at Keyes, Oklahoma) was opened—apparently led the government to undertake a study of the long-range potential availability of adequate helium supplies. The new plant at Keyes was to recover helium from a hydrocarbon gas field found in 1943, and it was sobering that every gas field discovered in the interim 15-year period had contained much less helium. During those same 15 years a great deal of natural gas exploration and discovery had taken place, and the Bureau of Mines had regularly tested samples from new wells for helium content.

The study was also stimulated by projections of large increases in demand for helium over the three or four decades ahead. The projections in part reflected the surge in demand at that time but were also reinforced by laboratory developments during the 1950s. It was assumed by scientists working with helium that some of the phenomena being studied then would result in engineering "payoffs" at some time in the distant, although foreseeable, future. These scientists foresaw new processes and products that would require very low temperatures, temperatures only available with the use of liquid helium.

Most of the known extensive sources of helium in the United States were (and are) in hydrocarbon natural gas that was already committed to distribution and use. Hence, unless the helium was extracted and stored before distribution, it would not be available later. Yearly demand at that time was almost 500 million cubic feet, while more than 5000 million cubic feet were being released into the air annually with the marketed helium-bearing natural gas.[3]

Two alternatives to recovery and storage (considered, but rejected in the 1950s) were (1) to let a conservation program take over the fields containing helium and operate them only when the helium was required or (2)

[3] When helium volumes are cited in cubic feet, this always refers to the volume at a temperature of 70° F and a pressure of 14.7 pounds per square inch.

to require the natural gas producers to recover and store or sell the helium in the gas before marketing the hydrocarbon gas in interstate commerce. Both notions were rejected because they were considered politically infeasible as well as unjustifiably expensive. It was believed that Congress, the Federal Power Commission, and both producers and consumers of natural gas would oppose the restraints on gas supplies implied by the alternatives.

The program that was proposed in 1960 and subsequently adopted had the following characteristics: (1) Under long-term contracts, the natural gas distributors whose source gas came from the extensive Hugoton and Panhandle fields—discovered before World War II in Kansas, Oklahoma, and Texas—were to build plants to recover the helium from the natural gas before the natural gas was distributed and to sell the helium to the federal government. (2) The government was to build a pipeline to transport the recovered helium from the recovery plants to a partially exhausted gas field (the Cliffside structure, near Amarillo, Texas) that the government had controlled since the 1920s. (See Figure 2.) (This structure had been used to store excess helium during the late 1940s; then when demand increased very rapidly in the mid-1950s, 85 million cubic feet of helium were retrieved from storage.) (3) Finally, a pricing policy for the sale of helium to consumers was designed with the intention of recovering, over a period of 25 to 35 years, all of the costs of the conservation program.

The government's selling price of helium, which at the time was $15.50 per thousand cubic feet for government agencies and $19 per thousand cubic feet for everyone else, was to be raised to $35 per thousand cubic feet for everybody, and the increased receipts from sales were to go to pay off the costs incurred in the purchasing of helium for conservation.[4] Although initially the costs of purchases for conservation were expected to exceed revenues from sales, and the difference was to be made up by borrowing from the United States Treasury, later (the crossover date was estimated to be around 1970), revenues from sales were expected to exceed costs, including interest on the borrowed funds. The entire program was to be administered by the Bureau of Mines, the agency (now within the Department of the Interior) that had been responsible for helium production since 1925.

When helium conservation legislation was before Congress, numerous amendments were added. The thrust of the amendments was to encourage the entry of private industry into the helium market. Yet the original concept of financing the program out of government sales of helium was retained. The amended legislation, with general bipartisan support, was passed in September 1960, and in August 1961 the program was authorized to spend $47.5 million per year for a 22-year period.

By November 1961 contracts for the purchase of helium from five helium-extraction plants (which were to be constructed) were negotiated with four private firms. Under these contracts the government expected to purchase up to 3500 million cubic feet of helium a year for 20 years. (This rep-

---

[4] At the time, the cost of production at Keyes was about $8 per thousand cubic feet, and the cost was higher elsewhere.

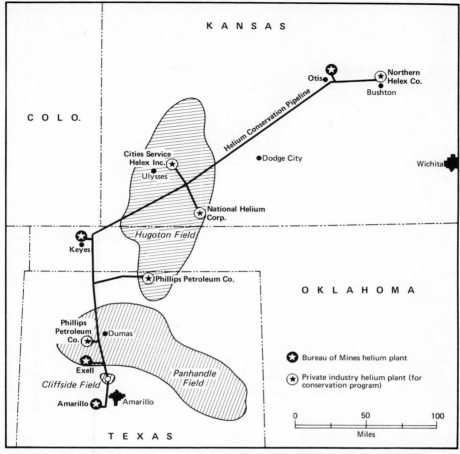

FIGURE 2 Location of the major helium extraction and storage facilities
that have been associated with the federal government's helium activities. The Bureau of Mines plants, built before the Helium Conservation
Program began, produce helium for sale by the government. By 1971
only the Keyes and Exell plants were open. The private industry helium
plants shown here produce helium for the conservation program. The
Hugoton and Panhandle gas fields are also shown. (U.S. Bureau of
Mines)

resents about four times the rate at which helium has been used in the last
few years and 60 percent of the helium that was then being wasted by not
being extracted from marketed natural gas.) The contracts set an initial
purchase price between $10 and $12 per thousand cubic feet, and provided
annual dollar ceilings on how much the government agreed to purchase
from each supplier. In letting out these contracts all of the authorized
funds were committed. The Bureau of Mines has since asked for authoriza-
tion to spend up to $65 million per year in order to increase the amount of
helium it could buy for storage, but more funding has not been granted.

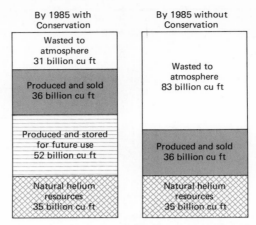

FIGURE 3 Bureau of Mines projections, made in 1960, comparing what would happen to United States helium with and without the Helium Conservation Program. (U.S. Bureau of Mines)

Figure 3 shows quantitatively how the Helium Conservation Program was expected to work. In 1960 the total known reserves of helium in helium-bearing natural gas amounted to 154 billion cubic feet. The amount of helium predicted to be available after 1985 with and without the program was 87 billion versus 35 billion cubic feet. (The difference, 52 billion cubic feet, would go into the air in marketed natural gas without the program.)

Initially the program worked as planned. By mid-1969, 22,000 million cubic feet of helium had been stored in the Cliffside field, and the demand for helium was rising according to expectations through 1967. Problems then arose in the financing of the program, problems which were rooted in inconsistencies that had been built into the original legislation. These problems were exacerbated by the high interest rates (for Treasury borrowing) of the late 1960s.

Setting the price of government-sold helium artificially high was a successful way for the government to earn revenues only as long as the government held a monopoly on helium sales. However, because the enabling legislation had sought to encourage the entry of private industry into the helium market, and all of the planned Helium Conservation plants had not been built (leaving desirable helium-bearing natural gas still going to market), the government's monopoly had been undermined. It was inevitable that the high government selling price would encourage the development of competition from private industry. By 1966 three private plants were in operation, and these new plants were underselling the government.

By 1970 private industry accounted for over half of the helium sales. Some of the sales came from the same plants that were supplying the helium to the conservation program, because they were producing helium at a rate faster than the government was permitted to buy it. The reason why the Bureau of Mines was able to sell *any* helium was that federal agencies were required by law to make all major purchases from the government.

The courts had decided that private contractors working for the federal government were not bound by the same restrictions on government purchasing as the government agencies themselves, although the Bureau of Mines had attempted to get the opposite ruling. And on top of all of this, total helium consumption, which had reached a peak of 948 million cubic feet in 1966, had dropped to 760 million cubic feet in 1969, primarily as a result of cutbacks in the space program.

The total situation is summarized in Figure 4. In brief, the government was not securing the revenues from sales which the enabling legislation had required, precipitating a budgetary crisis which, at the time of this writing, has not been resolved.

Alternative Sources of Helium

It is clear that a program designed to conserve helium for use many decades into the future would not be a wise investment of government funds if there were a reasonable prospect of obtaining adequate quantities of helium in the future at the same or a lower cost from some other source. To make a quantitative comparison, one must find a way of computing the cost in the future of storing helium now. That cost, called the *implicit cost,* is the sum of the initial cost (about $12 per thousand cubic feet) and the interest on that cost, compounded over the period the unit of helium is in storage. Choosing the appropriate interest rate is a critical and controversial business. Over a few decades of accumulation, even one percentage point increase in this rate may spell the difference between a program that can or cannot be justified before Congress.[5]

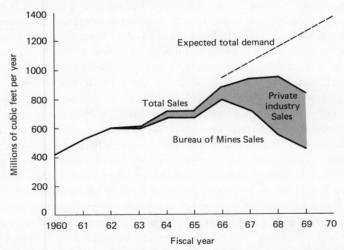

FIGURE 4 Sales of helium by private industry and by the United States Government since the beginning of the Helium Conservation Program. The dashed curve shows the demand for helium that was predicted in 1960. (U.S. Bureau of Mines)

[5] At 4 percent interest, a $12 expenditure becomes a $39 implicit cost in 30 years; at 10 percent interest it becomes a $209 implicit cost.

One must also estimate the costs associated with the sources of helium that will be available. The three likely sources are the air, possible power plants using nuclear fusion, and remaining deposits of natural gas.

Helium from Air

Helium and neon could be available as by-products of the liquefaction and separation of air, which is today a large-scale industrial process primarily designed to recover oxygen. Practical considerations indicate, however, that this source is inherently a minor one. At present levels of oxygen recovery in the United States, the helium *potentially* available (it is not now collected) is about 10 million cubic feet per year. There is no reason to believe oxygen demand will grow at many times the rate of growth of helium demand. Yet unless such a relative growth rate occurs, by-product helium could not be produced in quantities sufficient to supply even nonsubstitutable uses, say, 50 years from now.

In the process of recovering helium from air, three and one half times as much neon becomes available at the same time, since (footnote 1) neon is present in a concentration of 18 ppm versus 5 ppm for helium. Neon, the second lightest of the noble gases, would be cheaper than helium, presumably, and might be used as a substitute for helium in some functions. If helium alone were to be obtained from air with present technology, it would cost $10,000 per thousand cubic feet; by-product helium is likely to cost $600 to $800 per thousand cubic feet.

Fusion Power

Energy may ultimately be generated by nuclear fusion. If it is, helium will be a by-product, because the basic process involves the fusion of two hydrogen nuclei to make a helium nucleus. Any fusion process, however, is likely to require large magnetic forces; the source of the latter is very likely to be superconducting magnets at liquid helium temperatures. Engineering calculations seem not to have been made on whether a fusion power plant will be a net user of helium or a net supplier. The critical question seems to be the extent of helium losses within the system. It is unlikely, however, that fusion reactions will ever be a significant source of helium. Even if a fusion process is developed which does not require superconducting magnets, and even if all of the helium is recovered and available for other uses, the total helium production is small. Only 10 million cubic feet per year would be produced if an amount of electricity equal to the total current U.S. production of electricity were generated by the fusion process in plants converting nuclear energy to electrical energy at 20 percent efficiency. Less helium would be available if the plants were more efficient. (See Comment: *Helium production from nuclear fusion.*)

Comment: Helium production from nuclear fusion

Controlled nuclear fusion may become the most important source of energy for man. We are interested here in verifying that nonetheless it will not be an important source of helium.

In a fusion reaction, two nuclei combine to make either one or two different nuclei, in such a way that the total mass of the final nuclei is less than the total mass of the original nuclei. The "lost" mass reappears as the energy released in the reaction.

Numerous nuclear reactions are candidates for "the" fusion reaction. We will take for definiteness the reaction

$$D + T \rightarrow He^4 + n + 18 \, MeV$$

in which a deuterium nucleus (D) and a tritium nucleus (T) collide and a helium-4 nucleus, a neutron, and 18 MeV (million electron volts) of energy are produced. (Deuterium and tritium are isotopes of hydrogen.) The helium-4 nucleus immediately picks up electrons to become a helium atom. If power plants use this process to create electrical energy at 20 percent efficiency, then one helium atom will be made with every 3.6 MeV of electrical energy.

The problem at hand is to calculate (with the help of Appendix 3) the amount of helium that would be produced each year if this were the process by which all of the United States electricity were produced. Using the 1969 figure that electricity was consumed at 0.75 kilowatts per person, we get an electrical energy consumption in the country in a year (200 million people, 3×10^7 seconds in a year) of 4.5×10^{18} joules, or 3×10^{37} electron volts (eV). Since 1 MeV $= 10^6$ eV, we find that 8×10^{30} helium atoms would be produced. Now, 6×10^{23} atoms (Avogadro's number of atoms) of gas at O°C and 1 atmosphere of pressure occupy 22.4 liters (or $22.4/28.32 = 0.79$ cubic feet), so 8×10^{30} atoms will occupy 1.0×10^7 cubic feet, that is, 10 million cubic feet. This number agrees with the number on page 80, the correction for the volume of gas at the temperature and pressure of footnote 3 being less than 10 percent. [One could have obtained answers differing by a factor of 2 or so by choosing a nuclear reaction other than reaction (1).] Thus the amount of helium currently used in the United States in about four days would be produced in a year.

J.H. and R.H.S.

"Leaner" Deposits of Natural Gas

The cheapest alternative source of helium will probably be leaner deposits of natural gas. It is likely that improvements in technology will reduce the costs of extracting helium present in very low concentrations in natural gas. A crucial question, however, is whether substantial quantities of natural gas will still be recovered 40 years or more hence. A recent National Academy of Sciences study predicted that natural gas sources in the United States, and perhaps the rest of the world, will be nearing exhaustion during the first decade or so of the twenty-first century.[6] If that estimate is in fact accurate, comparing the cost of extraction from "leaner deposits" with the implicit cost of stored helium 40 to 50 years from now becomes an academic exercise.

[6] M. King Hubbert, "Energy Resources," in *Resources and Man*, published by Freeman and Co. for the National Academy of Sciences, 1969 (pp. 187ff).

Prospects for the Conservation Program

Any detailed comparison of the implicit cost of stored helium at some date in the future with the alternative cost of producing helium at the same date directly (by any process) obviously involves one in guesswork. One must make assumptions about both future production and future consumption. One must choose an interest rate to calculate implicit cost, and one must estimate future production costs for alternative processes. The author did nearly a score of estimates, using various assumptions about future demand and production, all of which indicated that the implicit cost of helium stored by the conservation program would be significantly less than the cost of by-product recovery from air, during the entire period that the stored helium would last.[7]

However, another economist, Professor Lee Preston, of the State University of New York at Buffalo, has testified before Congress that the helium conservation program cannot be justified on economic grounds.[8] His conclusions are so different from mine primarily because he is willing to be much more optimistic than I am concerning the future discovery of now unknown natural gas fields. He is also counting more than I on new technology which will be able to lower the cost of extraction of helium from deposits substantially leaner than those with 0.3 percent concentration. Finally, he is not expecting demand for helium to rise significantly from current levels. Professor Preston's conclusions are being used by some members of Congress and in the Bureau of the Budget to support a proposal to the effect that the Bureau of Mines should immediately abandon the acquisition of helium for conservation.[9]

The author does not believe that the underlying premises of the conservation program have been invalidated by the developments of a financial nature of the last two years. The way to solve a financial problem is with alterations in the structure of the financing. One way to continue the helium conservation program under redesigned financing would be to tax all sales of helium a fixed amount (say $10 per thousand cubic feet), using the revenue to pay for the conservation program. Another way would be to postpone the time at which the program is required to repay its loans from the U.S. Treasury until such time as the helium in storage is sold.

If the conservation program is allowed to function as was originally

[7] Charlotte Alber Price, "The Helium Industry: A Study of a Federal Government Monopoly." Ph.D. dissertation, Columbia University, New York, 1967.

[8] Hearings before the Subcommittee on Economy in Government of the Joint Economic Committee, Congress of the United States, 91st Congress, 1st Session, Part 2 (September 1969), pp. 363–371 and 420–423.

[9] In January 1971 the Department of the Interior notified the four private firms supplying helium for conservation that their 22-year contracts were to be terminated "60 days after the receipt of the letters." The President's budget proposal for fiscal 1972 anticipates abandonment of the Helium Conservation Program. (Note added February 1971. C.A.P.)

proposed, with some new financing provision, then by 1983, when the program ends, approximately 62,000 million cubic feet of helium will have been recovered instead of wasted. How much of this would then be in storage would depend on whether demand had required using some of what would otherwise have been stored. There would seem to be every reason to extend the program beyond 1983 and to continue to attempt to conserve as much as possible of the helium still in the ground at that time.

It was 40 years ago (May 1930) that a congressional committee held special hearings on the new government-owned Amarillo Helium Plant and its complementary helium-bearing Cliffside gas field. During the course of testimony it was explained that the gas field had enough helium to last for 200 to 300 years. At the consumption rate of that period, 8 million cubic feet per year, that was a proper estimate. In recent years consumption of helium has been about 100 times that of four decades ago, but now the conservation program is being attacked on the grounds that current known resources plus helium already in the Cliffside structure (by virtue of the helium conservation program) are more than adequate for the next 40 years. This is the kind of prediction which, if wrong and followed, represents an irreversible decision: not to recover additional helium while it is available from natural gas means one cannot go back 40 years if the forecast is wrong. It would seem prudent to base natural resource policy on an approach that, if it errs at all, errs on the side of overconservation.

POSTSCRIPT: Helium Waste and Entropy

John Harte and Robert H. Socolow

Helium waste appears to be one of the more esoteric examples of the loss of a valuable resource. The possible scarcity of other more basic resources such as oil, wood, clean water, or rich soil are, deservedly, more in the spotlight today. Yet the helium story needs to be understood, for it illustrates in an extremely simple and clear-cut manner the principle underlying the *irreversible* nature of the loss of a nonrenewable resource.

When a pocket of helium is wasted, that is, dissipated into the atmosphere, it undergoes a transition from a relatively concentrated and accessible state to a diffuse state, where it is far harder to retrieve. It is in this sense that the loss is irreversible.

The phenomenon of the irreversible loss of helium illustrates a fundamental law of nature—the second law of thermodynamics. Let us look first at a different example that illustrates this oft-quoted law and then attempt to state the law more generally and explore its implications.

Suppose you are given a metal rod at some moment in time, and it happens that the temperature of the metal at one end is greater than at the other. In other words, the molecules of metal at that time are, on the average, moving faster at one end than at the other. Suppose the rod is isolated from all external influences so that, for example, it cannot exchange heat with its surroundings. What do you expect to find if you measure the temperature at each end of the rod at some subsequent time? The end that was hotter may be even more hot, it may remain at the same temperature, or it may cool. Conservation of energy (the first law of thermodynamics) simply tells us that if the hotter end cools, the cooler end will heat, and vice versa, for the total quantity of heat in the isolated rod is constant. The second law provides more information, for it assures us that the two ends of the rod will *almost invariably* approach each other in temperature; heat will flow from the hot end to the cool end.

The basis of this law is statistical and that is why we say *almost* invariably. The odds are much much greater that the hot end will cool rather than heat up. This can be understood by considering matter at the molecular level. To any given macroscopic state of the rod, characterized by a temperature distribution along the length of the rod, there exist many microscopic states characterized by the positions and velocities of each of the molecules comprising the rod. (You might think, by way of analogy, of a party of people in a closed room. If you "measure" the macroscopic state of the party by a device that detects the total noise level, then many different combinations of the individual voices will correspond to the macroscopic state.) If you consider two *different* macroscopic states of the rod, in general, more microscopic states will correspond to one macroscopic state than to the other. (A silent room full of people can only occur if all voices are silent, so there is only *one* microscopic state corresponding to the silent state; whereas a macroscopic state with a nonzero noise level can arise from many different combinations of voices.)

In order to relate these concepts to the second law, we have to make a fundamental assumption: *all microscopic states are equally probable*. Granting this assumption, it follows that a system is most likely to be found with a macroscopic state to which the largest number of microscopic states correspond. Hence, if the system happens to be found in a macroscopic state A to which few microscopic states correspond, and if the system could exist in a macroscopic state B with which a greater number of microscopic states are associated, then it is likely that the system will evolve from A to B.

It is possible to show, using the same kind of mathematical thinking which leads to the conclusion that of all the two-child families twice as many have one boy and one girl as have two girls, that an overwhelmingly greater number of molecular states are available for a uniformly heated rod than for a rod in which one end is hotter than the other. Thus the probability is overwhelmingly great that the nonuniformly heated isolated rod will evolve in time into a uniformly heated rod.

There is a high degree of *order* inherent in a rod in which the molecules are moving faster at one end than at the other. If the molecular velocities are random or unordered throughout the rod, then its temperature will be uniform. According to the second law, our original rod will evolve in time toward the state of least order, or greatest randomness.

Physicists introduce the term *entropy* to discuss the disorder of a system. Our original rod was in a state of low entropy (high order), and it will progress, with overwhelming odds, to a state with higher entropy. The second law of thermodynamics states that the entropy of an isolated macroscopic system will, with overwhelming probability, either increase or stay the same.°

Although we have used a specific example of a metal bar to introduce the basic ideas, the second law of thermodynamics is a completely general result. With the single statement that the entropy of an isolated system will almost invariably increase or stay the same, the second law of thermodynamics describes how *every* isolated system will evolve.

We can now understand why certain changes are irre-

° Our example of the bar illustrates why we have to allow the possibility that the entropy remain constant. Once the temperature is uniform, no macroscopic state is available which is more disordered, and the entropy will thenceforth remain constant.

versible. If an isolated system increases its entropy, it is virtually impossible for it to return to its original state; for its entropy would have to decrease to do so, and that is what the second law forbids. On the other hand, changes in a system which keep the entropy constant (like rotating a metal rod that is at a uniform temperature) are reversible.

If the system is not isolated, then its entropy *can* decrease. To go back to our example, we can heat one end of a rod at uniform temperature, and the entropy of the rod will decrease. But the entropy of the *larger* system, consisting of the rod plus the heat source, will, if *it* is isolated, either increase or remain constant.

Returning now to helium, the underground reserves are in a state of relatively low entropy compared to the disordered state in which they are diffused into the atmosphere. Having helium molecules concentrated somewhere, rather than spread throughout the atmosphere, is analogous to having the fast-moving molecules in the metal bar concentrated at one end. If helium is allowed to dissipate into the atmosphere, the loss is thus practically irreversible. Only considerable external energy could retrieve the initial situation.

There are examples of the use of energy to retrieve a resource which has evolved, as a consequence of the second law of thermodynamics, into a high-entropy state. If we pour salt into fresh water, we have mixed, or disordered, the two ingredients. To retrieve fresh water, we must expend energy. For example, we can boil the salt water and obtain fresh water and salt in a state of lower entropy than that of the seawater.°

Another low entropy state derived from seawater is a fish. The source of energy which allowed the slow transition from seawater to fish is the sun.

There appears to be inherent value in order or low entropy. A low entropy ecosystem with a wide variety of interdependent species may be more resilient to stress than a high entropy ecosystem with less diversity. Similarly, a uniform mixture of paint affords no aesthetic pleasure, whereas a low entropy pattern of paints may be a masterpiece.

The degree of order of a system can be thought of as a nonrenewable resource; when it is wasted, the second law of thermodynamics tells·us it will inevitably cost energy to bring the system back to what it was. Man is already producing energy on a scale that is causing reverberations throughout his environment.† Actions of society which needlessly increase the present or future requirements for energy must be avoided; the second law of thermodynamics helps us to spotlight some of these actions.

° The essay on Water contains a practical discussion of this problem.

† See the essay on Energy.

C. INITIATIVE IN THE SCIENTIFIC COMMUNITY

There is always drama in a trial. When the life of the prosecutor, the defendant, and the judge may be at stake, the drama is heightened. So it was, in Wisconsin, when DDT was on trial. The story of this trial, and an explanation of the evidence, is the subject of the essay by Professor Loucks.

Numerous verdicts are being handed down against DDT at this time, and the use of DDT in agriculture in the United States is falling off sharply. But elsewhere in the world, particularly in the underdeveloped countries, its use is not decreasing. DDT is still the most commonly used chemical weapon against the insects that spread malaria and other terrible human diseases. It is also a component of various programs designed to grow more food. Yet, as Professor Loucks explains, no matter where DDT is introduced into the global ecosystem, it can turn up everywhere else, because it is readily transported over thousands of miles by wind and rain. The banishing of DDT from the planet awaits the development of *safe* alternatives and the decision of the prosperous nations to subsidize the production of these alternatives should they prove more costly in dollars.

Professor Eipper at Cornell forged a coalition of scientists and citizens as broad as the one that fought and won in Wisconsin. The scientific arguments they developed have had wide application; the explanation of how deep lakes pulse with the rhythm of the seasons and of how easily man can shatter that harmony has enlarged our understanding of thermal pollution.

Behind Professor Eipper's discussion of a particular power plant, there loom larger issues of energy policy. There is, for example, the question of whether power plants should be located near populated areas or on lakes as small and as vulnerable as Cayuga Lake. Perhaps only the open oceans are appropriate locations for large new power plants. The vast size of the oceans would greatly reduce thermal pollution effects, and the direct hazards posed by radioactive and other wastes might be minimized. A policy that assigned new power plants preferentially to the oceans could only be implemented at a national or, perhaps, international level.

Now that the environmental side effects of energy production have been recognized, other questions follow naturally: Can existing technologies be modified to use energy more efficiently without detriment elsewhere? Can demands for energy be discouraged, especially those which have been artificially stimulated?

87

7. THE TRIAL OF DDT IN WISCONSIN

Orie L. Loucks

The DDT story begins with the ready acceptance of this insecticide almost three decades ago. The use of DDT rose on the strength of unprecedented service to man on two fronts: first, it all but ended the threat of malaria wherever it was used in the tropics during World War II and in other areas since; second, it allowed an expansion of agricultural production in the postwar years that did much to prevent the world famines which had been predicted for that period. The acknowledged side effects or costs of using DDT were the mortality of nontarget insect and bird species. These were judged insignificant compared with the great benefit to man.

The story of the formal opposition to DDT begins only a few years after the war with suggestions that there might be a buildup of the concentration of DDT in the food webs supporting birds. Leading the opposition was Professor George J. Wallace, an entomologist at Michigan State University, who chastised the public for its indifference to the decimation of robins each spring, and described the violent tremors that preceded their death.

Other isolated expressions of opposition to the use of DDT appeared during the 1950s, but the first extensive indictment came in 1962 with publication of the book *Silent Spring* by Rachel Carson. Looking back now, there is no doubt but that Miss Carson's summary of the evidence against DDT at that time was a major milestone. But leaders of the agricultural and chemical industry were able to neutralize the effect of the book by alleging that it contained distortions and misrepresentations.

In the long run the real contribution of *Silent Spring* seems to have been the stimulation of a more intensive research effort on the secondary effects of persistent pesticides. Until the early 1960s, the opposition to DDT was based almost entirely on qualitative judgments linking bird mortality with these pesticides. The attack on *Silent Spring* led to a redoubled effort to demonstrate precisely the mechanisms by which DDT applied to crops or trees in one area could affect fish, birds, and possibly man, hundreds of miles away. The intuitions of even the best-informed conservationists had no chance of being included in the cost-benefit analysis that weighed in favor of DDT at that time, but with a more complete record of the cumulative indirect costs to society in their hands, the conservationists could expect to be taken more seriously.

By the late 1960s incontrovertible evidence of the persistence of DDT in the environment and of its sublethal toxic effects on many species was becoming available. The stage was being set for a more formal and more complete evaluation of DDT than had ever before been possible.

The story is ending now after some 20 years of costly scientific and legal effort. A critical turning point occurred at the marathon hearings in Wisconsin on a petition to declare DDT a water pollutant under Wisconsin law. This essay is specifically the case history of the Wisconsin hearings on DDT, but in a more general sense the hearings themselves told the case history of environmental degradation by DDT. The record produced by these hearings may well be a preview of case histories for many other persistent toxic compounds being introduced into the environment.

Organizing to Oppose DDT

At the same time that scientists were responding to the charge of inadequate evidence against DDT, citizen groups were organizing to continue the kind of opposition Rachel Carson had voiced. Inevitably, the scientists and citizens would meet and join forces. Two such unions took place in 1966, one in Wisconsin, and one in New York. Two years later, in 1968, the two groups formed from these unions collaborated to bring about the most comprehensive hearings ever undertaken on the impact of an uncontrollable chemical on the environment.

One of the two groups, the Environmental Defense Fund (EDF) of New York, began as an informal coalition of laymen and scientists including a dynamic lawyer, Victor John Yannacone, Jr. The group brought its first court litigation opposing the use of DDT in 1966, a citizens' class action against the Suffolk County Mosquito Control Commission on Long Island.° The success of that action encouraged the group to incorporate the following year, 1967, and to file suit against the Michigan Department of Agriculture and municipalities in Michigan using DDT for the control of Dutch elm disease. Although the Department of Agriculture was declared immune from suit, court orders were obtained against 53 municipalities.

The second group, the Citizens' Natural Resources Association (CNRA) of Wisconsin, was established in the early 1950s. A coalition of scientists and laymen, it lobbied in the Wisconsin legislature against exploitative resource development legislation and in favor of constructive conservation law. Partly in response to the strong attacks on Rachel Car-

---

° A citizens' class action is a lawsuit brought by injured citizens on behalf of the class of all citizens to whom the same law and fact apply; common relief is sought for the entire class. [J.H. and R.H.S.]

son's book *Silent Spring* the CNRA organized a conference titled "Pesticides—a Special Review" in April 1966 in Madison, Wisconsin. At the conference, local citizens heard scientists discuss changing insect control measures, DDT chemistry, genetic effects, the nature of ecosystem structure, and the potential of biological control. In 1967 the CNRA and the Environmental Defense Fund began considering the possibility of a major case against DDT in Wisconsin.

The ingredients for a reexamination of the desirability of using DDT were now coming together: strong scientific evidence of its persistence and damage, a substantial group of involved scientists, a small but determined citizens' group willing to raise the financial support necessary to wage a court battle with the agricultural and chemical industry, and a lawyer willing to immerse himself in the complex scientific and legal questions. By 1968, however, it was apparent that the judicial forum needed for questions as difficult as DDT contamination of the environment would have to be chosen carefully. Local and federal courts ordinarily will not take jurisdiction of an essentially scientific problem. In Michigan the courts ruled that the decisions about pesticide use made by specialists in the agency with assigned responsibility, the State Department of Agriculture, are not subject to judicial review, because they fall within the discretionary authority of that agency; in effect, this means that the Department cannot be prosecuted. Another forum, the legislative hearing, is unacceptable for two reasons: (1) the legislators or their representatives are not in a position to judge complex scientific testimony and (2) there is no established right of cross-examination by which to test the competence and relevance of evidence offered for or against an issue.

That the hearings on DDT took place in Wisconsin is primarily the result of some unique features of Wisconsin's administrative law. The most important of these features allows a citizens' group of at least six to petition a state agency for a declaratory ruling on whether a law that the agency administers applies to a particular question. The water quality legislation in Wisconsin gives authority to the Department of Natural Resources to control pollutants introduced into Wisconsin's lakes and streams. A pollutant is defined as any material that may be damaging to fish, game, or wildlife in state waters. The basis for the Wisconsin hearing, therefore, was a petition to the Department of Natural Resources alleging that DDT was producing a deleterious effect on the fish, game, and wildlife of Wisconsin waters, and that DDT should therefore be declared a pollutant under Wisconsin law.

The Wisconsin administrative law, furthermore, provides for a quasi-judicial hearing in petitions for a declaratory ruling. The rules of evidence, requiring relevance, materiality, and competence of the testimony, must be observed as in a court of law, and all testimony must be subject to complete cross-examination. The role of judge is filled by the hearing examiner, a man who is expected to be as competent in analysis of complex scientific testimony as he is in questions of law.

On October 28, 1968, the CNRA of Wisconsin (and three days later the Wisconsin Division, Isaak Walton League of America) filed a petition with the Wisconsin Department of Natural Resources. Paraphrased from legal form, the petition alleged as follows:

1. That DDT is an inherently dangerous, highly toxic, broad spectrum, persistent chemical biocide capable of cycling throughout the biosphere and increasing in ecological effect in the process of biological concentration;

2. That the use of an inherently dangerous, persistent chemical biocide such as DDT will result in the direct mortality of many nontarget organisms and reduction of biological diversity throughout the ecosystem within which it is applied;

3. That the people of the State of Wisconsin are entitled to the full benefit, use, and enjoyment of the national natural resource that is the Wisconsin Regional Ecosystem without diminution from the application of the chemical biocide DDT, and that the application of DDT will cause *serious, permanent,* and *irreparable* damage to this Wisconsin resource;

4. That the Wisconsin Regional Ecosystem is influenced by a number of environmental factors including the occurrence of DDT. The application of DDT to any part of the Wisconsin ecosystem will diminish the economic, recreational, educational, social and cultural value of the system and deprive the people of the State of the benefit, use, and enjoyment of this regional ecosystem; and

5. That as a result of said DDT application for the treatment of Dutch elm disease and other uses, the waters of the State of Wisconsin have been polluted, contaminated, rendered unclean and impure and further made injurious to public health, harmful for commercial or recreational use, and deleterious to fish, bird, animal and plant life.

On the basis of these statements, which the CNRA and EDF had to be prepared to prove in court, the petition went on to request a hearing under Wisconsin law. It asked the Wisconsin Department of Natural Resources to declare the use of DDT restricted to such uses wherein the DDT applied cannot enter the ecological cycles of the biosphere. They asked further that the department declare that DDT causes pollution of the lakes and rivers of the state and violates the provisions of the Wisconsin water quality statutes; and further, that the department order all persons, commercial businesses, municipalities, and governmental agencies to cease and desist from any further use of DDT in the state of Wisconsin.

On the receipt of the petition the Department of Natural Resources could have taken the position that there was no cause to hold the hearing. They did not, and scheduled the hearing to begin December 2, 1968, in the

Legislative Assembly chamber of the State Capitol in Madison before Chief Examiner Maurice H. Van Sustern.

Strategy of the Petitioners

The agreement between the CNRA of Wisconsin and EDF of New York was that EDF scientists and lawyers would prepare and present the technical evidence at the hearing on behalf of the CNRA. Thus began the next major milestone after *Silent Spring:* the attempt to prove in a court proceeding what Rachel Carson had asserted in her book. The damage alleged in the petition would have to be proved with substantial, uncontroverted evidence that would stand up to intensive, on-the-spot cross-examination. Furthermore, proof of the allegations would require evidence, not only of the laboratory effects of DDT, but of the actual mechanisms by which DDT moves from the site of application to the fish and game in remote waters of the state. To provide this, the entire environmental system to which DDT was being added would have to be described; then the chemical characteristics of DDT and its primary breakdown product, DDE, would have to be documented; and finally the response of the system *as a whole* to these chemicals would have to be analyzed. The following sections describe the evidence introduced on these three closely interwoven aspects of the DDT story.

Movement of Materials through an Ecosystem

The first steps by EDF were to describe the Wisconsin Regional Ecosystem, its internal relationships, and its interconnections with the surrounding states and the Great Lakes. Many of the most important studies on DDT had been done in other states, and parallels between other ecosystems and that of Wisconsin had to be established if these data were to be accepted in the hearing as relevant.

As an environmental scientist, I was among the first to testify. I introduced studies of the transport and redistribution of DDT which showed that water passing into the atmosphere could be a potential carrier of DDT. I also explained that moisture in the soil moves into the atmosphere continuously, from both soil and plants, and that contaminants may be transported with this moisture to downwind areas where they are precipitated back into surface waters and streams. I cited studies of the movement of materials such as tree pollen, gaseous industrial wastes, and even organisms as large as insects, to indicate the potential of the atmospheric transport mechanisms to carry a material such as DDT.

Also early in the hearing, Professor H. H. Iltis, a botanist at the University of Wisconsin, explained some of the ways in which plants and animals are interdependent. There are very delicate relationships between a given plant species and the specific insect which is required to complete pollination in the plant life cycle, and in many cases these insects have been affected by DDT. However, he concluded, it is not surprising that we

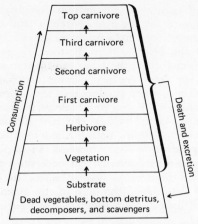

FIGURE 1 Diagrammatic illustration of the flow of nutrients in a food chain.

have not yet observed any decrease in the abundance of the corresponding plants, because most of these plants live a relatively long time, whereas DDT has been in widespread use for only about 30 years.

A third aspect of the regional environment, the food chain (trophic structure is the term used by ecologists), was documented through testimony and scientific papers introduced by several witnesses. The trophic structure in an ecosystem is illustrated in Figure 1. Each layer in the food pyramid except the substrate is called a trophic level. Every trophic level is characterized by an instantaneous weight of living organisms (biomass) and by the total biomass produced in a year. Of the biomass produced in one level 10 to 20 percent will be eaten by the organisms of one of the higher levels. At each level from 0 to 30 percent of the biomass that is eaten is retained by the organisms, depending on age and a great many other factors. Furthermore, at each level the biomass is roughly one-tenth as large as at the level below. (See Comment: *A look at the links of the food chain.*) It was shown that DDT is present in organisms at every trophic level, being highest in the upper levels. As plants or animals die, the DDT they contain is returned to the substrate, where it reenters the system.

Comment: A look at the links of the food chain

Three relationships between adjacent trophic levels are stated in the text: (1) the ratio of the biomass at one level to the biomass at the next lower level, (2) the fraction of the biomass production at one level which is eaten each year by organisms at a higher level, and (3) the fraction of the biomass eaten which is retained in the organisms. These relations are always mutually compatible, and they lead to further relations that are fun to explore.

To gain some familiarity with these ideas let us work out a simple example: a two-level world consisting of cows and grass in a pasture. We will assume that only cows eat the grass and that grass is all that cows eat. For definiteness, let the field contain a total biomass of 200 tons of grass and assume that the field produces 400 tons of grass in a year. Assume also that the field contains 20 tons of cows (10 percent of the biomass of the grass), that the cows eat 20 percent of the grass production, and that the cows retain 5 percent of what they eat. The cows, then, will eat 80 tons of grass and retain 4 tons of grass each year.

These 4 tons are the new cow biomass produced in the field and, in the steady state, exactly replace the cow biomass lost through cows which die. But now we know both the cow biomass in the field (20 tons) and the rate of production of cow biomass (4 tons per year).

The reader can easily verify that once the ratio of grass biomass to the rate of grass production was specified, the three relations given above uniquely determined the ratio of cow biomass to the rate of cow production. The first ratio, in our example, was picked to be

$$\frac{200 \text{ tons}}{400 \text{ tons}/\text{year}} = \tfrac{1}{2} \text{ year}$$

and the second turned out to be

$$\frac{20 \text{ tons}}{4 \text{ tons}/\text{year}} = 5 \text{ years}$$

What is the significance of the ratio of the biomass, M, to the rate of biomass production, P, in a population? Let us define this ratio to be T,

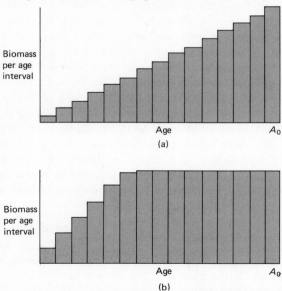

FIGURE A. A representation of two possible ways in which ages of organisms of a species could be distributed. The height of each bar is proportional to the fraction of the biomass of the total population that is accounted for by organisms in the age bracket spanned by the width of the bar.

so that $M = PT$. Clearly T is the time interval during which the new growth of biomass equals the total standing biomass M. To interpret T we have to know something about the mass versus age dependence of the population. Figure A shows typical distributions for two different populations. The total biomass within a single age interval is represented in the figure by the area of one of the thin vertical strips. The total biomass, M, is therefore the area under the entire graph.

We assume the population is in a steady state, by which we mean that the total mass in each age interval does not change with time. We further assume, for simplicity, that all individuals in the population die at the same age, A_0. In the steady state, the annual growth of biomass, $P \times$ (one year), must equal the biomass which dies annually.

The reader can show by simple geometry that for mass distribution (a), T is equal to $A_0/2$, and for mass distribution (b), T is approximately equal to A_0.

The two graphs in the figure were chosen for a purpose. The first describes in a crude way the mass versus age dependence of many plant species. The second describes many species of birds and mammals which grow rapidly until, at an age relatively young compared to the lifetime, the growth tapers off. For the example worked out above, we find that the lifetime of a blade of grass is

$$2 \times \frac{(200 \text{ tons})}{(400 \text{ tons}/\text{year})} = 1 \text{ year}$$

and the lifetime of a cow is approximately

$$\frac{20 \text{ tons}}{4 \text{ tons}/\text{year}} = 5 \text{ years}$$

From this simple example, we can also see how the magnification of DDT up the food chain occurs. Assuming that the grass contains a concentration of d parts per million of DDT, and that cows retain all the DDT they eat, the cows will have a concentration of DDT of $20d$ parts per million, because (1) everything the cows are was once grass and (2) they only retain one-twentieth of the *grass* they eat. (Of course, some DDT is lost at each trophic level by excretion, and this complicates the picture. Nevertheless, this is the essential mechanism of biological concentration.)

J.H. and R.H.S.

Chemical Properties of DDT

Against this background Dr. Charles Wurster, a biochemist and biologist, presented the physical, chemical, and biological properties of DDT. The behavior of DDT in the environment, he argued, was a predictable function of four key physical and chemical properties: (1) chemical stability, leading to persistence in the environment; (2) mobility in various environmental transport systems; (3) biological activity beyond pest species; and (4) low solubility in water combined with great solubility in lipids.

Wurster testified for only one hour on direct examination and was cross-examined for almost three days.

DDT was first discovered in Switzerland in 1870, but its insecticidal properties were not recognized until 1939. Although DDT itself is very stable, it can be degraded in sunlight, or by metabolism in certain organisms, to the closely related chemicals DDD and DDE, sometimes called metabolites. Other isomers and other metabolic breakdown products occur, but DDE is the most abundant and exhibits chemical and biological properties most similar to that of DDT. It is thought to be somewhat less toxic. The degradation of DDE is not as well understood.°

Wurster illustrated the persistence of DDT in the environment by citing a study in which 39 percent of the DDT applied in a field in Maryland was present 17 years later.[1] A half-life for decomposition of about 10 years would be implied if all the DDT that disappeared had decomposed, but there was no measurement of how much DDT was degraded compared with how much was transported elsewhere, so that the half-life could actually be much longer. It was conceded that the half-life of DDT can be much shorter in DDT-resistant insects such as cockroaches, in certain soil organisms, and in full sunlight, but it was noted that these conditions apply to only a small part of the DDT in the biosphere.

Wurster testified further on the great mobility of DDT and its metabolites, and he noted that, with time, it can leave the site of application and be transported by air and water over great distances. The tendency of DDT to form suspensions in water or to be adsorbed on particulate matter in water greatly increases the capacity of water to transport DDT, since DDT has a very low solubility in water (1.2 parts per billion).

Three mechanisms allow DDT to move into the atmosphere and be transported around the world. First, many application procedures result in the formation of tiny droplets that travel great distances from the application site as a suspension in the air. Second, DDT residues adsorbed on soil or plant particles also may be picked up by the wind and carried in the atmosphere. High concentrations have been found on dust far out over the oceans. Finally, DDT passes into the air by codistillation with water as it evaporates.

The third property, the wide range of organisms affected by DDT, particularly nontarget organisms, was documented by other witnesses who explained how epidemic outbreaks of an insect pest result from use of DDT, through its effect on the natural enemies of that insect pest. This is an upset of the population relationships among species in whole ecosystems and will be treated more fully below.

The fourth key property of DDT is its low solubility in water combined with its high solubility in lipids (fats). The solubility of DDT in

° The structural formulas for DDT, DDE, and DDD, along with an introduction to the nomenclature of organic chemistry, are presented in a Postscript to this essay. [J.H. and R.H.S.]

[1] R. G. Nash and E. A. Woolson, "Persistence of Chlorinated Hydrocarbon Insecticides in Soils," *Science*, vol. 157, pp. 924–926 (1967).

water is 1.2 parts per billion, but it is more than a million times more soluble in the lipids of plants and animals. Living organisms therefore "scrub" DDT from their environment by providing an almost unlimited sink for dissolving it. In addition, DDT and DDE remain very stable as they pass from one organism to another, carried in the flow of nutrient through a food web. Only a limited portion of food ingested by these organisms is excreted, and therefore relatively little of the DDT is excreted. Most of the consumed nutrient is burned, but the DDT—which is mainly in the fat of the consumed organism—is retained, and accumulates.

Dr. Joseph J. Hickey, a wildlife ecologist from the University of Wisconsin, provided evidence of higher and higher concentrations of DDE in the food pyramid of a lake ecosystem. He testified that DDE averaged 3 parts per billion in the bottom sediments of Lake Michigan at depths of 33 to 96 feet. In the small invertebrate animals (*Pontoporeia*) living in the mud, DDE averaged a hundredfold increase, 200 to 300 parts per billion. In the fish that normally feed on these invertebrates, DDE ranged from 1.8 parts per million in alewives through 2.3 parts per million in whitefish muscle, a further tenfold increase. The muscle tissues of adult oldsquaw ducks that fed on these invertebrates contained 4.8 parts per million of DDE, and in the muscle of adult herring gulls that fed on the fish, DDE averaged 80 parts per million. The result of these unique solubility characteristics therefore is magnification of DDE concentrations by more than 10,000 times in the food chain of aquatic ecosystems.

Effects on the Physiology of Whole Organisms

Several different physiological effects of DDT were documented in further testimony. It was noted that chlorinated hydrocarbons, such as DDT, can affect the enzyme balance in an organism. DDT induces the synthesis of certain liver enzymes in several species of test animals, and these induced enzymes have a broad activity that includes hydroxylation of the steroid sex hormones testosterone, progesterone, and estrogen. The significance of the estrogen breakdown by DDT-induced enzymes lies in the fact that estrogens affect calcium metabolism in birds. High estrogen hydroxylation could result in less calcium for the formation of eggshells. Museum collections of birds' eggshells gathered through past decades provided a surprising source of information on calcium formation. The eggshells of several species of birds studied by Ratcliffe were found to be of constant thickness from the early 1900s up until the late 1940s, when they became greatly reduced in thickness in both Europe and North America.[2] The precipitous drop in the ratio of weight to size, an index of eggshell thickness, is shown for peregrine falcon eggs in Figure 2.

Direct experimental evidence linking DDE to the observed decrease in eggshell thickness was described by Lucille F. Stickel, Bureau of Sport Fisheries and Wildlife, U.S. Department of the Interior. Eggs from mallards receiving 3 ppm of DDE, when compared with controls, had shells

[2] D. A. Ratcliffe, "Decrease in Eggshell Weight in Certain Birds of Prey," *Nature*, Vol. 215, pp. 208–210 (1967).

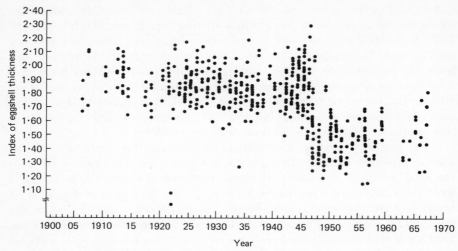

FIGURE 2 Change in the index of eggshell thickness (ratio of weight to size) for peregrine falcon eggs in Great Britain between 1900 and 1968. [D. A. Ratcliffe, "Decreases in Eggshell Weight in Certain Birds of Prey" *Nature*, Vol. 215, pp. 208–210 (1967)]

that were 13.5 percent thinner. They cracked or were broken six times as often, and produced less than 50 percent as many healthy ducklings. She testified that DDT also gave results that were comparable to those of DDE.

A further sublethal physiological effect produced by DDT was discussed in testimony by Dr. Alan Steinbach, a neurophysiologist. He described DDT as a neurotoxin, exerting a direct effect on the movements of ions across the nerve membrane. DDT causes the channel for sodium ion transfer through the nerve, which ordinarily closes after a nerve firing, to remain open for a longer time than is normal. Prolonging the active state of the nerve may permit one nerve impulse to follow another without further stimulus.[3] One probable effect, according to Steinbach's testimony, is the complete failure of the transmission, and another is uncontrolled repetitive firing. The fluttering of robins dying from DDT might be accounted for on this basis.

The Response of Entire Populations

After testimony on the chemical characteristics of DDT and DDE, and the molecular and physiological effects these chemicals induce in living organisms, the strategy of the petitioners was to show that whole populations of fish and bird species respond directly to DDT or its metabolites.

[3] Toshio Narahashi and Hans G. Haas, "DDT: Interaction with Nerve Membrane Conductance Changes;" *Science*, Vol. 157, pp. 1438–1440 (1967); Toshio Narahashi and Hans G. Haas. "Interaction of DDT with the Components of Lobster Nerve Membrane Conductance," *The Journal of General Physiology*, Vol. 51, pp. 177–198 (1968); Bertil Hille, "Pharmacological Modifications of the Sodium Channels of Frog Nerve," *Journal of General Physiology*, Vol. 51, pp. 199–219 (1968).

The DDT concentrations reported in these organisms were high enough to account for the observed disturbances of the enzyme and nerve systems; now it was necessary to explain how these sublethal disturbances in the physiology of native species could affect the populations of these species far from the site of DDT application.

Dr. Kenneth J. Macek, Bureau of Sport Fisheries of the U.S. Department of the Interior, presented evidence that DDT reduces reproduction in fish by inducing mortality of fry shortly after they hatch. Macek fed DDT to groups of brook trout and then studied their progeny. He found that mortality was higher if either parent had been fed DDT.

Dr. Robert Risebrough, a zoologist at the University of California, went on to explain that contaminated species include commercially important resources of the Pacific Ocean such as anchovies and tuna. Since the level of contamination in these populations approaches that at which fry were killed in freshwater systems, Risebrough concluded that important fisheries may be threatened.

Probably the most significant population effect identified as attributable to DDT and its metabolite DDE is the catastrophic decline of certain carnivorous birds (specifically the raptors) in both Europe and North America during the past 20 years. The peregrine falcon had experienced slow, long-term population fluctuations in the past, but in England the population varied only within 8 percent of its mean in the present century. Because of public interest in falconry, almost all nesting sites were known over centuries, and the nesting populations counted annually. Dr. Hickey testified that the decline in peregrine populations that began around 1950 was international. In Finland the nesting population has now declined approximately 95 percent; in West Germany, 77 percent; in Great Britain, about 60 percent. During this same period, the nesting population of peregrines in the eastern United States declined 100 percent and in the western United States 80 or 90 percent.

Dr. Hickey went on to testify that although DDT residues were suspected as a cause of the population decline, an explanation of just what was happening to produce the decline had been lacking. A breakthrough occurred in 1967 when Ratcliffe's study (recall Figure 2) showed that an extraordinary change had occurred in the weight and thickness of the shells of eggs of several top carnivore birds. It appeared that DDT-induced enzymes were upsetting the calcium metabolism, so that these species were laying thin-shelled eggs. Several additional studies corroborating this view, as well as photographs such as that of the bald eagle nest (see page 101) were introduced as evidence. Thus for the first time there was conclusive proof of the potential for DDT to bring about extinction of a species without ever spraying one individual directly.

Integrated Effects on the Regional Ecosystem as a Whole

The testimony which explained the movement and impact of DDT, as we have seen, ranged from transport mechanisms, through the chemical properties, to physiological effects on individual organisms, and finally to

Left: Children in a Guatemalan village are sprayed with DDT by a United Nations mobile field unit in order to control diseases such as malaria and typhus which are carried by insects. The insects die rapidly, but the children show no ill effects; this selective toxicity has given DDT a favored place in the public health arsenal. *Below:* The elm trees in this Michigan town are sprayed with DDT in an attempt to combat Dutch elm disease. The effectiveness of such spraying, carried out in many towns in the United States, is a subject of contention. (UNATIONS; Michigan Department of Natural Resources)

No chick will hatch from the crumbling bald eagle egg in the foreground. There is strong evidence that DDT was the cause of this and many other similar examples of breeding failure. The fish in the background, partially eaten by the adult eagle, is one step in the food chain which reaches up to the carnivorous eagle. The concentration of DDT in the eagle is higher than that in the fish, and the concentration in the fish is higher than in the food supply of the fish. Breeding failure among the birds of prey has resulted in sharp declines in their population in recent years. Knowledge of these declines is obtained from the field records of birdwatchers throughout the world. (Michigan Department of Natural Resources)

the upset of whole populations of certain species. In order to make the argument more complete we presented an assessment of the effects of DDT on the entire Wisconsin ecosystem. Such an assessment required what is known as a systems analysis: a mathematical model of a complex system emphasizing the interrelationships among the component parts of the system. To prepare this study, we required skills in systems engineering, mathematics, and statistics to integrate the chemical and physiological knowledge about the effects of DDT on whole populations of native species. The research studies needed to prepare this preliminary study of DDT flows were carried on while the hearing progressed.

The study incorporated the information about the exchanges of DDT through the Wisconsin regional ecosystem, which we have already described, as well as information about the ways in which DDT can be stored in the ecosystem. We employed a series of differential equations relating inputs, losses, and storages at a given trophic level to (1) the concentrations of pesticide in the next lower trophic level (see Figure 1); (2) the life-span of the longest-lived organisms in that trophic level; and (3) the mass of living organisms in that level.[4] Our equations incorporated the fact that the concentrations of DDT and DDE will be higher, the higher the trophic level, the more long-lived the species, and the smaller the biomasss in the trophic level. The results were consistent with data on DDT concentrations in fish in Lake Michigan, which, as mentioned above, had revealed a magnification of pesticide concentrations of over 10,000 times during the transport of DDT from the lower to the upper trophic levels of the ecosystem.

The inescapable conclusion of this study was that the known carrier and transformation mechanisms, the large storage terms in several parts of the system, and the reported low efficiency of the degrading mechanisms, all conspire to make for a rate of breakdown of DDT which is very slow compared with the annual worldwide additions during the 1950s and 1960s. As a result DDT will continue to accumulate in the fatty tissue of organisms living in the system.

The analysis of the flow of DDT upward from one trophic level to the next, and its return (by decomposition of the host) into the organisms at the base of the food chain, revealed that the average time required for the concentration of DDT to reach equilibrium in the top level of the system is roughly equal to the life span of the longest-lived species in the chain. Because DDT use continued on a large scale in the United States until the end of the 1960s, and is continuing around the world even now, this means that unless breakdown mechanisms become more efficient at higher DDT concentrations, the DDT and DDE concentrations in the world biosphere system will continue to increase. In particular, concentrations at the upper trophic levels and in the long-lived species can be expected to increase for

[4] H. L. Harrison, O. L. Loucks, J. W. Mitchell, D. F. Parkhurst, C. R. Tracy, D. G. Watts, and V. J. Yannacone, Jr., "Systems Studies of DDT Transports," *Science*, Vol. 170, pp. 503–508 (1970).

some years to come. An evaluation of the total impact of DDT on the system cannot be complete until a full life-span *after* the *last* introduction of DDT into the world transport system.

A final conclusion drawn from the systems analysis was that the decline in the predator populations would have a major degrading effect on the stability and quality of the ecosystem as a whole. This effect would occur because predators exert a natural control on their prey. Models of the expected increase in prey resulting from the removal of a predator have been known for years, and formed the background to this part of our case. Epidemic outbreaks of pest insects, for example, have occurred when important insect species that prey on these pests have been killed by DDT. Witnesses pointed out that an alternative method of insect control, biological control, may be more effective and offers much less risk of upsetting the whole system. An example of a method of biological control would be the introduction into agricultural areas of large numbers of sterile members of the pest species. The systems analysis showed that even at the higher trophic levels small changes in the effectiveness of a predator species in removing prey can deprive the system of essential natural control mechanisms. It was shown that the removal or decrease in the numbers of predators in the system by the presence of DDT and DDE can lead to exploding populations of the prey species. This imbalance is an indirect result of the persistent pesticide, but the result, we claimed, would have to be acknowledged as a significant degradation of the regional ecosystem.

Strategy of the Agro-Chemical Task Force

The testimony on behalf of the continued use of DDT, presented by a special task force of the National Agricultural Chemicals Association, began by crediting DDT with the worldwide control of malaria and other diseases in the post-World War II period. A widespread ban on DDT in the United States, it was claimed, would have a disastrous effect on its use in the rest of the world, and epidemics would ensue.

Members of the Task Force testified that there was no conclusive evidence that DDT is at all unsafe for man. Dr. Wayland J. Hayes, Jr., cited studies he supervised as chief of the toxicology section of the U.S. Public Health Services.[5] Convict volunteers who were fed 200 times the amount of DDT then ingested by the average person had DDT concentrations in their body fat which reached 234 to 281 parts per million after 21 months of feeding. Levels for the general population ranged from 2.3 to 4 ppm at that time. Factory workers in DDT manufacturing plants were also studied. They ingested 17 to 18 milligrams per day, some of them for as long as 19 years, compared with the average amount ingested per person in the population as a whole, 0.028 milligram per day. Yet, DDT was found to

[5] W. J. Hayes, Jr., W. F. Durham, and C. Cueto, Jr., "Effect of Known Repeated Oral Doses of Chlorophenothane (DDT) in Man." *J. Amer. Med. Assoc.*, Vol. 162, p. 890 (1956).

have no apparent effect on the health or work performance of either the convicts or the factory workers.

To counter the testimony by Dr. Hayes that DDT is safe, a Swedish toxicologist, S. Goran Lofroth, was brought from Stockholm and introduced as a witness of the Wisconsin State Attorney General's Office. Lofroth testified that studies of DDT toxicity in humans should have included observations on women and children, rather than male convicts and laborers. He stated that many breast-fed babies around the world are getting every day 0.02 milligram of DDT per kilogram of body weight—twice the maximum daily dose recommended by the World Health Organization. This rate of intake is in the range where laboratory animals show physiological changes, and one cannot predict the consequences if these and similar changes occur in man. Adults in the United States take in, as remarked above, an average of 0.028 milligram per day (roughly fifty times less DDT per kilogram of body weight than breast-fed babies) and today have an average of roughly 12 parts per million DDT in their body fat, with great variation from one part of the country to another. What the future holds for persons exposed to these levels of DDT, over a 60- to 70-year life span, is still unknown.

Further, the Agro-Chemical Task Force pointed out that in gas chromatography tests for pesticide residues, PCB (polychlorinated biphenyl), a plasticizing agent that is common in the environment in industrial areas, may be identified as DDT or DDD. A recheck of coho salmon samples, one witness claimed, indicated that 70 percent of the DDT-DDD residue might have been PCB. However, PCB and DDE are unlikely to be confused, and the argument of the Task Force was weakened when it became clear that the data introduced by the Environmental Defense Fund had been for concentrations of DDE only.

To challenge the earlier testimony that DDE has caused drastic reproductive failure among birds of prey by disrupting their calcium metabolism, the Task Force called Dr. Frank Cherms, a University of Wisconsin poultry specialist. Cherms reported that he had fed DDT to five generations of Japanese quail in hopes of developing a resistant strain, but found no sign of either thin eggshells or of changes in their reproduction. He testified that eggshell thickness could be affected by a host of factors— heredity, nutrition, environmental stress (fright and excessive heat), and disease. However, he acknowledged that there were great differences in calcium metabolism between carnivorous birds like the peregrine falcon and grain-eating birds like the quail.

Another member of the Task Force, Dr. Bailey Pepper, an entomologist from Rutgers University, testified that DDT continues to be needed for mosquito control in New Jersey, where mosquito-borne encephalitis persists. In addition, he pointed out that the cotton bollworm and several other insects were extremely hard to control without resorting to DDT, and he argued that the new insect control methods have not proved dependable against some crop pests.

Testimony on the economic cost of banning DDT came from Dr. Keith Chapman, a University of Wisconsin entomologist. He predicted that Wisconsin farmers would lose $2 to $3 million if deprived of DDT. Hardest hit would be the carrot growers, who reap a $2.3 million crop in Wisconsin.

The Outcome

Even as the hearings closed in May 1969, it was recognized that months would pass before a formal decision could be announced by the hearing examiner or the Secretary of the Department of Natural Resources. A transcript of over 2800 pages would have to be prepared and distributed to each party, then summary briefs were to be submitted by each counsel, and finally the examiner would have to complete a painstaking analysis of all testimony. It was June 1970 before the examiner's ruling was released, supporting virtually all of the petitioner's arguments. But over the intervening months, the national publicity that resulted from the testimony at the hearings did much to set in motion events that are now bringing the DDT story to a close.

During the early months of 1969, while the hearings were in recess, legislation to ban the sale of DDT was introduced in several states, including Wisconsin. Large quantities of canned coho salmon from Lake Michigan waters were seized by the Food and Drug Administration because of dangerous levels of DDT contamination. The FDA began an extensive procedure for establishing acceptable tolerance of DDT and its metabolites in fish marketed for human use. Eventually a standard of 5 ppm was established despite the opposition of states whose commercial fishing was threatened.

In July 1969 the U.S. Department of Agriculture ordered an immediate 30-day suspension of all pest control programs supported by the department, at both state and federal levels, in which any of the nine persistent chlorinated-hydrocarbon pesticides were being used. The suspension provided for immediate review of the programs and of alternative control methods. In November Secretary of Agriculture Clifford Hardin announced that in 30 days he would cancel the registration of DDT for use on pests in shade trees, gardens, aquatic areas, and tobacco fields. The department also filed a notice of intent to cancel all other DDT uses except those needed to control diseases for which no DDT substitute is available.

Early in November 1969 Prime Minister Trudeau announced that Canada would eliminate 90 percent of the DDT use in that country during 1970. One week later, U.S. Health, Education, and Welfare Secretary Robert Finch announced endorsement of a report recommending that major use of DDT be brought to an end in the United States within two years. Finch could hardly take stronger action, because authority to curb most uses of DDT rests with the Secretary of Agriculture.

In May 1970 Chief Examiner Maurice Van Sustern of the Wisconsin

Department of Natural Resources released his findings from the Wisconsin hearings. Van Sustern summarized the substantial evidence in a 25-page report and issued a 1-page ruling which states in part:

> DDT, including one or more of its metabolites in any concentration or in combination with other chemicals at any level, within any tolerances, or in any amounts, is harmful to humans and found to be of public health significance. . . . DDT and its analogs are therefore environmental pollutants within the definitions of Sections 144.01 (11) and 144.30 (9), Wisconsin Statutes, by contaminating and rendering unclean and impure the air, land and waters of the state and making the same injurious to public health and deleterious to fish, bird and animal life.

By mid-1970 the legislation to outlaw most uses of DDT had been passed in a few states and for the first time stood a chance of passage in the United States Congress. Conservation groups, following the initiative of the Environmental Defense Fund, continued to press for more rapid action, particularly at the national level. In June 1970 the U.S. Court of Appeals for the District of Columbia ordered the Secretary of Agriculture to suspend within 30 days the registration of DDT for interstate shipment, or to give reasons for not doing so in sufficient detail to permit prompt and effective review. The Department of Agriculture responded by presenting its case for continued limited use of DDT. On January 7, 1971, the U.S. Court of Appeals, District of Columbia, ordered the Environmental Protection Agency (that had acquired jurisdiction in the interim from the Department of Agriculture) to issue immediately notices of cancellation of all uses of DDT. Since DDT could continue to be used during an appeal of such cancellations, the Court also ordered EPA to determine whether DDT was "an imminent hazard to public health," in which case all interstate shipments of DDT would be suspended immediately pending the outcome of cancellation appeals.

More than any decision by government, however, the outcome of the hearings is represented by a greatly increased level of public understanding. The national demand for responsible action is far greater than when the DDT hearings closed, and greater still than a year earlier when the hearings were the only hope of a few determined people in New York and Wisconsin.

Implications for Scientists

A significant aspect of these court proceedings was the unusual number of scientific disciplines required to tell the story in adequate detail. The complete transcript is over 2800 pages, and only the most significant highlights have been touched on here. The supporting sciences in the DDT hearings included population biology, ornithology, several types of chemistry, molecular biology, biochemical pharmacology, statistics, and systems engineering.

The DDT story could not have been told without all of these disciplines. Giving the detailed record of other contaminants in the environment will require equally broad responses by laymen and by the scientific community. A breadth of specializations far beyond the interests and capability of any single individual must be assembled if we are to make satisfactory estimates of long-range effects.

I have reviewed the DDT hearing, however, not only because of what was achieved there, but also because it is representative of the new era in science which many scientists and students have been looking forward to. The nature of our government and university structure, and of research support procedures, has compelled us in the past to study problems that one person or one closely knit group could undertake. Scientists, including ecologists, have systematically skirted some of the most vital problems of our time because one laboratory simply could not undertake them. Even worse, the researcher in his laboratory has generally believed that to become involved in unrelated problems being studied in other laboratories would compromise his independence.

The DDT hearings herald the beginning of a significant change: the end of the era in which progress in science has meant to subdivide and specialize, and the beginning of a time when science and society will demand an unprecedented level of synthesis and insight. The hearings rekindled a genuine sense of community between disciplines and between scientists and laymen that will not be forgotten.

John Harte and Robert H. Socolow

We can describe an organic molecule at many levels of detail. The most elementary way is to use a set of letters, like DDT, which abbreviate the full name the organic chemist has assigned to the molecule. When the full name is written out, it is often very long, and the chemists have invented a pictorial way of representing the same expression, called a structural formula. Both the full name and the structural formula express the fact that a complicated organic molecule can be usefully thought of as a set of simple molecules linked together.

In the case of DDT the full name is *1-1-1-trichloro-2-2-bis(p-chlorophenyl)-ethane* ° and the structural formula is

The intent of this postscript is to penetrate a little of the nomenclature of organic chemistry which permits the initiated to draw the structural formula given the full name, or vice versa. We will use the formula for DDT as our first example.

The full name of DDT is built beginning with the word at the right of the formula, ethane, an extremely simple molecule whose structural formula is

Note that each carbon atom is attached by four "bonds" to other atoms.

The six hydrogens of ethane are numbered by organic chemists: the three attached to one carbon atom are given the number 1 and those attached to the other carbon atom are given the number 2. In ethane derivatives, other atoms or molecules are substituted for the hydrogen atoms, and the nomenclature tells which hydrogens have been displaced.

° The abbreviation DDT stands for *dichlorodiphenyltrichloro-ethane*, a name not sufficiently complex to specify the molecule fully, but nonetheless often used.

Thus 1-1-1-trichloroethane is the molecule whose structural formula is

$$
\begin{array}{c}
Cl \\
| \\
Cl - C - Cl \\
| \\
H - C - H \\
| \\
H
\end{array}
$$

The word-fragment *bis* means that what follows is "taken twice" and thus the class of organic molecules 1-1-1-trichloro-2-2-bis(X)-ethane, where X is still to be made explicit, has the structural formula

$$
\begin{array}{c}
Cl \\
| \\
Cl - C - Cl \\
| \\
X - C - X \\
| \\
H
\end{array}
$$

We see that five of the six hydrogens of ethane have been replaced, the three "1-hydrogens" by chlorine atoms and two of the three "2-hydrogens" by X.

In the full name for DDT we see (above) that X is p-chlorophenyl. Had X just been phenyl this would have meant that a benzene ring was attached. Benzene, the most basic "ring" molecule of organic chemistry, has six carbons and six hydrogens in a ring, with the following structural formula:

(benzene)

The two parallel straight lines between carbon atoms represent the "double bond," an unsaturated linkage of enormous importance in chemical reactions. To attach a benzene ring to another molecule, say, Y, one hydrogen must be substituted, obtaining which is called phenyl-Y. Ordinarily, one does not show explicitly the carbons and hydrogens in the benzene ring, instead drawing phenyl-Y as the following:

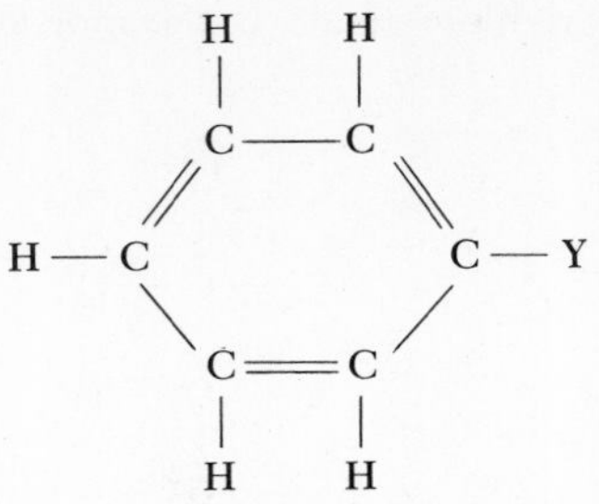

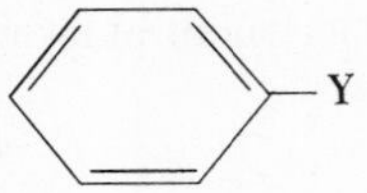

The final step is to learn how to describe the further modification of the benzene ring. Relative to the position of the atom or molecule labeled Y in phenyl-Y, three inequivalent hydrogen atoms can be replaced: one, two, and three carbons away. The nomenclature describing which H has been replaced can be discerned from the following examples:

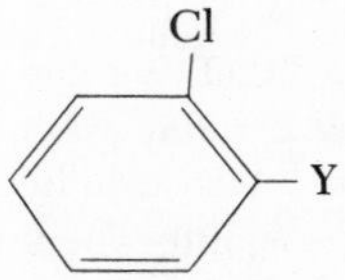

(o-chloro-phenyl-Y)

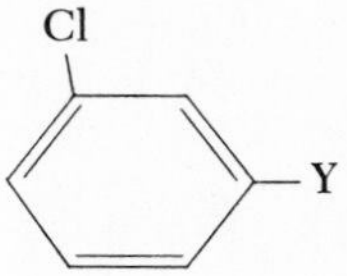

(m-chloro-phenyl-Y)

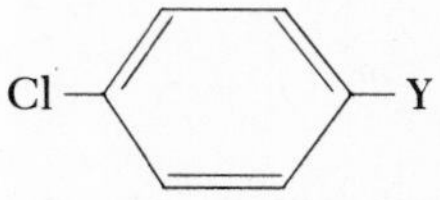

(p-chloro-phenyl-Y)

Here o, m, and p stand for ortho, meta, and para, respectively. The three molecules shown here are isomers of one another, that is, molecules with different structures but the same chemical composition.

The reader should now be able to put the whole formula for DDT back together. He will also find it straightforward to associate the full name for one of many isomers of DDD, 1-1-dichloro-2-2-bis(p-chlorophenyl)-ethane, with the following structural formula:

(DDD)

Finally, upon learning that *ethylene* has the structural formula

(ethylene)

the reader will be able to associate the full name for DDE, 1-1-dichloro-2-2-bis(p-chlorophenyl)-ethylene, with the structural formula

(DDE)

In Figure 1 of Dr. Galston's essay (on page 138), the reader will find additional structural formulas, including those for 2-4-D and 2-4-5-T, two toxic herbicides. In comparing these formulas with their full names (written beside them), the reader will verify what he surely suspected—that we have not given all the rules of naming organic molecules in the previous pages. There are a few more conventions, and the one by which the numbers 2,4,5 are arrived at is one of them. There are, in addition, *many* more names of basic building blocks, acetic acid among them.

The reader ought not to forget that even though the structural formulas convey much more information than a name like DDT does, they are still highly simplified representations of dynamic molecules. A higher level of description, in which the dynamic properties are made explicit, is now available in much of chemistry, thanks to the development of quantum mechanics.

8. NUCLEAR POWER ON CAYUGA LAKE

Alfred W. Eipper

Controversy over threats to Cayuga Lake, New York, from a proposed electric power plant has aroused national interest. This attests to the concern over the increasing seriousness and difficulty of allocating fixed water resources among demands that are proliferating, both in kind and in quantity. Escalating population and competition for resources are forcing us to a new awareness of concepts like sharing, compromise, and multiple use. The Cayuga Lake case is a prime example of the ecological problems involved in a nuclear power plant proposal, and the complex of scientific, social, and political questions closely associated with them. It demonstrates the variety of interests and attitudes that are involved in today's natural resource controversies, and some of the impediments to framing legislation that would promote equitable solutions. The complex decisions about uses of a natural resource must be public decisions in the truest political sense, and can no longer be made exclusively by any one special-interest group, be it a citizens' committee, a regulatory agency, or a "public" utility.

Demands on natural resources are increasing much faster than the population. In the case of water, per capita use in the United States, excluding transportation and recreation uses, appears to be doubling every 40 years.[1] This means that in the 60 years it is expected to take our United States population to double, *total* water use will increase sixfold. According to the U.S. Department of the Interior, under existing use patterns, the total amount of water needed just to sustain the present United States population for the remainder of their lives is greater than all the water that has been used by all people who have occupied the earth to date.[2] And although we tend to think of water supplies as fixed,

[1] U.S. Water Resources Council, *The Nation's Water Resources*, Washington, D.C.: U.S. Government Printing Office, 1968.

[2] U.S. Department of the Interior, *Quest for Quality*, U.S.D.I. Conservation Yearbook, Washington, D.C.: U.S. Government Printing Office, 1965, p. 10.

increasing amounts of water are being rendered unusable through various forms of pollution.

Steam electric stations pose some of the most crucial problems to future allocation of water resources in the United States. Electrical needs in this country are doubling every 10 years at present.[3] This is the result of population increase combined with rapid increases in per capita consumption of electricity. These needs will be met largely by construction of steam electric stations now that most sites suitable for hydroelectric generation have been utilized.

Generation of Electric Power

Steam electric stations operate by turning water into high-pressure steam to spin a turbine that drives an electric generator. In a fossil-fueled plant the combustion of coal, oil, or gas produces the heat that creates the steam. In a nuclear plant, the heat is derived from a self-sustaining nuclear fission chain reaction in the fuel (usually enriched uranium) of the reactor. In any steam electric plant the spent steam must be cooled immediately beyond the turbine to lower the back-pressure on it; the difference between the pressures on the two sides of the turbine is what makes the turbine spin. As the steam is cooled, it condenses to water, and the water is cycled back to the boiler (see Figure 1). The spent steam is cooled in a condenser by coming in contact with pipes containing cold water (coolant). During this process the coolant is heated 10° to 30° F, depending on plant design.

All steam electric plants require large amounts of coolant water for the condensers. The colder this water is, the more efficient the plant's oper-

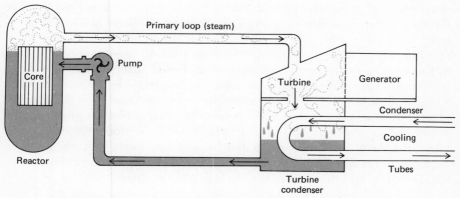

FIGURE 1 A schematic diagram of the steam and cooling systems in a boiling water reactor.

[3] Joint Committee on Atomic Energy, *Selected Materials on Environmental Effects of Producing Electric Power,* Joint Committee Print, 91st Congress, 1st Session, Washington, D.C.: U.S. Government Printing Office, 1969.

ation. There are basically three ways of obtaining cooling water for the plant: one is to take it from some natural body of water. The problem here is that returning the coolant to the water body some 20° F warmer than when it left can produce various kinds of ecological upsets in the receiving water. To avoid such problems the warmed water can be cooled again before returning it to the river, lake, or estuary from which it was taken, using one of the cooling devices described in the next paragraph. A second source of condenser cooling water is a cooling pond—a small "lake" impounded for the purpose. Cooling water is withdrawn from the impoundment, and discharged directly back into it without cooling. Here cooling is achieved largely by evaporation from the pond surface, and about 1½ surface acres are required for each megawatt of electricity generated. A third source of condenser cooling water is a "closed circuit" system in which the coolant, after leaving the condensers, passes through a cooling device, thence back to the condenser, and so on, in a cycle analagous to the cooling system of an automobile engine.

Cooling devices are usually towerlike structures in which the water is spread out and cooled by radiation or evaporation. In "dry" towers the water is dispersed in pipes that are cooled by fans. Dry towers are the most expensive to operate, but water loss is negligible. In evaporative cooling towers the water trickles down through series of baffles, and cools by evaporation. Natural draft evaporative towers rely on air convection produced by their lamp-chimney shape, and may be 200 to 400 feet high. They are expensive to build but relatively inexpensive to operate. In mechanical (induced) draft evaporative towers, air is forced up through the baffles by fans. These "towers" are typically very low and arranged in series. They are much less conspicuous and are cheaper to build than natural draft towers, but more expensive to operate. An exceptionally lucid and detailed description of steam electric stations and cooling methods is given by Tor Kolflat.[4] Zeller *et al.* have surveyed environmental effects of cooling facilities.[5]

Roughly two-thirds of a steam electric station's total energy output is in the form of waste heat. Because nuclear-fueled plants must be large to be economically efficient and because no appreciable amount of their waste heat is dissipated to the atmosphere (in contrast to a fossil-fueled plant), nuclear plants pose more serious thermal pollution problems for water bodies. Furthermore, nuclear plants produce more waste heat per kilowatt-hour of electricity generated because, for safety reasons, they are required to operate at lower steam pressures and temperatures than fossil-fueled plants, and this lowers the nuclear plant's efficiency.[6]

[4] *Thermal Pollution—1968*, Part I. Hearings before the Subcommittee of the Committee on Public Works, U.S. Senate, 90th Congress, 2nd Session, February 1968, Washington, D.C.: U.S. Government Printing Office, 1968.

[5] R. W. Zeller *et al.*, *A Survey of Thermal Power Plant Cooling Facilities*, Pollution Control Council, Pacific Northwest Area, Portland, Oregon, 1969.

[6] Joint Committee on Atomic Energy, *op. cit.*

Nuclear power plants have certain definite advantages. They do not pollute the atmosphere with particulate matter or oxides of sulfur and nitrogen, as do plants using fossil fuels such as natural gas, coal, and oil. Because they operate for months without refueling, they are especially useful in areas where fossil fuel would have to be transported great distances to the plant. They *may* also provide more economical production than fossil fuel plants. On the other hand, nuclear power plants routinely release small amounts of radionuclides to the water and to the atmosphere.

Although nuclear power accounted for less than 1 percent of the nation's total electrical output in 1967, it is expected to provide over half our total capacity by the year 2000, at which time the total cooling water requirements of the United States' steam electric industry are expected to approximate one-third of the country's entire yearly supply of runoff water.[7] °

Cayuga Lake

Cayuga and several similar Finger Lakes nearby are quite unusual among lakes of the United States. Exceptionally long, narrow, and deep, they were formed during the Pleistocene when glacial scouring straightened and deepened a preexisting group of adjacent valleys in soft shale. Cayuga is 38 miles long, has a mean width of 1.7 miles, and a surface area of about 66 square miles.[8] It has a maximum depth of 435 feet, a mean depth of 179 feet, and a volume of 331 billion cubic feet. The mean flushing time ("average" length of time required for a drop of water to move all the way through the lake) is 9 years or more.

During summer months the lake is thermally stratified: the upper layer (epilimnion) and lower layer (hypolimnion) differ so much in temperature—and hence in density—that they do not mix, even under the influence of strong winds. During the period of stratification (May to November) temperatures in the epilimnion range from 50° to 73° F; temperatures in the hypolimnion range from 40° to 43° F, just above the temperature at which water reaches its maximum density (about 39° F). The epilimnion layer becomes thicker as summer progresses, usually extending to depths of 35 to 50 feet below the surface. In autumn this layer begins to cool and consequently to increase in density. This process continues until late October or November, when density of the water in the epilimnion has increased to a point where it is near enough that of the hypolimnion so

---

[7] J. F. Hogerton, "The Arrival of Nuclear Power," *Scientific American,* Vol. 218, No. 2, pp. 21–31 (1968); J. E. McKee, "The Impact of Nuclear Power on Air and Water Resources," *Engineering and Science,* Vol. 31, No. 9, pp. 19–22, 31–32 (1968); J. R. Clark, "Thermal Pollution and Aquatic Life," *Scientific American,* Vol. 220, No. 3, pp. 18–27 (1969).

[8] The following description of Cayuga Lake is slightly modified from A. W. Eipper *et al., Thermal Pollution of Cayuga Lake by a Proposed Power Plant,* Ithaca, N.Y.: Authors, Fernow Hall, Cornell University, 1968.

° The assumptions underlying this estimate are developed in part I of the essay on Water in this book. [J.H. and R.H.S.]

Above: Early morning fishing on Cayuga Lake. *Right:* This lush growth at the north end of Cayuga Lake is the result of eutrophication. Were eutrophication to accelerate, the entire shoreline of the lake could become this choked with plants and its recreational value would greatly diminish. (Elmer S. Phillips; Daniel Jackson)

that the next strong wind causes the two layers to mix. The lake then becomes essentially the same temperature throughout and remains so (generally without ice cover) until May, when warming and density reduction of the surface layer reach a point where stratification is reestablished.

During the period of stratification, biological production, particularly of single-celled algae, is largely confined to the epilimnion, where light is available in combination with nutrients and warmer temperatures. Dead plant cells and planktonic animals sink to the hypolimnion. While stratification persists, there is no source of additional oxygen for the hypolimnion. Animals confined to the hypolimnion—such as lake trout (a major resource in Cayuga)—use the oxygen there in respiration, and additional oxygen is consumed by bacterial decay of the dead plant and animal matter continually sifting down from above. Thus oxygen in the hypolimnion decreases throughout the summer until that time in the fall when the epilimnion cools to a point at which winds can mix the entire lake again, and rejuvenate oxygen supplies in the deeper water.

Most lakes gradually become more fertile with the passage of time by a process known as eutrophication. Nutrients, including nitrogen and phosphorus in various forms, are continually being added from the surrounding watershed. As the concentration of nutrients increases, so does biological production. Also, as a lake becomes greener from the proliferation of single-celled algae, it absorbs more solar radiation and reflects less. This creates higher temperatures, and added heat speeds biological production. Thus biological production tends to be self-accelerating.

Lakes vary in their rates of aging, depending on nutrient load, temperature, area, and depth. The activities of man, moreover, can speed eutrophication considerably. Lake Erie is one of the most recent and extreme examples of the acceleration of natural eutrophication with all its attendant problems. Greater use of the Cayuga Lake watershed for housing, industry, and agriculture will continue to increase the influx of nutrients and to advance eutrophication. Heating the water will hasten these changes and possibly produce some others.

The Utility Company Plan

In the summer of 1967 New York State Electric and Gas Corporation (NYSEG), a private utility with some half million customers in the south-central area of the state, announced plans to construct an 830-megawatt nuclear-fueled steam electric station next to the 290-megawatt coal-fired plant (Milliken Station) they had been operating on the lake for about 10 years. The new installation was to be named Bell Station. Clearing and excavation of the site, 12 miles northwest of Ithaca (and Cornell University), was scheduled to begin in April 1968, and the plant was expected to begin operating in mid-1973.

In October of 1967 several Cornell staff members and one representative of the State Conservation Department were invited to an informal meeting by several officials of the utility and a representative of the construction engineering firm that was to build the plant. At this meeting the utility company's representatives briefly described to us their needs and plans for developing the site, and solicited comments. The State Conservation Department representative discussed the possibility that by using some of the warm water from the plant, it might be possible to operate a large fish hatchery for coho salmon on the lake, and utility personnel replied that such a program might have great public relations benefits for the company. Cornell staff members asked the company representatives what possible ill-effects the new power plant's operation might have on the lake ecosystem as a whole, and were told that the company had retained a biological consultant to study some of those aspects, and hoped to engage additional consultants from the Cornell staff. Apparently the company was at this time working to compile data for its voluminous Bell Station Preliminary Safety Analysis Report (1968), an Atomic Energy Commission prerequisite to filing for an AEC construction permit. At this meeting it was also suggested to the company that, before actual work on the site was started, it would be desirable to hold public hearings or the equivalent to settle any possible differences or misunderstandings in advance.

In February 1968 the Cornell Water Resources Center held a meeting for all interested University staff members at which NYSEG officials and consultants made an elaborate presentation of their plans for Bell Station. They described the company's need for more electrical power, and the desirability of locating the plant adjacent to the present Milliken Station. Advantages included a centralized location in the company's power grid, capability of using already owned transmission facilities and rights-of-way, and an abundant year-round supply of cold water. In the company plan, 45° F cooling water would be obtained from a depth of about 100 feet, warmed 20° to 25° in the nuclear plant's condensers, and returned to the surface waters at a rate of 1100 cubic feet per second (500,000 gallons per minute), year-round. (See Comment 1: *How much is water heated in the condensers?*) About one-tenth of the lake's total volume would thus be "processed" each year. Heat from the plant would add approximately 6 billion Btu per hour to the water, over and above the reported 1.3 billion Btu per hour heat rejection from Milliken.

A number of people at this meeting raised questions about possible ill-effects on the lake's ecology from the operation proposed by the utility. Although discharging water to the lake surface at 60° to 75° F might seem harmless, in fact, adding the predicted amount of heat could reasonably be expected to produce changes in any aquatic ecosystem, and there was good reason to expect that some if not all such changes would be deleterious (these changes will be discussed later). Not surprisingly there was a wide divergence of opinion between the questioners and the utility personnel regarding what sorts of possible changes constituted harmful changes. When company officials were asked what attention they had given to use of available technology for cooling the condenser water before returning it to the lake, it was obvious—from the amount of "fielding" which that particular query received within the company team present—that the matter had received little if any serious consideration. This meeting left many of us unsatisfied. It served a very useful purpose in raising and defining questions about possible effects of the proposed operation on the lake. But the utility's answers to some of these questions, and nonanswers to others, were distinctly disquieting.

Comment 1: How much is water heated in the condensers?

The water that cools a power plant by passing through its condensers will obviously get heated up. It is a quite straightforward matter to estimate the amount that the water is heated, given three things: (1) the capacity of the power plant (usually stated in megawatts), (2) the thermal efficiency of the power plant (the ratio of the electric power output to the total power input), and (3) the rate at which cooling water passes through the condensers (usually stated in cubic feet per second). We shall do this calculation for the proposed Bell Station nuclear power plant. A capacity of 830 megawatts and a rate of flow of cooling water of 1100 cubic feet per second are planned (see text). The thermal efficiency of nuclear power plants being built at this time is typically 32 percent; thus only 32 percent of the total energy becomes electrical energy (830 megawatts) and 68 percent becomes heat.

Almost all the heat goes into the water passing through the condensers. Hence, at Bell Station, while the water flows by at 1100 cubic feet per second, it will be heated at a rate of $(68/32) \times 830 = 1780$ megawatts. The production of one megawatt of power for one second is identical to the production of 947 Btu of energy, so $1780 \times 947 = 1,680,000$ Btu of energy are put into 1100 cubic feet of water every second. Finally, since 1 Btu is defined as that amount of heat that will raise the temperature of one pound of water by $1°F$, and 1 cubic foot of water weighs 62.4 pounds, the water will be heated $1,680,000/(62.4 \times 1100) = 25°$ F while passing through the condenser.

The calculation is almost identical for a fossil fuel plant, except that (1) typical efficiencies are higher, generally about 40 percent, and (2) about one-fourth of the heat escapes through the stacks in the form of hot gases, instead of going out through the condensers.

One of the loose statements that one sees these days is that such and such a river or lake will be heated, say, $25°$ F by a proposed power plant. What is meant, of course, is that the water is being heated $25°$ F *in the condensers*. The amount of heat being added to the body of water, rather than the amount the water heated in the condensers, is the crucial quantity for most considerations, since, for example, by doubling the rate at which water is passed through the condensers, the same power plant will heat the condenser water only half as much, while heating the lake or river almost exactly as before. However, if you were a limnologist trying to keep the nutrients in the hypolimnion from being transferred to the biologically productive epilimnion, or a fishery biologist trying to keep the number of small aquatic organisms entering the condensers down to a minimum, you would prefer the smaller condenser, which transfers less water (and fewer organisms) from the bottom to the top of the lake. On the other hand, the greater temperature change will be more devastating to those aquatic organisms which *are* transported through the condensers. Moreover the smaller the condensers, the higher the *average* temperature of the coolant while passing through the condensers. With a higher temperature coolant the plant will operate at a lower efficiency, and more heat will be put into the lake for each megawatt of electric power.

J. H. and R. H. S.

Scientists' Concerns Take Shape

By March 1968 we had learned that in addition to studies by its private consultant, NYSEG was planning to delegate investigations of the lake to independent research organizations. One of these was a heat budget study to be conducted by engineers of Cornell Aeronautical Laboratory at Buffalo, New York. The other was a study of the ecology of Cayuga Lake to be conducted independently by certain biologists and limnologists (freshwater ecologists) at Cornell appointed by—and solely responsible to —the Cornell Water Resources Center. Several of these were scientists who had previously conducted research on the lake.

In the weeks following the utility company's "briefing" for the Cornell community, the nagging questions left unanswered at that meeting intensified and proliferated among some two dozen of us who had been present. More unsettling still were the growing indications that these questions were to be left in limbo, with most of the action at Cornell being directed toward development of the Water Resource Center's research program for the lake. By this time a few of the more detached scientists were even referring to the desirability of studying ecological effects of one or more large power plants on Cayuga Lake, and using its adjacent "twin," Seneca Lake, as a control.

By March 1968 increasing numbers of us were articulating our unanswered questions about the power plant's effects on the lake in the form of letters to local editors and legislators, the utility was preparing to break ground for Bell Station (as planned), and it appeared unmistakably clear that a unilateral industrial decision about use of a needed resource was beginning to be implemented.

In reaction to this, about 20 of us met in early April to discuss the possibility of collaborating on a position paper delineating our questions about Bell Station's effects on Cayuga Lake and available technology for avoiding these threats. We felt impelled to give concrete expression to our conviction that natural resource decisions of this sort must have input from more than one special-interest group. We hoped that publication of a position paper might provide the focal point or catalyst needed by those in the area who were unsatisfied with the narrow course being followed in the lake issue but did not know how to modify it. After three successive drafts, each one of which was worked on by nearly all of the 21 coauthors involved, we submitted a review draft of the paper to officials of some state and federal regulatory agencies, the University, and NYSEG. All reactions were favorable except (not surprisingly) those of the utility. Company officials, it appeared, were particularly concerned that 4 of the 21 authors were men also involved in the Water Resources Center's research project that NYSEG was funding. Presumably the company took the position that it would give too much the appearance of conflict of interest for a scientist to appear as an author of our position paper and, later, of a research report for the utility. At any rate, the 4 authors in question withdrew their names, and we went to press with 17 remaining. Eight fields were represented by these authors: limnology—3, fishery biology—6, aquatic botany —2, natural resources—2, and 1 each from aquatic microbiology, water resources engineering, electrical engineering, and geology. We obtained a loan from the Environmental Defense Fund to pay for the printing of 5000 copies of our position paper and by June 1968 had mailed copies to a list of several hundred individuals and organizations who we hoped might be interested in the issue and/or influential in disseminating the paper or its thoughts more widely. By the end of the summer we had received enough voluntary contributions to repay the Environmental Defense Fund, and our supply of the paper was nearly exhausted.

The position paper described Cayuga Lake and drew attention to limnological data, gathered periodically since 1928, which suggested that Cayuga Lake was becoming measurably more eutrophic, particularly in recent years. The paper discussed NYSEG's proposed operation on Cayuga Lake (see also Figure 2):

> The total volume of water withdrawn from the hypolimnion during a 6-month period of summer stratification would be 18.5 billion cubic feet, or roughly 10 percent of its average volume during the May–October period.
>
> Every 24 hours, year-round, 100 million cubic feet of heated water (65°–70° F) would be added to the epilimnion. Spread uniformly over the entire lake, this would be an addition of about one-half inch per day to its 66-square-mile surface.
>
> The large, continuous addition of heated water throughout the stratification period would increase the epilimnion's normal October volume some 20 percent.

The paper stressed certain effects of the power plant operation on the lake —effects which were entirely predictable from company data and limnological data, although the precise magnitude of these effects was less predictable.

> 1. The onset of thermal stratification will occur earlier in the spring and, because volume of the epilimnion will be increased during the course of the summer, stratification will extend longer into the fall.
>
> 2. The length of the growing season for plants and animals in the upper layer will, therefore, be extended.

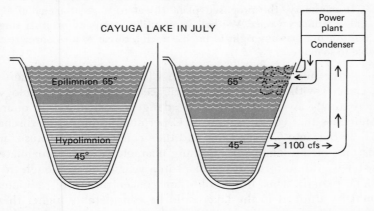

FIGURE 2 If water to cool Bell Station were taken from the hypolimnion of Cayuga Lake and were then returned directly to the epilimnion, the effect would be to change the relative volumes of the two regions, as shown schematically in the comparison of the left and right portions of the figure. The temperatures in the two regions would not change substantially, because in the process of passing through the condensers at 1100 cubic feet per second, the water would be heated approximately 20° F. This circulation system would also enrich the epilimnion with nutrients removed from the hypolimnion.

3. Water brought up from the lower layer and flushed into the surface where most biological production (growth and reproduction of plants and animals) takes place, will contain nutrients previously unavailable to plants in the lighted portion of the upper layer.

4. A longer growing season and more nutrients in the surface layer of the lake will result in greater capacity for biological production.

5. Prolonged stratification will extend the period of oxygen depletion in the large underlying layer of cooler water, where trout live, during the summer. Thus oxygen levels will become lower than they do at present, before being replenished by the delayed fall mixing of the upper and lower layers.

The position paper then described the applicable state and federal permit-granting procedures, and the different sorts of cooling systems available and in use, noting that operation of the power plant on the lake, but with appropriate cooling technology, represented a reasonable and very inexpensive compromise.[9] Our report concluded:

> Our society has often indicated its ability and willingness to pay for maintaining environmental quality, therefore a negligible power cost increase to meet the higher construction and operation expenses of closed-circuit cooling should be acceptable. The momentum of public opinion increasingly obliges those individuals or firms who are polluting our air, water, and land to include as part of their operating cost satisfactory solutions to their waste disposal problems. This should be the case with thermal pollution of Cayuga Lake at Bell Station.
>
> In summary, the proposed plant threatens a great many of Cayuga Lake's primary values. We believe no utility company or other single-interest group has any right to impose such a threat to a resource so valuable to so many.

During the summer of 1968 Professor Clarence Carlson, a fishery biologist at Cornell with training and experience in aquatic radioecology, became increasingly aware of the possible dangers from the small quantities of radionuclides that could (within AEC permissible limits) be discharged from Bell Station into Cayuga Lake and then concentrated because, among other reasons, of the lake's slow flushing characteristics (approximately as much water would go through the plant in a year as goes through the lake in the same period). Thus the equilibrium concentrations of radionuclides that would be attained in the lake could be considerably higher than the concentrations of those radionuclides in the water entering the lake from the plant at the beginning of plant operations. This is unlike the situation

---

[9] The Alabama Power Company recently announced its intention to install induced-draft cooling towers at its proposed 829-megawatt SEALA nuclear generating plant at a cost of about $4 million (just over 2.4 percent of the total plant cost). The Georgia Power Company has announced that it will use closed-circuit cooling at its planned nuclear power stations on the Altamaha River and elsewhere. (R. H. Stroud, *Nuclear-Fuel Stream-Electric Stations*, Bulletin No. 202, pp. 1–2, Washington, D.C.: Sport Fishing Institute, 1969).

in a river or tidal estuary, where the radionuclide concentration in the power plant effluent would be higher than that in the adjacent receiving waters. Dr. Carlson was joined in this concern by 11 other scientists (including others conversant with aquatic radioecology, several physicists, a medical doctor, and 5 authors of the thermal pollution paper, myself included), who published in November 1968, a second position paper, "Radioactivity and a Proposed Power Plant on Cayuga Lake." [10] Using available data on radioactivity decay rates and the rates at which particles would be added to and flushed out of the lake, Dr. Carlson, aided by nuclear engineers and limnologists, calculated equilibrium concentrations and total amounts of various radionuclides that would be accumulated in Cayuga Lake if "typical expected" releases and if AEC maximum permissible concentrations were continuously discharged to the lake.[*]

In addition to concentration of radionuclides caused by Cayuga Lake's slow flushing characteristics, Dr. Carlson and his colleagues stressed the following:

> There will . . . be concentration of radionuclides in aquatic organisms. Many plants and animals concentrate specific radionuclides in certain organs or tissues. For example, iodine is concentrated in the thyroids of higher animals and strontium in bones, scales, and shells. . . . The extent to which different radionuclides are concentrated under various conditions by different organisms varies widely. The scientific literature contains reports of concentration factors for strontium-90 by freshwater organisms up to 500,000 for filamentous green algae, 100,000 for insect larvae, and 20,000 to 30,000 for fishes. Each element can be expected to behave differently in different organisms. Though radionuclide levels in human diets may not be significantly increased or exceed "permissible" levels, no one can accurately predict the effects such accumulations might have on aquatic organisms. Additions of radionuclides and the resulting increase in radiation exposure may be particularly damaging to aquatic organisms, because they are normally subjected to relatively small amounts of ionizing radiation.

The authors made special efforts to avoid invoking nuclear hobgoblins or similar scare tactics, and tried to be as factual as possible. It is interesting to note that the authors succeeded in avoiding an emotional reaction to the radionuclide issue in this case. Indeed, the general public's reaction to the radioactivity paper was far more subdued than it was to the thermal pollution bulletin. Additional points made in summarizing the radioactivity paper were the following:

> 1. The proposed Bell Station is an "experimental" nuclear power station in the sense that no plant of such large capacity or of the precise design

---

[10] C. A. Carlson *et al., Radioactivity and a Proposed Power Plant on Cayuga Lake*, Ithaca, N.Y.: Authors, Fernow Hall, Cornell University, 1968.

[*] The buildup of concentration is treated in a simplified model in the Postscript to this essay. An extensive discussion of radiation and its effects is presented in the essay on Radiation. [J.H. and R.H.S.]

proposed here has ever been operated elsewhere. No nuclear power station has been sited on or discharged liquid radioactive wastes to a relatively small, slow-flushing lake like Cayuga.

2. The quantity of radioactive waste discharges from the proposed plant cannot be accurately predicted in advance.

3. Radiation exposure of every person who will use the lake will be slightly increased as a result of normal plant operation.

4. An accident in the plant which could result in greatly increased exposure of the local population is possible, though highly unlikely.

5. Any exposure to radiation involves some biological risk to ourselves and our descendants.

6. If operating procedures other than those proposed by the company were used, it would be possible to reduce or prevent routine discharges of radionuclides to the environment.

7. The local public has not been informed of the risks to itself and its environment inherent in the operation of Bell Station or of the cost necessary to avoid at least some of those risks.

Public Response

Reactions to the two position papers ranged from anger through apathy to acclaim. This spectrum of response was manifest within the University as well as outside it, although among the academicians there was less neutrality as anyone who has ever attended a faculty meeting could guess. Identifiable disapproval of our activities included feelings that we were doing something that violated propriety, adademic decorum, and/or the scientist's pristine role. Others felt that we were against nuclear power, progress, and new concepts, despite our continuing and completely sincere contentions that we did not oppose the plant—only the proposed method of operating it. In short, we were damned by some for being radicals, by others for being reactionaries, and still others likened us to "the little old lady in tennis shoes who never goes near the lake, but wants to know in her heart that it is pure."

Although there was certainly no wild outburst of public acclaim when our first position paper appeared, as the weeks went by it became obvious that a growing number of people in the Cayuga Lake area felt that the questions we had raised should be widely considered and discussed before a decision was made to let the utility operate the power plant with a once-through cooling system. Requests for copies of the position paper gradually accelerated, and increasing numbers of these came from other states—even a few from other countries. By late July 1968 there was at least one request per week to speak to some sportsmen's, fraternal, or college organization around the lake. We were especially gratified to learn that two remarkably conscientious local legislators, Assemblywoman

Constance Cook and Senator Theodore Day, and also the League of Women Voters, were devoting serious study to the issues that had been raised.

NYSEG's blasting and excavation for Bell Station proceeded rapidly through the summer of 1968, until by November the company had over 2 million dollars invested in the site, which was clearly visible from points on the lake three miles away. Company press releases tended to be bland, reassuring—and repetitive. I was receiving increasingly strong and frequent inquiries about the possibility of forming some sort of citizens group to provide a vehicle for unified expression of the mounting number of individual concerns about the plant. Finally, in early August we held a meeting in a local grocery store that was attended by some 30 members of the community, including several authors of the first position paper. I reviewed the current status of the controversy, handed out packets of background information, and tried to explain my view that the role of the scientists in such an issue is to make information available on pollution hazards and alternatives that will help the voter in making his own decision, without telling him how to vote. Such decisions must be broadly based public decisions. Furthermore, from the standpoint of political strategy (and ethics) the most important factor is not *who* speaks, but *how many*. I therefore felt that if people in the community wished to take some sort of action on this issue, they must make the decision on whether—and how—any action group should be organized, but it should not be by the people who wrote the position papers. After the meeting was over, a young man I didn't know—a Mr. David Comey—introduced himself and said he might be able to help get some sort of organization started, or at least determine if one was wanted. Dr. Carlson and I gave him our files of reprints, mailing lists, and correspondence on the Cayuga Lake case.

Mr. Comey proved to be an unusually dynamic individual, and he became deeply committed to this issue. He rapidly attained an astonishing level of expertise in scientific and technical areas relating to the power plant controversy, and his organizational activity was prodigious. Within a few weeks he formed the Citizens' Committee to Save Cayuga Lake. The number of paid members grew to 300 within three months and over 800 within six months, and affiliation with the Cayuga Lake Preservation Association, sportsmen's clubs, and property-owner associations brought in more than 2000 associate members.

Informative, hard-hitting newsletters soon appeared, state and federal legislators were made aware of the problem, and news releases went out to a variety of newspapers and radio stations. The Citizens' Committee printed the position paper on radioactivity, and reprinted the thermal pollution paper. In October Mr. Comey and his capable committee arranged a large civic luncheon which featured a powerful address by Barry Commoner entitled "The Hidden Costs of Nuclear Power." The local radio station rebroadcast it several times.

The Utility's Response

NYSEG's actions in the Cayuga Lake case seemed to follow a behavior pattern fairly common in resource preemptions by industry. The strategy was to announce the proposal *after* plans for implementing it were already well under way, and to keep things moving ahead rapidly thereafter. Company officials consistently refused invitations to debate their critics in public. The company's numerous publicity releases stressed progress, electrical needs, a tradition of good neighborliness, electric rates that had not been raised in 15 years, monetary benefits of the plant to the community, and the enticing possibility of a large salmon hatchery on the lake—courtesy of NYSEG and the New York State Conservation Department.

The company would never allow its second plant to "harm" the lake, it reiterated, and was conducting contract research programs which it expected would demonstrate that its operations would not damage the lake.[11] That the company did indeed contract at least two lake studies to highly qualified independent research teams is to its credit. Needless to add, the researchers did not share the company's preconception of what the results might show. Company officials' reactions to the position papers, as given in various press releases, indicated that the authors were "jumping to conclusions before the facts are in." They stated that plans for design of the plant were still flexible, pending conclusion of the company-sponsored research on the lake.

The utility's posture that it was not already committed to a once-through flushing design for the plant's condenser cooling water seemed a little ridiculous to the Citizens' Committee. In the summer of 1968 a company official had told a large audience that if the company were required to employ a system for cooling the water before returning it to the lake, financial considerations might force the company to consider moving its plant elsewhere. Since several million dollars had already been invested in this project, it seemed obvious that the company planned on *not* moving elsewhere, and hence planned on not being required to employ a cooling system in the plant design. We subsequently learned that NYSEG had already contracted to sell about two-thirds of Bell Station's power output to Consolidated Edison in New York City—this while "research to assure no harming of the lake by the plant" was under way and before a half year's data were available for analysis.[12] Law Professor Harold P. Green has caricatured the attitude of such utilities (and some regulatory bodies) toward risk-taking as a four-step process:

> (i) we do not have enough scientific knowledge to tell us whether or not the risks are really significant, but our present judgment is that the risks are insignificantly small; (ii) as the project goes forward, further re-

---

[11] This approach was documented in the company's Preliminary Safety Analysis Report to the Atomic Energy Commission.

[12] From the January 16, 1969, issue of the trade journal *Nucleonics Week*.

search will be undertaken to verify our judgment that the risks are insignificantly small; (iii) whatever risks do exist can be reduced to tolerable dimensions through technological devices; (iv) if the risks indeed are found to be, and remain, significant, the program will of course be abandoned or drastically restricted or controlled to protect the public interest. QED.[13]

Legislators' Response

By September 1968 citizens' expressions of concern to their legislators about Bell Station had become a rising groundswell. They realized that the AEC would not, as a matter of policy, take thermal effects into account and based its licensing solely on radioactivity criteria. Moreover, as observed above, even these criteria were designed for flushing situations, rather than essentially nonflushing waters like Cayuga. The New York State Health Department was the only permit-granting agency that could by law take thermal effects into account, and the many citizens who wrote the Health Department about thermal problems were not reassured by the standard reply they received, which read in part:

> The State Health Department looks on Cayuga Lake as a valuable asset to the State and, therefore, is interested in protecting it. On the other side of the coin, the State Health Department is involved with industrial development of the state.

Doubts about the Health Department's true role were subsequently reinforced by this announcement in the December 19, 1968, Ithaca *Journal*:

> Dr. Hollis S. Ingraham, the State Health Commissioner [Head of the Health Department] has been elected chairman of the New York State Atomic Energy Council. ... The council is charged with encouraging the development of atomic energy in the State and at the same time protecting the health and safety of the public.

It appeared the state had installed a "closed-circuit cooling system" of its own.

Accordingly State Assemblywoman Constance Cook and State Senator Theodore Day took actions that led the Joint Legislative Committee on Conservation, Natural Resources, and Scenic Beauty to hold an all-day hearing in Ithaca on November 22, 1968, to help determine whether existing state licensing criteria and their current interpretation by the agencies involved were adequate to protect Cayuga Lake from damage by the power plant's proposed operation.

The hearing opened with a number of presentations by utility officials. They described their activities to date and their intentions. These were followed by presentations from five of the authors of the position papers, Drs. Clifford Berg, Clarence Carlson, Lawrence Hamilton, John Kingsbury, and

[13] Quoted by L. J. Carter, "Technology Assessment," *Science*, Vol. 166, No. 3907, pp. 848–852 (1968).

myself, each speaking about threats of the proposed operation to Cayuga Lake from the standpoint of his own particular discipline. David Comey added information on cooling devices now being used by more progressive electric companies in the Northeast and the deficiency of current regulations to protect the citizens' interests here.

In the afternoon Professor Leonard Dworsky, director of the Cornell Water Resources Center, stated his conviction that current legislation and limitations on functions of regulatory agencies were such that "under these circumstances we believe that all available technology should be utilized in connection with the Bell Station plant so as to eliminate any further deterioration of the Cayuga Lake environment." It was obvious that the utility people were taken aback by Professor Dworsky's testimony. That they clearly had expected him to testify more on their side implied to us that they had failed to grasp the full meaning of scientific integrity. The afternoon ended with the legislators asking the utility vice-president to reiterate, for the record, his earlier estimate (subsequently found high by a factor of about 2) that technology to eliminate most possibilities of thermal or radionuclide pollution to Cayuga Lake would add not more than 50 cents to the monthly household electric bill of the average NYSEG customer.[14] In retrospect it was a clear victory for the Citizens' Committee and its allies, although at the end of that long day many of us were too exhausted to be sure.

In late January 1969 the Citizens' Committee announced its intention to seek legal designation as an intervenor in hearings that would be prerequisite to an AEC permit for construction of Bell Station, stated that it had engaged Harold P. Green, a Washington attorney of national prominence in such matters, to act on its behalf, and set out to try to raise $10,000 to cover costs of this proceeding.

In early March Mrs. Cook, supported by Mr. Day and other legislators, introduced three bills to provide environmental safeguards for lakes in nuclear power plant cases such as the one pending on Cayuga Lake. One of these bills provided restrictions on the place and aggregate amount of heat additions to the epilimnion, another restrained a utility from making any expenditures on a site (after simply acquiring it) until the New York State Health Department discharge permit and Atomic Energy Commission construction permit had been obtained, and the third bill limited the

---

[14] In a more detailed analysis on January 20, 1969, the company vice-president, Albert Tuttle, estimated the total construction and operating costs for cooling the plant's thermal discharge by natural draft evaporative towers at $21.3 million or (prorated) $2.4 million per year. This would amount to a cost of slightly less than $5 per year to each of the company's half million customers. Subsequently a careful analysis by Cornell economist Jeffrey Romm indicated actual cost to be $12.8 million ($1.4 million per year). Mr. Romm's estimates are in good agreement with cost estimates for comparable facilities at comparable-size plants elsewhere. (Jeffrey Romm, "The Cost of Cooling Towers for Bell Station," Manuscript, mimeo., Ithaca, N.Y.: Department of Agricultural Economics, Cornell University, 1969).

concentration of radioactive waste discharges from a plant to not more than 50 percent higher than the intended maximum discharge stated in the plant's Preliminary Safety Analysis Report. These bills seemed to us eminently sound, reasonable, fair, and not unduly restrictive. All three of them passed the State Assembly—and the first two the Senate also—unanimously.

In this same period the State Health Department proposed a set of thermal discharge criteria for New York waters and held a series of hearings on them. These criteria were sharply criticized by the Cornell Water Resources Center administrators, biologists working on its Cayuga Lake study, Cornell engineers working independently on the problem, the Citizens' Committee, and others because they failed to provide any scientifically sound basis for regulating thermal discharges in stratified lakes. The criteria were also criticized by Associated Industries of New York State as being too restrictive—an interesting sidelight.

On April 11, 1969, NYSEG announced an indefinite postponement in company plans for construction of Bell Station "to provide more time for additional research on cooling systems for thermal discharge from the plant, and for consideration of the economic effect of such systems." Presumably the favorable outlook for Mrs. Cook's three bills, and uncertainty as to what thermal discharge criteria the state would adopt, were factors in the company's decision. Commenting on this action, the trade journal *Nucleonics Week* stated: "Observers suggest, also, that the postponement decision may have been influenced by NYSEG aversion to a court battle over AEC licensing." [15]

The scientists, the citizens, and the legislators had acted conscientiously in their appropriate roles. This is the way resource allocation problems should be solved, and the outlook for a reasonable decision in the Cayuga Lake case was bright.

Disappointments from the State Capitol

One of Mrs. Cook's three bills (the one limiting radionuclide discharges to 50 percent more than the stated concentration) was killed in the Rules Committee of the State Senate as a result of heavy pressure from utilities in the state. Then in May 1969 Governor Rockefeller vetoed the other two bills. In turning down the thermal discharge bill, he stated legislative action at this time was premature because the State Water Resources Commission was now developing thermal discharge criteria which would supply the needed safeguards. In vetoing the bill to restrict heavy investment in site development until permits had been granted, he said: "It would not be reasonable to expect the Public Service Commission to evaluate and balance against one another the myriad of factors involved in selecting sites. . . . [The bill] would be detrimental to the State's power program and would seriously retard the economic growth of the State."

[15] Vol. 10, No. 6, p. 2 (April 17, 1969).

A second blow came on July 25, 1969, when the State Water Resources Commission approved the much-criticized thermal discharge criteria in their originally proposed form. It seemed that the lengthy series of state-wide hearings on these proposals had been a meaningless exercise. The criteria were forwarded to Washington for evaluation by the Federal Water Pollution Control Administration (now called the Federal Water Quality Administration).

Later Developments

In November 1969 the Water Resource Center's study of Cayuga Lake was published. It covered a 9½ month sampling period. The data comprised a valuable increment to the store of information on the lake. However, as the director of the Center pointed out, the data from such a short-term study did not—and could not have been expected to—provide definitive answers about the safety of the company's proposal for using Cayuga Lake.

In August 1970, NYSEG president William A. Lyons indicated that construction of Bell Station was contingent in part on FWQA comments on New York State's thermal discharge criteria. He stated: "It now appears that an optimistic estimate of the earliest possible on-line date for a nuclear plant at our site on Cayuga Lake would be late in 1977. It more likely would not be until 1978." The utility recently sold the reactor vessel for the proposed Bell Station to the Long Island Lighting Company.

In December 1970, NYSEG stated that it would push for the development of two nuclear generating plants—one of these on Cayuga Lake —as soon as the state legislature established site selection procedures.

It is now 2½ years since the first position paper on Cayuga Lake was written, and at this writing (the end of 1970), Bell Station remains dormant—a multimillion-dollar excavation on Cayuga's shoreline. The increase in cost of the completed plant that has taken place through price rises during the two years the company has been resisting compromises has been estimated at $100 million. One wonders how the company can reconcile this with the relatively small additional cost (perhaps $15 million total) of accepting these compromises in the first place.°

° The "compromises" Professor Eipper has evaluated for Cayuga Lake especially emphasize cooling towers.

The future role of cooling towers in electric power generation is not easy to predict at the moment. As with many other technological devices that reduce an environmental problem, there will be side effects that must be assessed. For example, cooling towers will increase local fog formation; whether at any particular site the extent of that effect is negligible or not will require research which takes into account regional climatic conditions. Cooling towers should be considered as one among a spectrum of possible solutions to the problem of thermal pollution. Other alternatives, in addition to those suggested by Professor Eipper, are (1) to build power plants in the open ocean where the heat increase to the water may be ecologically insignificant, (2) to develop ways of utilizing the "waste" heat

The Citizens' Committee remains viable and is expanding its activities. In the Cayuga Lake region, in Cornell University—and indeed throughout the nation—there is noticeable increased concern with environmental quality, the management of natural resources, and the development of controls to prevent pollution before it happens.

---

from the power plant for practical purposes such as heating homes, melting snow on roads, prolonging the growing season of irrigated crops, or accelerating the rate of sewage treatment processes, and (3) to halt the growth of electric power consumption and stop building additional power generating plants. Each of these alternatives involves costs and benefits which must be evaluated. [J.H. and R.H.S.]

POSTSCRIPT: How Radioactive Waste Builds Up in a Lake

John Harte and Robert H. Socolow

If a factory or power plant discharges a fixed amount of waste every day into a lake, the concentration of that waste in the lake will build up to an equilibrium level at which the waste leaves the lake at the same rate as the waste enters the lake. The concentration of the waste at equilibrium and the length of time required to reach, say, one-half the equilibrium concentration, both depend on the rate at which the lake is flushed. Other things being equal, the slower the lake is flushed, the longer the time during which the concentration of waste keeps increasing substantially and the higher the final equilibrium concentration.

Wastes can leave a lake in ways other than by being flushed out. For example, nutrients can react chemically to form insoluble salts, falling to the bottom of the lake, or a waste may pass through the food chain into birds or land animals. All of these arguments apply whether or not the particular waste is radioactive. If the waste *is* radioactive, this provides another mechanism by which the substance can leave the lake, because a radioactive substance decays. The waste may be said to have "left" the lake if, as is usually the case, the decay products are not radioactive.

The decay of a radioactive waste and the flushing of a uniformly mixed waste out of a lake have one property in common: for each process, the rate at which waste leaves the lake at any instant is proportional to the amount of waste in the lake at that instant. When there is more than one way ("channel") for waste to escape, there will be a characteristic rate associated with each exit channel, and if these channels are *independent*, these rates will *add* to give the total exit rate. In the case of radioactive waste, the two most important exit channels, radioactive decay and getting flushed out, are clearly independent.

Let $N(t)$ be the amount of waste in the lake at time t. Then the rate at which waste leaves the lake by flushing is $N(t)/L$, where L, the mean flushing time for the lake, is the time by which a volume of water equal to the volume of the entire lake will have flowed through the lake. The rate at which the radioactive waste decays is $N(t)/R$, where R is the mean-life of the radioactive isotope in question.°

If we let

$$\frac{1}{T} = \frac{1}{R} + \frac{1}{L} \tag{1}$$

° The reader who is not familiar with these concepts may find Appendix 1, Halving and Doubling, helpful.

132

then, because the rates add, the total rate at which waste leaves the lake is $N(t)/T$; T is the mean time for waste to exit from the lake.

Now suppose that a waste product is being added to the lake at a constant rate B. Then at equilibrium, the rate at which waste is removed, N/T, equals the rate at which it is added, B, and so the equilibrium amount is $N = BT$. In words, BT is the amount of nutrient which will enter the lake at the rate B within the mean time T. Other things being equal, a slow-flushing lake means a large value of the mean flushing time L, hence a large value of the mean exit time T, and hence large equilibrium concentrations.

For Cayuga Lake, L is about 10 years. The time R, of course, depends on the isotope; for strontium-90, with a half-life of 28.1 years, R is 40.6 years, so that $T = 8.0$ years. This means that the equilibrium amount of strontium-90 in Cayuga Lake will equal the total amount of strontium-90 discharged (at a constant rate) into the lake in 8.0 years. The equilibrium *concentration* will equal this amount divided by the volume of the lake.

The detailed way in which the amount of waste product builds up, under these assumptions, is given by the solution of the differential equation

$$\frac{dN}{dt} = -\frac{N(t)}{T} + B \tag{2}$$

Each side of this equation is an expression for the net rate of change of waste in the lake. If the discharge of waste begins at a time $t = 0$, then the solution is

$$N(t) = BT(1 - e^{-t/T})$$

After a time T, the amount of waste product has reached $(1 - 1/e) = 0.63$ of its equilibrium value. A graph of $N(t)$ versus t is shown in Figure A. In equilibrium, $dN/dt = 0$.

In view of this discussion, it is clear that proper standards for radioactive waste disposal from nuclear power plants built on lakes must be based on equilibrium concentrations, rather than on the concentration of radioactive waste in the water passing into the lake from the condensers. The latter type of standard, however, is easier to administer and has the advantage of being independent of the properties of the body of water being used for cooling. An example of this kind of standard is the AEC upper limit for strontium-90 discharge, set at 3×10^{-7} *microcuries per cubic centimeter of cooling water*.° Let us work out an example that illustrates the relation between these two types of standards.

For the proposed Bell Station plant, where 1100 cubic

° A more complete discussion of radioactivity and an explanation of the units in which it is measured are found in the essay on Radiation in this book.

feet (or 3.1×10^7 cubic centimeters) of cooling water are emitted every second, the AEC standard just quoted permits 9.3 microcuries to enter Lake Cayuga every second. The equilibrium amount of strontium-90 in Lake Cayuga is the amount entering the lake in 8.0 years, which works out to be 2.3×10^9 microcuries. Since the volume of Lake Cayuga is 9.4×10^{15} cubic centimeters, the equilibrium *concentration* in the lake is 2.5×10^{-7} microcuries per cubic centimeter. By a set of coincidences which can be summed up by saying that water is drawn through the Bell Station condensers at nearly the same rate that it is flushed through Cayuga Lake, this equilibrium concentration is very close to the concentration of isotope which is first emitted in the condenser cooling water. The reader will note, however, that the *concentration* of the radioactive waste in the cooling water had nothing directly to do with the equilibrium concentration; rather, it is the *amount* of radioactive waste discharged per second that enters into the calculation.

The reader should be reminded that an equilibrium concentration in no sense implies a uniform distribution of radioactive waste in the lake; in particular, biological concentration of particular wastes in particular organisms must be included in any complete assessment of the hazards involved in a particular rate of release of radioactive waste to a lake.

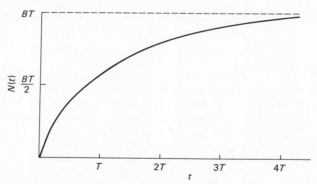

FIGURE A. This graph, plotting equation (3), shows how the quantity of waste product in a lake builds up toward its equilibrium value, given that waste enters the lake at a constant rate and leaves the lake at a rate proportional to the amount present.

D. CONFRONTING THE MILITARY

In the two essays that follow, it is possible to be outraged at man's assault on nature even without thinking about what men are doing—or getting ready to do—to men. Professor Galston tells how herbicides might be turning parts of Vietnam's soil into brick and making the mangrove forests so infirm that they can be washed into the sea. Professor Shifferd describes how northern Wisconsin will be defaced if the Navy carries out its proposal to dig up thousands of square miles of woods to bury a radiotransmitter that will be able to send messages to distant American submarines deep under water.

Both Professor Galston and Professor Shifferd have had to contend with layers of secrecy in trying to ascertain the ecological consequences of these military projects. Because neither the locations that have been sprayed with herbicides nor the physical dimensions of the radio antenna are matters of national security, the military's lack of openness on these matters has led to speculation about motives that probably often underestimates both the competence and the good will of those responsible for the projects.

Starting from certain premises, it can be argued that spraying herbicides and building the radio transmitter are not absurd things to be doing. If men are to be on patrol in Vietnam, then they ought to be protected from ambush; clearing the leaves from the jungle will help do this. American and Soviet submarines—carrying missiles with nuclear warheads and in secure communication with their home base—are essential elements in the most sophisticated system yet devised to make sure that each super-power is afraid of the other. Sanguine is to be the "button" for the American system. Such a defense strategy leads relentlessly to projects like Sanguine.

Environmental destruction by the military will continue so long as warring camps persist on our planet. In a way we are all like passengers on a plane carrying hijackers. The hijackers have let us know that they may have to have a shootout among themselves to determine where the plane ought to go, and they have asked us to relax and enjoy the trip.

135

9. WARFARE WITH HERBICIDES IN VIETNAM

Arthur W. Galston

There was a time, not long ago, when to be a botanist meant to engage in work that could lead only to the improvement of the condition of mankind. Botanists could point proudly to their roles as discoverers of new economically useful plants, as breeders of higher and higher yielding crops, as pathologists providing better and better means of warding off predatory fungi and insects, as agronomists rationalizing and optimizing agricultural productivity, as physiologists discovering chemical and physical means for controlling the developmental cycles of plants, and as biochemists uncovering new pharmacologically useful compounds in plants. Botanists were entitled to feel that all men everywhere derived benefits flowing from the botanist's special knowledge of and relationship to the green plant.

Even discoveries made to foster the welfare of mankind are frequently perverted into instruments of destruction and of war. Alfred Nobel, the inventor of dynamite, learned this to his sorrow when explosives were put to bigger and bigger wartime uses. The airplane, designed to aid man's transport from place to place became, in World War II, one of man's most terrible weapons of destruction. Nuclear energy, which promises to give man virtually boundless sources of energy for his increasingly technologized society, powered the bombs that devastated Hiroshima and Nagasaki. And now the botanist, possibly the last of the scientific innocents, has become guilty of aiding the fabrication of weapons which create hunger rather than plenty, which kill rather than promote the growth of plants, and which may cause ecological catastrophes rather than the improvement of nature. Retrospectively, the botanist has become aware that some of his proudest accomplishments, whether designed for use in peace or war, have had unforeseen harmful side effects.

To understand the consequences of herbicide use in peacetime agriculture and as weapons of war, it is necessary to go back a bit into history and to recall the development of the plant hormone concept. About 40

136

years ago; it was discovered, both in the Orient and in Europe, that plants, like animals, are controlled in their behavior by minute quantities of mobile organic substances produced at specific locations in the body. These substances, called hormones, migrate all over the plant from their centers of production in localized areas like terminal buds. They can influence such processes as cell division, cell enlargement, cell differentiation, growth patterns of stems and leaves, and the production of reproductive organs, like flowers.

With the development of sophisticated biochemical methods for the separation of closely related organic materials, it became possible to isolate these hormones and to deduce their chemical structures. One type of growth hormone was given the generic name of *auxin;* the main auxin in most plants is believed to be the simple substance indole-3-acetic acid (IAA). This compound promotes the elongation of most plant cells and can also regulate a variety of developmental phenomena such as the inhibition of the growth of buds below the apex, the formation of new roots on cuttings, the promotion of flowering and fruiting in certain species, and the control of the relative growth of various organs. Organic chemists soon discovered that they could synthesize molecules in the laboratory which would act like IAA in producing these typical effects on the plant. Such synthetic compounds had the advantage of being cheaper and more stable than IAA. They are also more effective in the plant, which can degrade the natural IAA readily, but cannot cope as well with unnatural synthetic compounds. Among such compounds are substituted phenoxyacetic acids, of which 2,4-dichlorophenoxyacetic acid (2,4-D), and 2,4,5-trichlorophenoxyacetic acid (2,4,5-T) are examples (Figure 1).

These compounds have one additional property that has greatly favored their use in agriculture; they are selectively toxic to plants in the sense that when specific formulations are applied at a certain rate (for 2,4-D, one uses 0.5–3.0 pounds per acre in an oil-water emulsion) they will cause the death of some plants, generally of the dicotyledonous or broad-leaved type, but will not affect the growth of the more resistant monocotyledonous plants, such as grasses. This permits their formulation and use in agriculture as selective herbicides, or weed killers, which substitute chemical for mechanical means of cultivation. In fields of maize, for example, the spraying of 2,4-D will kill such broad-leaved nuisances as bindweed while permitting the maize to grow. The application of 2,4-D as an aerosol, either with ground-based equipment or from low-flying airplanes, has greatly facilitated the precise timing and mechanization of agricultural practices, and has saved many hours of work and much money that would have gone into less convenient, more destructive cultivation practices such as tractor-based hoe devices. Clearly, chemical herbicides, as originally designed for use in agriculture, must be classed as a boon, increasing mankind's food production and freeing him from much backbreaking labor.

This use of agricultural chemicals has caused a tremendous growth in the chemical industry synthesizing such compounds. It is now estimated

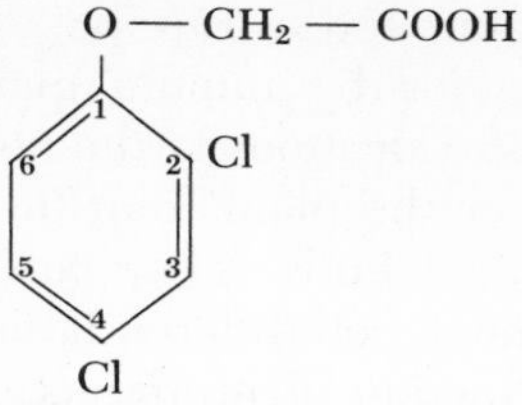

O — CH₂ — COOH

2,4-D (2,4-dichlorophenoxy-
acetic acid)

O — CH₂ — COOH

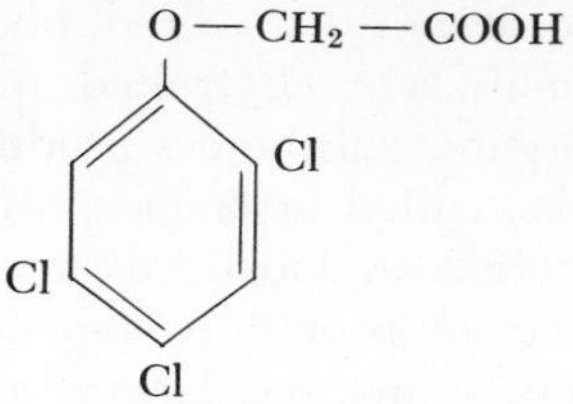

2,4,5-T (2,4,5-trichloro-
phenoxyacetic acid)

NH₂

Cl

Cl

Cl

COOH

N

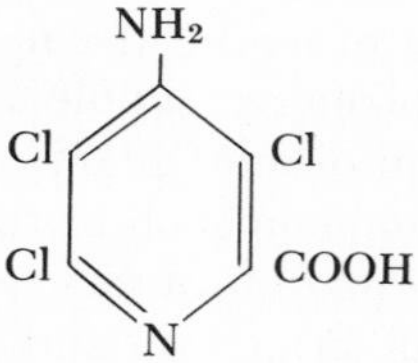

Picloram

CH₃

CH₃ — As — OH

O

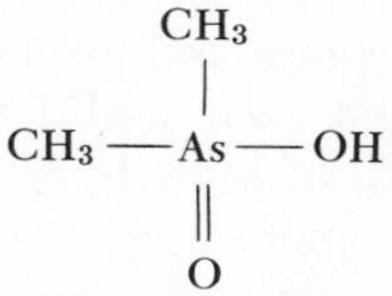

Cacodylic acid

FIGURE 1. Herbicidal materials used in Vietnam. (An explanation of some of the nomenclature of organic chemistry will be found in the Postscript to the essay by Professor Loucks.)

that a total of more than 800 million pounds of various chemicals worth more than $2 billion are used in agriculture in the United States each year, and more than half of this total belongs to the chemical herbicide class. The use of herbicides has been growing at a rate of about 16 percent per year recently, and on this basis will overtake fertilizers sometime in the mid-1970s. In 1950 farmers spent 25 cents per acre for pesticides (0.5 percent of the sale price for his produce); in 1969, it was $3.65 per acre (4.6 percent of the sale price for his produce). This prodigious increase is designed to blunt the effects of some 50,000 kinds of fungi, 30,000 kinds of weeds, 15,000 species of nematodes, 10,000 species of insects, and several species of rodents that together cause crop losses estimated at about $13 billion per year.

Chemical herbicides are not limited to the phenoxyacetic acids, since the competition between independent chemical companies, each with its own specialties and patents, leads to the formulation of an ever-increasing armamentarium of compounds effective against particular weedy pests. Thus Dow has recently introduced picloram (Figure 1), a wide-spectrum herbicide based on the common chemical picolinic acid; Du Pont has introduced antiphotosynthetic herbicides based on chemically substituted urea, and Monsanto has Avadex, an allylic compound that kills wild oats and other grasses.

In the closing years of World War II, the possible military use of herbicides was investigated by a group of plant physiologists located at Fort Detrick in Frederick, Maryland. In the early experiments it was clearly established that proper formulations of 2,4-D and related chemicals could be used to control the fall of leaves from trees. Where military operations had to be conducted in densely overgrown or jungle terrain, the localized application of selected herbicides could be used to cause the shedding of leaves, opening up the terrain to observation and lowering the possibility of ambush, infiltration, or the mounting of enemy offensives. World War II ended before military application could be made of the use of herbicides.

Use in Vietnam

In Vietnam herbicidal operations for military advantage have become very widespread and of great strategic importance.[1] An estimate of the areas treated with herbicides is shown in Table 1, and the formulations of the main preparations used are described in Table 2. It can be seen that the number of acres subjected to defoliation activities rose from 17,000 in 1962 to more than 1½ million in 1968, which seems to have been the peak year. Each acre is sprayed with about 3 gallons of mixed derivatives of 2,4-D and 2,4,5-T, to yield about 26 pounds of acid equivalent per acre, a figure about ten times as high as the recommended rate in the United States. To achieve longer-lasting effects on the sprayed areas, a derivative of picloram, which is a more resistant compound, is used, mixed with a 2,4-D salt; this is also applied at a rate of 3 gallons to the acre. The application deposits 6 pounds per acre of 2,4-D and 1½ pounds per acre of picloram. We shall have more to say about picloram later.

Column 3 of Table 1 details the acreages of crop lands destroyed. These acreages represent rice growing in isolated hinterland areas believed to be under Viet Cong control, the harvest of which is used to feed guerrilla groups. These acreages rose from 717 in 1962 to a maximum of almost a quarter of a million in 1967, then declined sharply since 1968. The chemical used to destroy rice is a sodium salt of cacodylic acid, an organic arsenical containing approximately 47 percent arsenic. It is used rather than 2,4-D because the latter type of compound is ineffective against narrow-leafed plants. As formulated, the cacodylic acid herbicide used in Vietnam is supplied at 3 gallons per acre, yielding 9.3 pounds of chemical per acre. The fate of this chemical in soil is not well understood; but whatever its pathway of degradation, much of the arsenic is there to stay and may accumulate in the soil or may possibly be converted to more toxic, chemically reduced derivatives such as arsine.

From the very beginning, the use of herbicides for the massive de-

---

[1] Reviews of U.S. herbicidal operations in Vietnam can be found in: F. H. Tschirley, *Science*, Vol. 163, pp. 779–786 (1969); G. Orians and E. W. Pfeiffer, *Science*, Vol. 168, pp. 544–554 (1970); *Chemical-Biological Warfare: U.S. Policies and International Effects.* Hearings before the Subcommittee on National Security Policy and Scientific Developments of the House Committee on Foreign Affairs (1970); Hearings on Military Posture . . . , House Armed Services Committee, Feb.–April 1970, p. vii following p. 8667.

TABLE 1. Estimated Areas Treated with Herbicides in Vietnam
(Acres sprayed per year)

Year	Defoliation	Crop Destruction	Total
1962	4,940	741	5,681
1963	24,700	247	24,947
1964	83,486	10,374	93,860
1965	155,610	65,949	221,559
1966	741,247	103,987	845,234
1967	1,486,446	221,312	1,707,758
1968	1,526,333	170,000	1,696,333
1969	1,404,333	115,233	1,519,567
Total	5,427,096	687,843	6,114,939

SOURCE: Constructed from data in the two Hearings cited in Reference 1.

TABLE 2. Herbicides Used in Vietnam

Agent [a]	Active Ingredients	Active Ingredient Present in 3 Gal/Acre Sprayed	Major Target
Orange	2,4-D (n-butyl ester)	12 lb/acre	Forest vegetation
	2,4,5-T (n-butyl ester)	13 lb/acre	
White	2,4-D (triisopropylamine salt)	6 lb/acre	Forest vegetation
	Picloram (triisopropylamine salt)	1.5 lb/acre	
Blue	Cacodylic acid and its sodium salt	9 lb/acre	Rice and other food crops

SOURCE: Constructed from data in the four sources cited in Reference 1.
[a] Names correspond to the color of the coded stripe around the canister in which the agent is packed.

struction of forested lands and of selected croplands in Vietnam met with the opposition of certain members of the American scientific community. Pressure applied on the Department of Defense by the American Association for the Advancement of Science in 1966 led the former to commission a literature survey of existing knowledge on the ecologic consequences of herbicide application.[2] At the annual meeting of the American Society of

[2] W. B. House *et al., Assessment of Ecological Effects of Extensive or Repeated Use of Herbicides,* Kansas City, Mo.: Midwest Research Institute, 1967.

Plant Physiologists in 1966, I proposed the issue in the form of a letter of inquiry, which was signed by several of my colleagues and sent to President Johnson. The letter said in part:

> The undersigned plant physiologists wish to make known to you their serious misgivings concerning the alleged use of chemical herbicides for the destruction of food crops and for defoliation operations in Vietnam. The use of such agents by United States forces was reported in the *New York Times* of December 21, 1965 and has never been denied by the Administration or by the leaders of our military operations. Our deliberations and our statements below are based on the assumption that this published report is true.
>
> We would assert in the first place that even the most specific herbicides known do not affect only a single type of plant. Thus, a chemical designed to defoliate trees might also be expected to have some side effects on other plants, including food crops. Secondly, the persistence of some of these chemicals in soil is such that productive agriculture may be prevented for some years. Thirdly, the toxicology of some herbicides is such that one cannot assert that there are no deleterious effects on human and domestic animal populations. It is safe to say that massive use of chemical herbicides can upset the ecology of an entire region, and in the absence of more definite information, such an upset could be catastrophic.
>
> Even if we assume that our military leaders have selected reasonably specific anti-rice herbicides, nontoxic to humans or to domestic animals for use in Vietnam, we must still be concerned with the effects of large-scale food deprivation on a mixed civilian-military population. As Prof. Jean Mayer of the Harvard School of Public Health pointed out in a letter to *Science* on April 15, 1966, the first and major victims of any food shortage or famine, caused by whatever agent, are inevitably children, especially those under five. This results mainly from their special nutritional needs and vulnerability to stress. Thus, the effect of our use of chemical herbicides may be to starve children and others in the population whom we least wish to harm.

The reply, which came from Assistant Secretary of State Dixon Donnelley, said the following:

> Chemical herbicides are being used in Vietnam to clear jungle growth and to reduce the hazards of ambush by Viet Cong forces. These chemicals are used extensively in most countries by both the Free World and the Communist Bloc for selective control of undesirable vegetation. They are not harmful to people, animals, soil or water.
>
> The elimination of leaves and brush in jungle areas enables our military forces, both on the ground and in the air, to spot the Viet Cong and to follow their movements, and to also avoid ambushes.
>
> Destruction of food crops is undertaken only in remote and thinly pop-

---

(Sponsored by Advanced Research Projects Agency of the Department of Defense. ARPA Order No. 1086.) See also a review of this document, by F. A. Egler, "Herbicides and Vegetation Management—Vietnam and Defoliation," *Ecology*, Vol. 49, No. 6 (Autumn 1968).

ulated areas under Viet Cong control, and where significant denial of food supplies can be effected by such destruction. This is done because in the Viet Cong redoubt areas food is as important to the Viet Cong as weapons. Civilians or non-combatants are warned of such action in advance. They are asked to leave the area and are provided food and good treatment by the Government of Viet-Nam in resettlement areas. [Dated Sept. 28, 1966.]

This was the first open admission by the administration of the aims and practices of the chemical warfare campaign against plant life in Vietnam. It has led to some searching confrontations between scientists and both the Johnson and Nixon administrations.

The admission by our administration that it was using chemical warfare agents in Vietnam raised several important and disturbing questions.[3] In the first place, the Geneva protocol of 1925 (Figure 2) prohibited the use of "asphyxiating, poisonous or other gases and all analogous liquids, materials or devices." Although the United States signed the protocol, our Senate never ratified it, and thus we were technically not bound by its terms. Nonetheless, the United States did ratify the United Nations resolution of December 5, 1966 (Figure 3), which calls for "strict observance by all states of the principles and objectives of the protocol signed at Geneva on June 17, 1925, and condemns all actions contrary to those objectives." It

---

Geneva Gas Protocol

Protocol for the Prohibition of the Use in War of Asphyxiating, Poisonous or Other Gases, and of Bacteriological Methods of Warfare,

Signed at Geneva on June 17, 1925

Whereas the use in war of asphyxiating, poisonous or other gases, and of all analogous liquids, materials or devices, has been justly condemned by the general opinion of the civilized world; and,

Whereas the prohibition of such use has been declared in Treaties to which the majority of Powers of the world are Parties; and,

To the end that this prohibition shall be universally accepted as a part of International Law, binding alike the conscience and the practice of nations;

Declare that the High Contracting Parties, so far as they are not already Parties to Treaties prohibiting such use, accept this prohibition, agree to extend this prohibition to the use of bacteriological methods of warfare and agree to be bound as between themselves according to the terms of this declaration.

FIGURE 2   Geneva Gas Protocol. (Only the substantive part of the protocol is reproduced here.)

[3] In addition to the chemical antiplant agents discussed above, the U.S. military forces have also used more than 14 million pounds of the CS-type tear gas (really a lung gas) against personnel in Vietnam. An analysis of the effects of this agent can be found in Steven Rose (ed.), *CBW: Chemical and Biological Warfare,* Boston: Beacon Press, 1969.

therefore seems that the United States government, through its use of chemical warfare agents in Vietnam, has acted contrary to the UN resolution.

The United States position on the use of herbicides and of riot-control chemical gases, which are also used in Vietnam, was summarized by Ambassador Nabrit of the United States shortly before passage of the UN resolution of December 1966, as follows:

> The Geneva Protocol of 1925 prohibits the use in war of asphyxiating and poisonous gas, and other similar gases and liquids with equally deadly effect. It was framed to meet the horrors of poison gas warfare in the first World War, and was intended to reduce suffering by prohibiting the use of poisonous gases, such as mustard gas and phosgene. It does not apply to all gases. It would be unreasonable to contend that any rule of international law prohibits the use in combat against an enemy for humanitarian purposes of agents that governments around the world commonly used to control riots by their own people. Similarly, the protocol does not apply to herbicides, which involve the same chemicals and have the same effects as those used domestically in the United States, Soviet

---

United Nations Resolution

Adopted by the General Assembly, December 5, 1966

The General Assembly,

Guided by the principles of the Charter of the United Nations and of international law,

Considering that weapons of mass destruction constitute a danger to all mankind and are incompatible with the accepted norms of civilization,

Affirming that the strict observance of the rules of international law on the conduct of warfare is in the interest of maintaining these standards of civilization,

Recalling that the Geneva Protocol for the Prohibition of the Use in War of Asphyxiating, Poisonous or Other Gases and of Bacteriological Methods of Warfare of 17 June 1925 has been signed and adopted and is recognized by many States.

Noting that the Conference of the Eighteen-Nation Committee on Disarmament has the task of seeking an agreement on the cessation of the development and production of chemical and bacteriological weapons and other weapons of mass destruction, and on the elimination of all such weapons from national arsenals, as called for in the draft proposals on general and complete disarmament now before the Conference,

1. *Calls* for strict observance by all States of the principles and objectives of the Protocol for the Prohibition of the Use in War of Asphyxiating, Poisonous or Other Gases, and of Bacteriological Methods of Warfare, signed at Geneva on 17 June 1925, and condemns all actions contrary to those objectives;

2. *Invites* all states to accede to the Geneva Protocol of 17 June 1925.

FIGURE 3   United Nations Resolution pertinent to the Geneva Gas Protocol.

Union, and many other countries to control weeds and other unwanted vegetation.[4]

Ambassador Nabrit fails to note the excessively high concentrations and quantities of the herbicides used in Vietnam and the use of the arsenical cacodylic acid on crop plants, a practice not permitted in the United States.[5]

Moreover, severe ecological effects of a massive herbicide defoliation program in Vietnam appeared possible. I wrote in 1967:

> When we spray a synthetic chemical from an airplane over a mixed population of exotic plants, growing under uninvestigated climatic conditions as in Vietnam, we are performing the most empirical of operations. We learn what the effects are only after we perform the experiment, and if these effects are larger, more complex, or otherwise different from what we expected, there is no way of restoring the original conditions.

I stated in conclusion:

> We are too ignorant of the interplay of forces in ecological problems to know how far reaching and how lasting will be the changes in ecology brought about by the widespread spraying of herbicides in Vietnam. These changes may include immediate harm to people in the sprayed areas, and may extend to serious and lasting damage to soil and agriculture, rendering more difficult South Vietnam's recovery from war, regardless of who is the victor.[6]

Effects on Plant Life

Recent observations, both in Vietnam and in the United States, have reinforced these guesses. Both Tschirley [7] and Orians and Pfeiffer [8] report that the mangrove associations lining the estuaries near Saigon and throughout the Mekong Delta area are, for reasons we do not understand, extraordinarily sensitive to Agent Orange, which contains 2,4-D and 2,4,5-T. A single spray with this mixture can kill an entire mangrove community, which consists of 20 different species, many in the *Rhizophoraceae* family, intermingled in an impenetrable mass. The estimated time for regeneration after killing, if it ever occurs, is a minimum of 20 to 25 years. In the mean-

[4] *Chemical and Biological Weapons: Some Possible Approaches for Lessening the Threat and Danger.* Special Subcommittee on the National Science Foundation of the Committee on Labor and Public Welfare, U.S. Senate, Washington, D.C.: U.S. Government Printing Office, May 1969.

[5] President Nixon announced his intention to resubmit the Geneva Protocol to the Senate in November 1969 (James M. Naughton, *New York Times*, November 26, 1969). Possibly because of the unresolved questions concerning the banning of herbicides and tear gas, the matter has not yet come up.

[6] A. W. Galston, *Scientist and Citizen*, Vol. 9, No. 7, pp. 123–129 (August–September 1967).

[7] F. H. Tschirley, *op. cit.*

[8] G. Orians and E. W. Pfeiffer, *op. cit.* See also P. M. Boffey, "Herbicides in Vietnam: AAAS Study Finds Widespread Devastation," *Science*, Vol. 171, No. 3966, pp. 43–47 (1971).

Two mangrove forests in South Vietnam in 1970. The one at the right was sprayed with chemical defoliants on a date that the United States government will not make public. (Arthur H. Westing)

time, many shellfish and migratory fish which depend on the mangrove root environment for completion of essential stages in their life cycles will have to go elsewhere.[9] A reduction in the fish and shellfish catch would deprive Vietnamese of much-needed protein with which to supplement their rice diet. Of the estimated 1.2 million acres of mangroves in South Vietnam, about one-fifth had been destroyed by mid-1968, and the figure has probably risen to over 30 percent by now because the Rung-Sat coastal zone southeast of Saigon was declared a "free-dump" area for herbicides. In view of this, it is difficult to be reassured by Tschirley's quotation [10] from official South Vietnamese government figures purporting to show that there has been no diminution of the official fish catch from 1965 to 1967. The ecologic effects of the dissipation of the mangrove community could take several years to become apparent because of the remarkable persistence of the tightly packed fibrous roots in the form of mangrove peat. Thus data for the years 1968–1970 are important to obtain. Then, too, much of the fish catch does not find its way into harbors for commercial weighing and recording but goes directly into use in nearby villages. Conscientious on-the-scene investigations would be needed to uncover such data.

Another harmful consequence of the killing of mangroves could be the destabilization of the estuarine waterway boundaries, and their ultimate collapse and erosion in the face of heavy rains and tidal movements. This

[9] J. P. Milton, in a speech on coastal conservation, Washington, D.C.: Conservation Foundation, 1968.
[10] F. H. Tschirley, *op. cit.*

could affect the ecology of the entire coastline in which such operations have been carried out, and the data will not be obtainable until several more years have elapsed.

What has been the effect of defoliation operations on the deciduous hardwood forests of Vietnam, which occupy about 25 million acres, or about 60 percent of the total area of that country? One spraying, causing defoliation, lasts for about 6 to 9 months, after which new growth is put forth. In the interval, the lower story and ground vegetation, which had been repressed by the failure of light to penetrate through the top canopy, will have grown vigorously. One plant which flourishes in this way, given half an opportunity, is bamboo. Once bamboo take over an area, there is no way to return it to trees except to bulldoze and burn and start all over; there is no herbicide that will do the job. Trees that have been sprayed two or more times may be killed entirely, and such areas are even more certainly taken over by the lower-lying vegetation.

These effects could be terribly serious for the economy of South Vietnam, since forest products are, potentially at least, the major source of export income (potentially $150 million annually) for that country. In spite of all the shellings, fires, and lack of security, as well as the defoliation, an estimated 317 sawmills were operating in 1970 (compared to 700 in 1961). Other forest-related industries are a wood preservative plant, a pulp and paper plant, a plywood factory, a particle-board plant, and numerous charcoal and woodworking plants. In 1968 these industries gave employment to about 80,000 people. If there has been serious debilitation of the forests due to the defoliation operations, then the economic consequences could be severe. One estimate is that more than 5 million acres, or about 20 percent of the total forest land, have been damaged or destroyed, mainly by herbicides.

The ability to grow crops of any sort depends, of course, on the fertility of the soil. Most of the soils of Vietnam are highly leached *podsols* (high in silicon, low in iron) or *laterites* (high in iron, low in silicon).[11] Both types of soil are dependent on their organic matter to hold minerals (the so-called base-exchange capacity) and to give the soil a physical quality that makes it possible for plant roots to grow well in it. If trees are defoliated and the exposed ground is heated by the sun, soil microbial activity speeds up, the organic matter is oxidized away, and the minerals formerly held are "dumped," leaching away in the next rain. Such soils are impoverished, and may no longer grow reasonable crops of any kind. Some lateritic soils so affected may compact irreversibly into hard, bricklike masses permanently unsuited for agriculture. That this phenomenon can occur in South East Asia is seen by the fact that part of the temple of Angkor Wat in Cambodia is constructed of bricks cut out of hardened lateritic soil. How much of this has gone on in Vietnam is hard to judge, but min-

[11] R. Maignien, "Review of Research on Laterites," *UNESCO Natural Resources Research*, Vol. IV (1966).

eral dumping has certainly occurred, to an extent which we may never be able to assess.

It is often assumed that the chemicals sprayed on the vegetation produce their effect quickly and then are rapidly degraded by soil microorganisms into innocuous end products. Neither of these assumptions is true. One of the herbicides used in Vietnam, picloram, has such an exceedingly long life in the soil that some botanists consider it as the herbicidal analog of the well-known persistent insecticide, DDT. The Dow Chemical Company's own research journal, *Down to Earth,* reports that this compound, which is a potent herbicide, disappeared to the extent of only 3.3 percent over 467 days in one dry soil in the United States.[12] The disappearance over the same period in other soils approximated 20 percent. The continued use of such a persistent material could build up the picloram content of the soil to the point where all plant growth could be seriously inhibited for years, since picloram is very toxic when absorbed by the roots. The persistence of picloram is made more graphic by the knowledge that there is no microorganism known which will degrade it directly. It has been estimated that approximately 10,000 to 100,000 pounds of soil organic matter must be broken down to effect the decomposition of 1 pound of picloram. For these reasons picloram has not been licensed for agricultural use in the United States. Its widespread use in Vietnam is therefore all the more indefensible.

Even 2, 4-D and 2,4,5-T, which are normally expected to disappear in several months, may accumulate in soils under relatively anaerobic conditions. Since most rice paddies are anaerobic for at least part of their growing cycle, some buildup of inadvertently sprayed or drifted phenoxyacetic acids could be expected. Such buildups could cause a crop to accumulate many substances, including nitrate in such quantities as to be toxic to farm animals or humans.[13] This toxicity of nitrate results from its conversion in animal cells to nitrite, which then causes methemoglobinemia, an aberration of the blood oxygen-transport system, caused by chemical modification of hemoglobin. This would be difficult to demonstrate in Vietnam, but evidence for such an effect should be sought, if we are to understand fully the consequences of our actions.

Effects on Man

Toxic effects in laboratory animals have recently been reported from 2,4,5-T, one of the major herbicides used in Vietnam.[14] When commercial preparations of this compound were injected subcutaneously or fed to

[12] C. R. Youngson *et al., Down to Earth,* Vol. 23, No. 2, pp. 3–11 (1967).

[13] L. M. Stahler and E. I. Whitehead, *Science,* Vol. 112, pp. 749–751 (1950).

[14] U.S. Department of Health, Education, and Welfare. Report of the Secretary's Commission on Pesticides and Their Relationship to Environmental Health, Parts I and II, December 1969 (chap. 8). See also K. D. Courtney *et al.,* "Teratogenic Evaluation of 2,4,5-T," *Science,* Vol. 168, pp. 864–866 (1970).

pregnant female laboratory mice and rats, the offspring produced had a wide range of abnormalities, ranging from cleft palates to cystic kidneys, to deformed nervous systems. In addition, many embryos aborted in utero. Doses as low as 4.6 milligrams per kilogram of body weight produced some effects; doses of 46 and 113 milligrams per kilogram of body weight produced litters in which up to 100 percent of all litters had abnormal individuals and up to 70 percent of all individuals were abnormal. If the toxicity rate in humans is the same as in mice and rats, then a 45-kilogram (100-pound) Vietnamese woman would need to ingest only about 200 milligrams of 2,4,5-T to be affected. This is the amount she would get from about 3 liters of water, assuming a spray of 26 pounds per acre of agent Orange, followed by a 1-inch rainfall in which all the 2,4,5-T dissolves and no dilution. Since much drinking water is gathered by collection of rain from rooftops or from very shallow wells, such an assumption is reasonable. Furthermore, there are reports in the Saigon press of new and unexplained birth abnormalities, starting late in 1967, when the spraying program became massive.[15] Definite medical evidence will be hard to obtain, but must be sought.

It now appears that these toxic effects of commercial preparations of 2,4,5-T may be due to an impurity called *dioxin,* which arises during the commercial synthesis from tetrachlorobenzene and methanol.[16] The original material tested, which caused fetal deformity in mice and rats, appears to have been contaminated with about 27 parts per million of dioxin, which is known to be highly toxic. The Dow Chemical Company, which synthesizes much of the 2,4,5-T used in Vietnam, says that its product can be made to have less than 1 part per million of dioxin. Tests are now being conducted by Dow, by the National Institutes of Health, and by the Food and Drug Administration to see whether the purified 2,4,5-T is or is not teratogenic (that is, injurious to the fetus); at the moment opinion appears to be divided, and one must await the results of carefully replicated teratogenicity tests before coming to any conclusion.

Even if pure 2,4,5-T turns out to have no harmful side effects in animals, its use will be suspect as long as dioxin is known to be teratogenic. One may ask, for example, whether metabolism of 2,4,5-T in the plant could lead to dioxin formation. To the biochemist, this is not impossible; all one needs is conversion of the 2,4,5-T to the corresponding phenol, oxidation of the phenol to a free radical and coupling of two such free radicals in a specific manner. Could this happen in the human body? In the soil? As the result of burning in a sprayed area? A recently activated research program, to be carried out in the research laboratories of the U.S. Department of Agriculture in Beltsville, Maryland, should provide some of the answers to these questions.

[15] T. Whiteside, *The New Yorker,* February 7, 1970, p. 32, and March 14, 1970, p. 124; reprinted with additional documents in *Defoliation,* New York: Ballantine/Friends of the Earth, 1970.

[16] *Medical World News,* Vol. 11, No. 9, pp. 15–17 (1970); *Nature,* Vol. 226, pp. 309–311 (1970).

As of now, while these issues are yet to be decided, the government has withdrawn 2,4,5-T from both military use abroad and agricultural use at home.[17] Registration of liquid forms of the herbicide has been suspended by the Department of Agriculture, making interstate sales illegal. Nonliquid forms, not considered imminently harmful, will remain on sale until hearings and possible appeals are completed. Various forms of 2,4,5-T will continue to be used for control of weeds on nonagricultural lands such as forests and pastures, and for miscellaneous purposes, not involving food crops, around homes. Since most of the 2,4,5-T sold in the domestic market finds its way into such uses, the ban will not completely halt the flow of this potentially dangerous chemical into an ecosystem. In 1968 about 42.5 million pounds of 2,4,5-T, worth about $30 million, were sold in the United States.

With respect to the deliberate killing of crops in order to deprive the Vietcong military of food, it should be remarked that whenever starvation is used as a weapon against an entire civilian population the main sufferers are inevitably the aged, the infirm, pregnant women, lactating women, and children under five years old.[18] The fighting man almost always gets enough food to sustain himself. Thus, in using hunger as a weapon, we are attacking that part of South Vietnamese society which is least involved in military operations and whom we would wish least to injure.

If about 650,000 acres of rice have been killed by agent Blue (cacodylic acid) over the last 7 years, then since one acre of rice in Vietnam will feed two people for a year, one may calculate that substantially more than 1 million man-years worth of edible rice has been destroyed. If one adds to this the widespread killing (sometimes inadvertent, through drifting spray) of jack fruit, cassava, papaya, and vegetables,[19] then the food denial program may have induced serious nutritional deficiencies, at least among restricted numbers of people, possibly Montagnards, in the remote upland areas mainly affected by the spraying program.

The advent of the use of chemical warfare weapons against plants raises the question of escalation and of control of this type of weapon. If we use 2,4,5-T, picloram, and cacodylic acid to kill jungle trees and rice crops, why should we not resort to the use of viruses, bacteria, or fungi which can produce epidemics in food-producing plants? It is well known that at Fort Detrick, Maryland, where most of America's research on chemical and biological warfare occurs, an especially virulent strain of the rice-blast fungus has been developed.[20] Under what conditions might this fungus be introduced into Vietnam or a future theater of war? The use of "big bang" weapons like nuclear bombs has been kept under control partly

[17] N. Gruchow, *Science*, Vol. 168, p. 453 (1970).

[18] J. Mayer, in Steven Rose (ed.), *CBW: Chemical and Biological Warfare*, Boston: Beacon Press, 1969.

[19] G. Orians and E. W. Pfeiffer, *op. cit.*

[20] President Nixon's announcement that the United States will relinquish its biological warfare capability will probably lead to the transfer of this facility from the Department of Defense to the National Institute of Health.

because of the ease of detection of such weapons, but the quiet, undetectable, and sinister way in which biological weapons can be introduced into warfare makes their control very much more difficult. Thus some biologists are objecting to the use of herbicides, partly on the basis that the use of such agents represents a new escalation of an already horrible war and opens the lid of Pandora's box just a bit wider.

In a letter dated September 29, 1967 to Don Price, then president of the American Association for the Advancement of Science, John S. Foster, Director of Research and Engineering for the Department of Defense, stated, "We have considered the possibility that the use of herbicides and defoliants might cause short or long term ecological impacts in the areas concerned. . . . Qualified scientists, both inside and outside our government . . . have judged that seriously adverse consequences will not occur. Unless we had confidence in these judgments, we should not continue to employ these materials." The announcement [21] of the phasing out of the defoliation program is a sign that this confidence has been shaken, if not shattered.

[21] White House statement of December 26, 1970. See also W. Buchel, *New York Times,* June 23, 1970, p. 1.

10. THE FIGHT AGAINST PROJECT SANGUINE

Kent Shifferd

The United States Navy is planning a massive environmental assault in Wisconsin. It intends to build "Project Sanguine," a giant radio transmitter, whose antenna will be buried in a grid pattern underneath the 26 northern counties of the state. This essay explains what this project is, its potential environmental hazards, and the organization and activities of the State Committee to Stop Sanguine.

What Is Sanguine?

The Sanguine transmitter is based on the phenomena associated with extra low frequency (ELF) radio waves. It will transmit at a frequency below 100 cycles per second, probably at either 45 or 75 cycles per second. At 75 cycles per second these waves are approximately 2500 miles long (since the product of the frequency and the wavelength is the speed of light, or 186,000 miles per second) and will girdle the globe; however, "to be efficient, an antenna must at least approach in size the wavelength it is to transmit."[1] Hence the requirement for the world's largest antenna. The Navy maintained, early in its publicity campaign, that it could take advantage of a piece of ancient rock—the Laurentian Shield. This rock is 3 to 30 miles thick underneath the northern third of the state and extends hundreds of miles into Canada. The Navy has varied the details of the project, and it has not been possible to obtain precise information about the system they want to build. One proposal, in approximate terms, is for a grid antenna consisting of some 6000 miles of cable, buried 2 to 6 feet underneath the surface of the soil over this rock. The Navy maintains that this dry rock formation, low in conductivity, would in effect become an extension of the antenna. Approximately 240, ten-acre transmitter stations would be built at the intersections of the grid. The system would require approximately 400 to 800 million watts of power, each of 60 cables carrying

[1] "How We'll Broadcast with Mystery Radio Waves," *Popular Science*, September 1969.

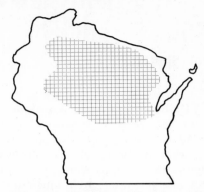

FIGURE 1 A schematic drawing of the Sanguine antenna, showing the rectangular grid of cables. The exact size and location of the antenna have not been chosen, but it would underlie a large portion of northern Wisconsin.

14,000 volts and a current of between 500 and 1000 amperes. Eight hundred million watts is equal to one-half the present power requirement for the entire state.

Figure 1 indicates the form Sanguine could take. The affected area is 21 million acres, according to this proposal.°

What is the purpose of this colossal project? It could be used to communicate with attack submarines below the surface of the ocean. Present communications systems require the submarines to rise near to the surface to receive messages; however, the ELF waves can penetrate the ocean to a depth of 1500 feet.

Project Sanguine was publicly announced in the spring of 1968 by Congressman Alvin E. O'Konski in whose Tenth District most of the antenna will lie. He made loud claims that it would stimulate an economic boom and give the north country a chance to serve the nation. During the previous year and subsequently a public relations team was sent around the state by the Navy, giving "briefings" to various county boards and other official groups including the state agencies and the governor and his cabinet. These sessions usually began with a film entitled *Sanguine: A New Voice for the North Woods.* Very little technical information about Sanguine was in the film, which was filled instead with pictures of aircraft carriers and rockets and accompanied by statements about the troubled world and the need for defense. Following the film Captain James Galloway, U.S. Navy, along with a public relations man, presented a talk complete with

---

° The Sanguine system has not been precisely defined, but there are definite constraints on the design. The strength of the antenna signal is proportional both to the square of the length of the antenna and to the square of the current in the cables, so the antenna can only be made smaller if the currents in the cables are made larger, and vice versa. The strength of the antenna signal is also inversely proportional to the conductivity of the underlying rock, so a comparable antenna would have to be bigger still if it were built over a rock formation with higher electrical conductivity. [J.H. and R.H.S.]

official-looking graphs and charts. At the end of their presentation they asked for questions. It was at one of these briefings (Ashland, Wisconsin, July 19, 1969), that opposition to Sanguine crystallized, although some concerned citizens had carried a resolution to the State Democratic Party convention one month before.

The briefing was attended by approximately 100 persons, some friendly to whatever the Navy would have to say, others extremely doubtful of what they had read in the newspapers—all packed into a small courtroom on a hot July afternoon. The presentation was designed to assure the audience that the Navy had the situation fully under control (the implication that some aspects of Sanguine had to be controlled was de-emphasized). But it was here that the Navy, in its desire to impress the audience with its foresight and technological competence, admitted that Sanguine did have certain "interference problems" associated with it. We learned, for example, that a Sanguine system interfered with telephones, television, railroad signals, and aircraft communications. We learned that the Navy would require "easements" on the land in order to build and then have access to the antenna should repairs or adjustments be necessary. Finally we were assured that the Navy had thought of all the problems and had contracted with various experts and corporations to carry out "mitigation" studies. Some of these studies were already under way at the Interference Mitigation Laboratory at Clam Lake, Wisconsin, in the heart of the Chequamegon National Forest 25 miles due south of Ashland. The audience was assured that "if Sanguine is not compatible with the environment it will not be built."

During the hour-long question period a number of issues were raised concerning power requirements, the need for atomic power plants, the likelihood of nuclear attack, the necessity of Sanguine, its cost, soil erosion, rock conductivity, and so on. In general there were three kinds of answers given by the Navy: (1) "We don't know," (2) "It's in the planning stage," and (3) "That may be bordering on the classified." At one point in the presentation the Navy maintained it did not know what the power requirements of Sanguine would be and a few minutes later claimed that it had contracted with a laboratory which was making studies of the effects on cows of ten times the Sanguine power level.

Organizing for Opposition

Following the briefing a small group of people in Ashland County began to consider the environmental hazards of Sanguine and to plan for organized opposition. We drew encouragement from the knowledge that Senator Gaylord Nelson, the dean of Senate conservationists, was concerned enough about Sanguine to communicate his support.

On September 21, 1969, approximately 40 persons met at Stevens Point, Wisconsin, and founded the State Committee to Stop Sanguine. These people represented conservation groups (for example, the Sierra

Club, the Wisconsin Ecological Society, the Audubon Society, the Wisconsin Resources Conservation Council), organized labor, the legal profession, and farm groups. They came from all parts of the state and represented both political parties. Local television stations had been alerted and were also present. At the meeting an executive board and a chairman were elected, a plan of action and goals formulated, and several resolutions passed.[2]

We decided to build as large a membership as possible by setting a low minimum contribution ($2), and to publicize our campaign through press releases, speeches, and later if funds materialized, through paid advertisements and television spots. We drafted two resolutions: One requested the governor to direct the state Department of Natural Resources to conduct an investigation into Sanguine and make the results public, and the other requested the attorney general of Wisconsin to take action in any and all courts to see that the environment of the state was protected as provided in our state constitution.

Soon thereafter we formed a Science Advisory Council made up of a number of scientists from the colleges and universities of Wisconsin to give us more complete technical analyses of the project and of Navy testing programs. In this way the potential environmental hazards have become more clearly identified.

Environmental Outrage

Since the lumber barons, there has been no greater assault on northern Wisconsin than that represented by Project Sanguine. That which we love may be destroyed. Northern Wisconsin is a unique area today. It is forest in an age of asphalt. Here are extensive stands of gentle aspen and white birch, thousands on thousands of acres of Norway pine and the rugged jack pine which alone can survive the forest fires. Wisconsin is cedar swamps and white trillium, trailing arbutus and bear berry in the spring. It is the quiet of undrained tamarack bogs. Here one can still know that slightly sinful, good feeling of being snowed in—a birch log fire on the hearth and outside, the black-capped chickadees picking for seeds in the drifts.

Northern Wisconsin, where Sanguine will intrude, is at this moment abundant animal life in an age of machine technology. It is deer eating fallen apples in the side yard, and the unmistakable print of the black bear

---

[2] The fight against Project Sanguine has involved hundreds of men and women of good will. Chief among these is Lowell Klessig of Madison, Wisconsin, whose organizational abilities, ecological insights, and tireless efforts deserve special commendation. Long hours, knowledge, and money have all been contributed by all the members of our board of directors: C. H. Stoddard, Fred Baumgartner, John Wilson, Louise Erickson, Jack Bohnen, Tom Ortman, and finally, Walter Thoreson, who took on the monumental task of running for Congress against the entrenched and long-time favorite incumbent. Walter did more than any of us to popularize our cause. My job as chairman was simply to keep up with these energetic crusaders for the environment.

in the sand on a wet spring morning. The snowshoe hare, the fisher, and the timber wolf are not zoo animals here; they are free. In our lakes and streams are the delicious walleye and rainbow and the hard-fighting northern pike, and in the air the marsh hawk, the tiny hummingbirds and sapsuckers; there are also sharptails in the woodland openings, and waterfowl beyond number.

This land is clean air in a society choking on its own wastes. Here people drink the water because it tastes good. It is the land of the last living of the Great Lakes—Superior, whose cold vastness reminds us that we are but men. Men work fifty weeks a year in Chicago and Milwaukee so they can spend two here. In a society racing toward environmental disaster it is one of the few places not already beyond saving.

Sanguine will change all this; the scale of its intrusion upon the environment is staggering. The instrusions come in three stages. First, even before the transmitter is installed, the forest will be despoiled by the enormous amount of ditching necessary to bury the antenna cables. To maintain permanent access to the cables, thousands of miles of 30-foot-wide corridors will be bulldozed through national, state, and county forests and private lands. This will leave northern Wisconsin carved up like a waffle, destroying its recreational value. It is not only that the physical appearance of the installation is contradictory to the wilderness aspects of the area, but also that its basic nature is contradictory to what this region is. In the controversy over an electric power facility at Storm King, New York, in the Hudson Valley, the principle of the aesthetic integrity of an area was effectively defined: No matter how much it may be possible to hide a large technological intrusion, the knowledge of its presence changes the character of the area. The North Woods will no longer be the North Woods if Sanguine, with all its twentieth-century connotations (including that of nuclear disaster) is here. And, of course, Sanguine cannot be hidden. In order to bury the cables during the construction phase thousands of acres of water-holding vegetation will be stripped away, resulting in the subsequent silting of the myriad streams and small lakes in the area. The degradation of this forest, moreover, will set a precedent for inroads by the military-industrial complex to exploit our other national forest and park lands for purposes other than those for which they were originally intended.

At a second stage, the power sources necessary for supplying Sanguine with 800 million watts will create serious environmental disturbances. The most likely location for power plants would be the Lake Superior shore, whether existing facilities already located there are expanded or, instead, new atomic plants are built. The thermal pollution from fossil-fuel and nuclear power plants and the air pollution from the former will be particularly unwelcome in an area which counts as its chief assets clean air and water and an ample supply of game fish. In addition, these large power sources would inevitably attract developers into the North Woods, making the wilderness vanish even more rapidly.

At a third stage, electromagnetic hazards may arise when the transmitter is turned on. The creation of a modified electromagnetic environment may have detrimental effects on soil bacteria, insects, birds, larger animals, and humans. In the presence of very large magnetic fields, much larger than Sanguine will create, numerous effects on growth rates and nervous system function have been observed in animals, and the directional growth of plants has been shown to be dependent on inhomogeneities in the surrounding magnetic field.[3] The problem of extrapolating these effects to weak magnetic fields is a difficult one. There has also long been serious discussion of the possibility that birds use the magnetic field of the earth to navigate,[4] and the magnetic fields of Sanguine are comparable in strength to that of the earth.

The oscillating electromagnetic fields produced by Sanguine will induce electric currents in conducting materials whenever these materials form closed paths. This can happen where wire fences, highway guardrails, railroad tracks, pipelines, and so on, either form closed paths themselves or are shorted by an unsuspecting animal. The Navy is sufficiently concerned that they are doing "mitigation studies." To mitigate a fence, they would have to cut it at intervals and insert insulating material; the induced currents which could then flow would be greatly reduced. To mitigate the whole North Woods, the Navy would have to find and cut every long fence. Yet the North Woods abound with miles of abandoned, partially downed fence lines (left over from an early attempt to establish an agricultural economy in the area) many of them out of sight, under a few inches of fallen leaves and pine needles.

The evidence of the effects of electromagnetic fields on biological organisms is insufficient to draw any hard conclusions. The point made by the State Committee, however, is that whoever wishes to alter the environment must prove beforehand that he will not harm it. The burden of proof should not lie with the public after the damage has been done, as was the case with DDT.°

---

[3] A scholarly review of these and other topics is to be found in Alexander Kolin, "Magnetic Fields in Biology," *Physics Today*, Vol. 21, No. 39 (1969).

[4] H. L. Yeagley, *Journal of Applied Physics*, Vol. 18, pp. 1035–1063 (1947), and Vol. 22, pp. 746–760 (1951).

° The electric and magnetic fields produced by Sanguine's cables are not much larger than those produced by an overhead power line, for the latter carries a 60 cycle per second alternating current of up to several hundred amperes. At a distance of d meters from a long cable carrying a current of I amperes, the magnitude of the magnetic field, B, measured in gauss, is given by the expression:

$$B = \frac{0.002}{d} I$$

Thus at the surface of the earth directly above the cable, if it is buried 2 meters deep and carries a current of 1000 amperes, the magnetic field strength will be 1 gauss. For comparison the magnitude of the earth's magnetic field is roughly 0.5 gauss. Sanguine's electric fields are also not very large, because the electric field is proportional to the frequency of the current, and these are very low frequency signals.

The Navy has interpreted the task of correcting the adverse effects of Project

Fighting the Project

In the months that the State Committee has been in existence we have built our membership to approximately 1200 individual and group affiliates. The individual members are as diverse a group as is the general population of Wisconsin. Some representative members are a man 92 years of age from the little hamlet of Cornucopia on the tip of the Bayfield peninsula who just plain distrusts "the government" from a kind of gut conservatism; a young liberal professor from Wisconsin State University at Eau Claire who simply opposes the military-industrial complex; a housewife from Cable, a resort town, who resents being called unintelligent by her congressman; a newspaper editor in Vilas County who smelled a rat from the beginning; a retired businessman from Chicago, now living in a little Wisconsin town, who worries about the nuclear attack aspect of Sanguine; a lady on social security; a wealthy Milwaukee businessman who owns valuable lake property in the affected area; the chairman of a local sportsman's club; a former Wisconsin resident now living in California who hopes to retire in his home state, and one living in Tennessee who simply retains her love for the lakes and woods of Wisconsin; a high school student doing a research project on Sanguine; a labor leader with headquarters in Chicago and locals in Wisconsin, whose now-affluent workers own their own camper trailers and boats.

Through our own efforts and through the efforts of Senator Nelson's office we alerted the press to our existence and our program. Their response was overwhelming: local, state, and national newspapers and television networks came to us for interviews. Although not minimizing the amount of work the executive committee has put into press relations, we had to conclude that we had become newsworthy because of the national scope of Sanguine, the increasing concern about the environment and about excessive military spending, and the human interest of the underdog citizens' committee battling the monstrous Pentagon. Stories have appeared in the *New York Times*, the Los Angeles *Times*, the San Francisco *Chronicle*, and in numerous other papers, as well as on the CBS Evening News. The two major Wisconsin papers (the *Capitol Times* and the Milwaukee *Journal*) have carried continual stories and have editorialized against Sanguine. Much of the effective work alerting local people to the project, and inci-

---

Sanguine in terms of mitigating the effects of the electric and magnetic fields. In so doing, its attention has been distracted from deeper issues that do not permit mitigation but instead require that choices be made among conflicting values. If the system is built, this will disturb the wilderness not only directly but also because the availability of plentiful electric power will spur development. The people of northern Wisconsin will be subject to the distress of becoming a nuclear target. And the country will have acquired a new component in its strategic deterrent, with complex effects on global arms control that have only been publicly discussed in a much oversimplified form. [J.H. and R.H.S.]

dentally raising membership for us, has come from the local and county papers of Wisconsin. For example, the Lakeland *Times* of Minocqua conducted an editorial campaign and reader survey, receiving over 1100 anti-Sanguine responses and only 8 in support.

We also identified a list of conservation clubs, women's clubs, garden clubs, and resort-owners associations and contacted their officers with our material. From these groups we have received invitations to "tell our side of the story." We have sent quantities of our literature and accepted speaking engagements. *In toto,* we have distributed 5000 copies of our ten-page pamphlet, called "What Is Sanguine," 10,000 flyers ("Let's Have the Truth About Sanguine"), and 1000 bumper stickers ("Stop Sanguine"). We are now in the process of distributing 20,000 four-page fact sheets. We found that the pamphlet was too long and the flyer too short. We have found it necessary and useful to prepare three model letters and mass-produce them to send to people who write asking, "What can I do?"

We have tried to identify the key decision-makers and bring our opposition before them. In our case this group includes Secretary of Defense Melvin Laird, Undersecretary David Packard, Deputy Director of Electronics Research Howard Bennington, the President's Council of Economic Advisors, headed by Roger Mayo, and the President's Council on Environmental Quality, headed by Russell Train.

Finally, we have gone into Federal court with a complaint that the Navy did not seek authority from the Federal Communications Commission and the Wisconsin Public Service Commission to build and operate their Phase I transmitter at the Clam Lake site, that this transmitter interferes with telephone service, and that Sanguine is a "public nuisance" on environmental grounds. Our legal counsels, Mr. Roy Tulane (former assistant to the attorney general of Wisconsin), and Mr. William Chatterton have generously donated their services, making our only costs the normal expenses of travel and court fees (approximately $1000 for the federal suit).

Our complaint was rejected, however, on the grounds that an insufficient amount of damage would be involved vis-a-vis the individuals directly named in the suit, a technicality which denies us the opportunity to present evidence and make argument. It is disappointing that a group committed to peaceful reform like ours is not given this chance to work through the system.

Our opponents have not been easy to handle. Our committee has actually been subject to a smear campaign, in which it has been said that we are comparable to Communists and Nazis, and that we "cohabit with the SDS." On the more sober side, the private interest groups, for example, the officials of the local power company have made public statements and testified before state legislative committees that Sanguine is harmless and will create an economic boom. Chambers of Commerce have made the same statements and have urged the Navy to build Sanguine.

On higher levels the politicians have made exaggerated claims for the number of jobs Sanguine will create. Tenth District Congressman Alvin O'Konski initially claimed that thousands of jobs would be created. Skeptical, we immediately wrote a letter to the Secretary of Defense and received a reply giving us the figure of several hundred, which we made public. We explained that most of these will be imported technicians, drawn mainly from the military personnel.

In fact, the economic consequences of Sanguine may well be detrimental. Sanguine will bring into the area a temporary, transient work force during the construction period. These people will put added pressure on schools, housing, public health services, highways, and police and fire protection; but because they will pay no property taxes, they will not share the cost of these increased services. The local property owners will bear the added cost. Sanguine will also waste tax dollars. Its total cost is estimated by the Navy at $1 billion. (Senator Nelson estimates $5 billion.) The Economic Development Administration in Wisconsin (a federal agency) is charged by law to prove that for every $10,000 spent they have created one permanent job.[5] At that rate Sanguine ought to create 100,000 permanent jobs rather than several hundred. The billions of dollars spent will not, in any case, ever enter the state. The list of contractors provided by the office of the comptroller general of the United States does not show one firm with head offices in this state. The economic impact may well fall most heavily on the current economic mainstay of the northern area, the tourist industry, which brings about $300 million annually to the residents.

Finally, the chief proponent has been the Navy. The Navy argues that Sanguine is required for our national defense, that it will be economically beneficial, and that it will be safe. The Navy gave a $178,000 contract to Hazelton Laboratories of Falls Church, Virginia, for an "environmental study." This included a literature search, a few laboratory tests on one species of plant, two cows, some rats, and fruit flies. Through the offices of Senator Nelson we obtained a copy of the research proposal Hazelton Laboratories submitted to the Navy and have criticized it to the press, in our literature, and at speaking engagements. We found out, for example, that no ecological survey of the affected area would be made, that only a few rudimentary on-site tests (at Clam Lake) would be conducted, and that ecological research was not a specialty of the Hazelton laboratory.°

In our view, the Sanguine system is indefensible even from a military standpoint: it is a "soft" system. Naturally any potential enemy would regard it to his advantage to neutralize the essential links in our military capability. Since the antenna cables will not, according to the Navy, be buried in hardened trenches and since the power supplies must be above ground, any aggressor would find it prudent to make Sanguine an early strike target.

[5] Walter Thoresen, former EDA director of northwest Wisconsin, testimony at state assembly hearing, Park Falls, Wisconsin, December 16, 1969.
° The reader can find descriptions of the biological research performed for the Navy in Appendix B, "Biological and Ecological Assurance," of the *Sanguine System Environmental Compatibility Assurance Program (ECAP) Status Report,* December 1970, available from the Sanguine Project Office, Naval Electronic Systems Command Headquarters, Washington, D. C. [J.H. and R.H.S.]

Above: The Namekagan River in the northern Wisconsin wilderness. *Right:* The Navy's Clam Lake station in northern Wisconsin. Test cables for the Sanguine system follow the clearing through the forest. (Wisconsin Natural Resources Department; U. S. Navy)

The Navy has continually changed the physical dimensions of the project, partly, we believe, as a tactical maneuver to confuse the opposition. The Navy has also sought to raise clouds of confusion by issuing a series of contradictory reports regarding the status of the project, including its geographic location, its power requirements, and even its fate. The latter is the most serious; in late October 1969 Congressman O'Konski announced in a lament to the people that Sanguine was apparently dead, killed by a handful of fanatics. This message was harmful—some conservation editors and individuals are surprised to find us still in active opposition to the project. This is an old and standard practice—the strategic withdrawal at the public level in the hope that the opposition will call off their activities. Military-industrial projects cannot be considered dead until they are killed by a vote in the Congress, their demise is announced by the Secretary of Defense, or (and the surest sign of all) no more funds are appropriated for them.

The Navy has contradicted itself in its references to the importance of the Laurentian Shield. Originally, the Navy maintained that Sanguine could be built only in northern Wisconsin because of the desirable conductivity and dryness of this rock formation. However, the Navy recently told the governor's investigating committee that the transmitter could be located elsewhere and that "other considerations" influenced this site choice.

What has been accomplished in the fight against Project Sanguine so far? We can take partial credit for the following six achievements. (1) We have raised a vocal segment of public opinion against Sanguine, thereby causing even its proponents to modify their positions from outright endorsement to a "wait-and see" attitude, or something stronger. By the autumn of 1970 Congressman O'Konski, in a campaign speech, was saying: "I would not shed a tear if they took Project Sanguine out of Wisconsin lock, stock, and barrel." (2) We have forced whatever testing has occurred, no matter how insignificant from a scientific point of view. When we began, the public officials displayed absolute confidence in the Navy's plan; now these same officials are calling for tests and accusing us of prejudging the project. (3) We have stimulated much significant editorial opinion against Sanguine. (4) We have aided the governor in his decision, announced early in December 1969, to appoint a blue-ribbon panel of scientists to investigate the project. (5) We have forced a legislative hearing on Sanguine. (6) We may perhaps have had a part in influencing the 75 percent cut made by the House Appropriations Committee in the Sanguine budget for Fiscal Year 1969–1970. The Navy had asked for $20 million in order to begin contract definition and some subsequent development. The House gave it $5 million and stipulated that the money be used for research into the environmental aspects. (One of the humorous incidents connected with this was the Navy's attempt to cover up this defeat by claiming, at the state legislative hearing, that it had *asked* the House committee to cut its request.) To summarize, we believe that we have slowed the momentum of Project Sanguine.

What have we learned in Wisconsin that might be valuable to other citizens' groups? With regard to organization, the structure of a citizens' group must remain simple—no elaborate constitutions are required, only a brief description of the organization and its goals for purposes of filing incorporation papers with the attorney general of the state (this is a legal necessity in order to raise funds and make expenditures, and it is a safeguard for the members in case of any legal action against the group). A relatively small working executive committee (in our case, nine individuals), actually carries out the administrative functions of the organization. Each must be charged with a responsibility for a clearly defined area of action: for example, fund-raising, press releases, building a research organization, maintaining a clipping file for duplicating material on request, contacting other environmental defense organizations, *et alia*. For our purposes we needed wide representation from across the state, but we have found this to be inconvenient for meetings. Beneath this small committee stands the broad base of the citizen members who, for a minimal fee, become members of the organization and in turn receive the literature, solicit others to join, and write letters to the press and public officials. These people are the heart of the organization; it is their voice that constitutes its clout. We have found the services of legal counsel indispensable, even before we started planning lawsuits.

We have learned to be politic about allies—to exclude no one but not to take in everyone. In this day of national polarization along ideological lines, any citizens' committee that expects to have an impact on the Congress and the bureaucracy must adopt a *via media*. We had made a conscious effort to gain the support of the so-called silent majority, the property owners and taxpayers, while at the same time publicly proclaiming that extremist groups of either the Left or Right are not welcome.

The impact we have had has been made on the basis of about $6000. We have found that the organization cannot live on the individual membership fees; if they are set high enough to provide substantial income, they will discourage membership, and if they are realistic, they may pay for the postage, but no more. Large contributors must be found. The executive committee must identify individuals and groups which have a stake in the affected area and make a case for significant contributions. (For example, absentee owners of large summer estates have no protection except that which the committee can provide. We have gone into two of the major feeder cities and encouraged the formation of fund-raising groups. Efforts of this nature are conducted on an informal, person-to-person basis with relatively low visibility.)

We have learned to identify the decision makers. This identification is important so that they can be bombarded with mail from the hundreds of individual members. A letter from the executive committee alone will not advance the cause.

We also learned to identify the existing environmentalist organizations and to work through them to gain individual members, group affiliations, and independent resolutions against the project. Finally, we have sought the support of businessmen's associations whose livelihood depends on an unspoiled environment, such as the tourist promotion organizations.

Project Sanguine represents a new kind of pollution, bordering on the realm of science fiction and having behind it the force of the military-industrial complex. More and more, however, the environment has behind it the citizens of the nation. We had underestimated the enormous amounts of time and energy that this kind of fight would demand of us, but we also underestimated our chances of success. Although we have been exhausted, we have been encouraged by the thousands of people from diverse backgrounds who have been willing to let differences lie dormant and unite to save the common environment. Our faith in citizen participation in democracy and in the ability of the average man to affect the conditions in which he lives has been reaffirmed.

E. PRESERVING QUIET CORNERS OF THE EARTH

Society bases criminal justice on laws, which in turn are derived from moral precepts that have developed over millennia. Upon what basis should land-use decisions be made? If there is a mountain and upon it you want to hike through unspoiled wilderness while I want to build a ski resort, how can we decide what to do? The vision of Woody Guthrie that "this land was made for you and me" haunts us at times as a sardonic reminder of our failure to cope with this problem. Those cultural and perhaps instinctive land ethics that most of us do share, such as the notion that "a man can do what he wants with his own land," have only contributed to this failure.

Many trends in our society act to homogenize the land and make every acre the equivalent of every other. But certainly there ought to be open land from which we will never see an airplane and mountain tops from which we will never see a ski tow. Just as a nation's economy or an ecosystem is more stable, the more it is diversified, so the stability of man may be linked to the diversity of his experiences which derive from the land. Perhaps variety, then, is a precept upon which we could begin to base a land policy that respects both land and man. Reversibility, too, seems important. Irreversible changes in our land are more devastating than reversible ones. Often this can be interpreted as a statement about the magnitude of man's activities on the land, for when certain limits are exceeded, there is no returning. And irreversible changes deprive future generations of the right to formulate *their* land policy.

The following two case studies deal with these issues, with the ecological consequences of ill-conceived land-use decisions, and with the politics of land-policy reform.

11. MINERAL KING: WILDERNESS VERSUS MASS RECREATION IN THE SIERRA

Michael McCloskey and Albert Hill

This country's ecological awakening is revealing the trap that is implicit in planning to meet projected demands and doing so with single-minded mission-oriented thinking. The environmental dilemmas posed by this kind of thinking are dramatically illustrated by the plans for a giant resort at Mineral King in California's Sierra. Mineral King is a glaring example of an approach to planning that becomes completely unacceptable once certain thresholds of size and impact are passed. In a number of quantum jumps, the project has grown to such a size that all other values are in danger of being overwhelmed.

The resort is proposed on a tract of national forest land that is tucked into an indentation on the south side of Sequoia National Park. (See Figure 1.) Feeling that Southern California needed a huge ski resort, and that the Forest Service needed a showpiece recreation complex, the Service issued a preliminary permit in 1965 to Walt Disney Productions to develop America's largest mountain resort. Conservationists, who felt that Mineral King belonged in Sequoia National Park, launched an immediate protest, and the matter is now entangled in litigation they have instigated. They claim that necessary studies have not been done to safeguard the area's fragile qualities, and that the Forest Service is blandly asserting it can handle the numerous problems it has hardly begun to investigate.

Today Mineral King is a largely unspoiled near-wilderness region. Largely surrounded by Sequoia National Park, the Mineral King valley and the surrounding high country form a truly spectacular scenic area. For over a century visitors have been making the trip to this remote area to enjoy its splendid scenery, its natural curiosities, and its quiet unhurried atmosphere.

Mineral King valley is the heart of this region—quite small, only about 300 acres—but more important than its size would indicate, since it is the natural scenic and ecological focal point of the area. The boundaries

165

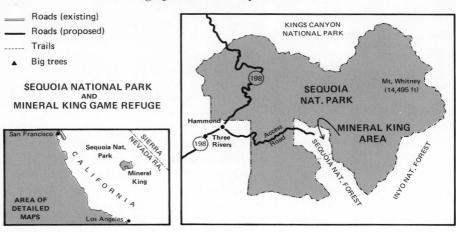

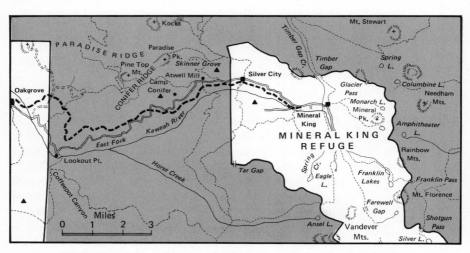

FIGURE 1 Maps showing the location and the detailed features of Mineral King Refuge.

of the region, except on the west, are a hydrographic divide, with Mineral King forming the headwaters of the East Fork of the Kaweah River, which flows out of the valley to the west across Sequoia National Park. The valley's floor is only at an elevation of 7800 to 7900 feet. However, it is open and alpine in appearance, largely because the persistent avalanches that sweep down the steep slopes periodically thin out any trees—or for that matter buildings—in their paths. In all directions high peaks, some up to 12,000 feet, present imposing views.

For its size it is overendowed with the things a visitor looks for in a mountain valley. Five major streams—Monarch, Crystal, Franklin, White Chief, and Eagle—and a number of smaller ones form cascades down the valley's walls. On their way to join the main stream of the valley, each side stream waters its own flower garden; tiny white orchids, lilies, gentians, sneeze weed, and countless others. These gardens merge into acres of meadows in the lower parts of the valley. Around its edges a scattering of cottonwood and aspen groves provides summer shade and sometimes, in season, bright spots of autumn gold.

Each of the major streams, and several others that join the Kaweah just outside the valley, flow from glacially formed hanging valleys, each one a separate lake basin. Of the approximately twenty-two lakes found in these high valleys, each one has its own character, its own unique setting; some are rimmed with tall pines, others are above the tree line, and still others are nestled up against great stone walls left from the time when the glaciers carved out these basins.

Much of what is interesting and beautiful in this high country is fragile, easily destroyed by foot or by the blasting and earth moving that would accompany development: the perfection of a group of flowers along a stream bank, or an area of polished rocks showing the marks made by the movement of the glaciers that long ago covered this High Sierra region. Although the Mineral King basin includes 15,000 acres, the impact of development will most likely be concentrated in the areas that are the most fragile and that form a relatively small proportion of the total: streamsides, lakesides, and meadows.

From atop the great walls at the tops of the cirques of these glacial valleys, you can look out across the Mineral King region at a panorama of contrasting geologies not often found in the Sierra Nevada. Turning around, you look into the wilderness high country of Sequoia National Park, and just to the east of Mineral King you look into one of the last great wildernesses, and among the most remote, of the entire Sierra.

There are many other things to see in Mineral King: bubbling soda springs, old mines, disappearing streams, and marble caves. Located at the end of a winding, narrow 25-mile-long road, Mineral King has been just far enough from the crush of visitation to preserve its fragile beauty from overuse and overdevelopment. One thing is certain: the visitor to Mineral King leaves behind smog, freeways, ringing telephones, crowded checkstand lines, and a great many other "benefits" of modern life, and finds blue skies, white clouds, clear lakes, green trees, rushing streams, and only the sounds of nature. Places like Mineral King grow fewer each year. It is all too easy for them to fall to ill-conceived developments, and it is almost impossible to reclaim them.

The Background of the Dispute

In 1926, in the same act which enlarged Sequoia National Park to twice its previous size, 15,000 acres of the Sequoia National Forest were designated the Sequoia National Game Refuge (popularly called the Mineral King Game Refuge). Previously an undesignated region of the National Forest, it apparently was excluded from the enlarged park because of mining interests still active in the area, remnants of the silver rush of 1872–1888. Since then, the mining operation and the few remaining structures associated with it have largely disappeared, and the principal value of the area is acknowledged to be for outdoor recreation.

As a culmination of interest originating in the 1930s, the U.S. Forest Service (a part of the Department of Agriculture) issued a prospectus in 1949 seeking private capital to develop the area for winter sports. Calling for a development that would be modest by today's standards, they were not able at that time to interest a developer because of the lack of an adequate all-weather highway from State Highway 198 to the valley. In 1965 the Forest Service issued a second prospectus, outlining a winter sports development projected to cost a minimum of $3 million. This time the Forest Service imposed a condition on the recipient of the 30-year operating permit that he secure construction of an all-weather highway to the valley before the permit could be issued. Six bids were received by the Forest Service that met the minimum qualifications. Four of these were for relatively modest developments, and two were for developments in the $35 to $40 million range. Ultimately the $35 million development proposal of Walt Disney Productions was accepted, and a three-year preliminary permit for further study was issued in December of 1965.

At that time no agreements had been reached with the National Park Service (a part of the Department of Interior) concerning highway construction across Sequoia National Park or access to the park from the Mineral King area. Yet the Forest Service issued a permit to a private developer, calling on him to develop a plan that would require access across the park and would undoubtedly have an impact on it.

Later, in 1968, the Forest Service's Regional Office promised to provide a period for the public to review the developer's plan before it was adopted. Conservationists were hopeful that major modifications would be made. In mid-December the Forest Service indicated that a limited impact study it was conducting had not yet been finished but "hopefully will be some time this winter."

Nonetheless in January of 1969, without notice or opportunity for review, a master plan differing in detail, but in outline essentially in agreement with the Disney Company's original proposal, was accepted by the Forest Service. However, the 30-year development permit was not issued at that time because agreement had not yet been reached over the access road crossing the park. The issuance of this permit has been further delayed by legal action initiated by the Sierra Club.

The Disney Development

The original Disney proposal for a $35 million project estimated that there would be 2.5 million visitors annually, 60 percent of them in the summer. Subsequently this figure has been reduced twice, first to 1.7 million and later to 980,000. The reasons for the reductions are not altogether clear, but they stem at least in part from the elimination of camping of various types in and near the valley. Even at this lower figure, and some expect it to be exceeded, the visitation density in the valley will still be 3 to 4 times as high as in overcrowded Yosemite Valley. A high proportion will still be visiting the area during the spring and summer seasons when the snow does not cover the ground and the spring growth is vulnerable.

The facilities planned to accommodate the visitors are truly enormous: an 8- to 10-story parking structure for 3600 vehicles including buses, a 1030-room hotel complex for visitors plus extensive accommodation for the nearly 1000 employees (some will be housed in the valley and others nearby), and a lift system employing between 22 and 27 lifts of various sizes and types extending up to 4 miles into the surrounding country.

In the vicinity of the parking structure will be an "auto reception center" with extensive areas for storage, food preparation, a wardrobe department, administrative offices, hospital, heliport, and shops. From the auto reception center the visitor can travel to the village on "colorful excursion trains," where he will encounter the hotel complex, numerous shops, restaurants, a convention center, theater, and an equestrian center and arena complex. Other major facilities include a complex located out of the valley near an area known as Miners' Ridge which will be for ski instruction and presumably other purposes, and a restaurant at a high elevation near an area known as Eagle's Crest.

The list of restaurants is impressive:

> Probable Mineral King food-service facilities in Year I will be a popular-priced buffeteria near the "family" rooms in the Village with 150 seats; a village center konditorei, or tea room and coffee shop, with a capacity of 100 seats; a 100-seat "teen center" near the "family" rooms; a gourmet restaurant with beverage service atop a village hotel, again with 100 seats; another popular-priced buffeteria, particularly attractive to one-day guests, at the automobile reception center with 200 seats; a quick-service snack bar immediately adjacent to the parking area buffeteria with a capacity of 150 seats; a 200-seat buffeteria at the Midway Gondola Terminal at Miners' Ridge, with beverage service, a facility which will convert to waitress service in the evening and for special occasions; a 150-seat coffee shop and snack service at the enclosed lift terminal on Eagle's crest, and a medium-priced waitress-serviced restaurant atop the village gondola station with beverage service, the capacity 150 seats.
>
> By Year 5 the buffeteria near the "family" rooms in the village would be expanded from 150 to 300 seats, and the reception center buffeteria from 200 to 300 seats. Additionally in Year 5 we will add a snack bar and sandwich shop at the Midway Gondola Terminal with a capacity of 150. A multi-themed 350-seat restaurant in the Village, with beverage service; a 300-seat, waitress-served restaurant just south of the automobile reception center, and a snack service facility in an area immediately adjacent to the centralized village ice-skating rink. [From Master Plan Presentation, Walt Disney Productions, January 8, 1970.]

It is important to remember that these facilities cannot stand alone, but require the development of a substantial water supply with associated water storage facilities, the development of sewage treatment plants, and local generation of power or its importation through transmission lines that would run across the park.

A development of this size will also require the construction of a substantial highway across Sequoia National Park with a capability of handling peak loads of at least 1200 cars per hour.

The Case against the Development

All of these facilities are needed to serve the huge numbers that will be attracted to the "mecca" that the Disney Corporation and the Forest Service plan for the area. The development of such an attraction goes far beyond their claims of meeting outdoor recreation needs. The difference in size between a $3 million development and a $35 million one was enough to alarm conservationists, but this difference does not fully reveal an equally, if not more alarming, underlying premise. This is the idea that you can vastly increase visitation and still maintain a reasonably natural condition by skillful design and cosmetics. It is specious to assert that you can build a hotel and entertainment complex, a so-called self-contained village, as large as in our major cities and still preserve a sense of naturalness and the integrity of a fragile ecology and its beauty.

In these high elevations the growing season is short and it is difficult for plants to become established. Any damage to the plant life may take many seasons to repair, and this disturbance, even if slight, may continue to be felt season after season. If it is heavy, as it will be in a high density recreation development, the plant life will ultimately disappear from the disturbed areas. Following disappearance of plant life, the soil can be more easily eroded. Once the soil is gone, there is no base for the reestablishment of vegetation, and it may take centuries for the soil to be rebuilt.

All of this development also is to take place in a game refuge. The most heavily impacted area, the valley, is a key part of the refuge because it is the only area of lush vegetation and is the summer range of deer and other animals that range along the East Fork of the Kaweah River. Under continual disturbance, many animals will simply leave an area, even if their habitat is not completely destroyed. No studies have been conducted to find out what the *total impact* will be on animal populations or on the plant life, some of which is unique, including the oldest known foxtail pines. On any large tract of National Forest land the public should be able to expect these matters to be looked into; in a refuge they can reasonably expect even greater care to be exercised. You cannot wait until after a project is completed to find out what damage will be done. By then the decision is largely irreversible.

A development of this size will unquestionably have a major impact on the ecology of Mineral King and on portions of Sequoia National Park.

Conservationists find it difficult to say exactly what this will be for two important reasons. First, needed ecological research has not yet been done in the area, and conservationists do not have the resources available to do it themselves, nor can this be expected of them. It is the government agencies that should have had these studies done, for they were needed as a basis on which to decide whether development should take place. Second, the master plan that has been approved by the Forest Service apparently allows for such great flexibility that major changes can take place, even within the sketchy outline of planning that is presented. It is impossible to analyze accurately what the impact will be when it is impossible to establish exactly what will be built and where.

It would seem that the public should have a right to know *exactly* what is going to be permitted before the project is authorized. Will a reservoir cover a large portion of the meadows in the valley? Will there be commercial helicopter service to the development across Sequoia National Park? Both of these have been, but reportedly are not currently, planned. There appears to be no guarantee, however, that both of these objectionable features could not be included at a later date, along with a number of others. The truth is that, except in a vague way, approval of the master plan by the Forest Service does not appear to specify what kind of development and operation will be built at Mineral King. The Disney organization defends this flexibility in the master plan by saying that it allows them to improve their plans up to the last minute before construction. But improve to whose standards? How large a role will economics play in setting these standards when perhaps other factors should apply? The Forest Service says it will have the final say before anything is done. Perhaps it will, but will the Service apply the same standards that have been applied to other ski areas, some of which are a blight on the landscape? These standards are not high enough for Mineral King or other areas. Actually, the Forest Service itself cannot tell what standards can be maintained without making the ecological studies that, as far as is known, have yet to be made.

In the alpine and near-alpine high country, where a harsh climate and thin soils are a constant challenge to plant life, how much trampling and disturbance can the meadows stand before the balance is tipped against them, and the area becomes barren? How will this use increase erosion? With 22 to 27 lifts, what grand views will be left without towers and cables? Here, in this clear atmosphere, things that are miles away appear close at hand, and construction scars will be clearly evident. In constructing the lifts and other facilities in the high country, will the materials and labor be brought in by helicopter, or will a road which will cause considerable damage be constructed? Will the continued maintenance and service of the lifts and restaurants be done by vehicular access or by air? Neither the Forest Service nor the Disney organization will commit itself as to which method will be used.

In its original proposal, the Disney Corporation indicated some idea of what it felt would be necessary:

Grooming and manicuring of most slopes without destroying the naturalness of the area particularly for intermediates will require extensive bulldozing and blasting in most lower areas and extensive rock removal at higher elevations. This is especially true for the Mosquito, Eagle, and White Chief Bowls and the lower Farewell Canyon area.

Other effects on portions of the area outside the valley will arise out of avalanche and flood control work. Debris dams will be constructed on the tributary streams to protect the main development from flood damage. An unknown amount of terrain will be modified in conjunction with snow fences that will probably be used to minimize avalanche hazards. A water supply and sewage disposal system will have to be developed for the remote facilities.

Although no agreement has been worked out with the National Park Service, visitors to the development will be dropped by lift within easy reach of Sequoia National Park. What facilities will be necessary to protect the park and to provide for the visitors' needs and safety? In the case of one ski lift, visitors will have to ski through a portion of the park to return to the Mineral King area.

The clearing for ski runs and other disturbances will increase runoff and the debris that will be carried in the streams. This will aggravate the highly disturbed conditions in the valley, which, even without additional disturbance, shows a strong tendency toward erosion. Increasing erosion will increase the danger of flooding and washout; artificial channels may be required for Monarch Creek and the East Fork of the Kaweah River to handle increased water flow and to protect the village.

In addition to the scarring already mentioned, the access road to serve the development would cut through unique portions of the Giant Sequoia groves, endangering a large proportion of the trees below the road. A study conducted for the State Division of Highways by Dr. R. J. Hartesveldt concerning the possible effects on the Sequoias raises these alarms:

> There is much evidence of soil materials on the move in these two groves and I believe that it merits the most serious engineering consideration before the road-building commences. . . . In all, there are 103 sequoia trees below the proposed roadway centerline, of which 45 could be affected in some manner by the construction of the road.

Later in the report he comments on a proposed bridge:

> In spite of plans to bridge this deep canyon because of the depth, accelerated erosion is very likely to be a major problem no matter what engineering activities occur there. . . . Of the 32 groves of giant sequoias familiar to the author, this drainage-way provides one of the most unique communities of plants he has seen in association with sequoias. This in itself makes the utmost protection desirable.

Below the valley, new problems are emerging along the proposed access road. A large private inholding within the Mineral King Refuge and adjacent to the National Park is the site of plans for a condominium-hotel complex that could bring additional millions of dollars of development to the area and would impose an additional load on the highway and available water supplies. So far, the county has rejected the developers' plans, but there is always the possibility in the future that Tulare County might find it economically expedient to approve them, much in the manner that it has encouraged the Disney project.

In Sequoia National Park the new road will be particularly damaging. The old road follows the terrain with much twisting and turning and does little damage to the canyon, which is steep and dissected by the smaller side canyons of tributary streams. On the other hand, the new road will have curves with a minimum radius of 850 feet, which will require extensive cuts and fills and bridging to handle higher speeds. The Clarkeson Engineering report, prepared for the National Park Service, raises strong doubts that this "two-lane-with-passing-lane" design will be adequate for the projected peak traffic flow of 1200 cars per hour. The design certainly does not provide for passing lanes adequate for winter use when there is the need to install and remove tire chains. There is a strong feeling that the current design is merely a foot in the door: it is less expensive and less damaging than what will ultimately be required.

This road does not serve any park purposes—indeed does not even fit into park plans; it will only serve to bring to the park the noise and air pollution of up to 1200 cars per hour and will overwhelm existing facilities. Who will find it desirable to camp at a park campground that is only a few yards from the edge of a highway with its day and night traffic?

It is clear that Sequoia National Park and Mineral King have an intimate relationship. Geographically and topographically Mineral King is isolated from the rest of Sequoia National Forest, but is surrounded by the park. Water that flows through the park originates in Mineral King and, reminiscent of the Everglades National Park situation, any change in its quantity or quality will affect life in the park. The only reasonable access to Mineral King is through the park. Increased access to the high country of Mineral King will also increase access to the high country of the park. The wildlife of the two regions is linked.

Conservationists seek a workable solution to the problem of using Mineral King's outdoor recreation potential while protecting the natural values of both the park and Mineral King. Such a solution does not appear to be forthcoming with separate administration of the two areas. There is no doubt about the fact that Mineral King is a region of National Park caliber and that the purposes to which it is best suited are compatible with National Park objectives. Coordinated planning and protection of the areas can best be achieved if Mineral King is added to the National Park.

The Narrow Vision of Bureaucracy

If Mineral King remains in the hands of the Forest Service, a workable solution seems unlikely. The Forest Service has always assumed that development at Mineral King should be pushed, that it was in the public interest. Building a resort was the assigned mission. This assumption grew out of studies done in the 1930s and 1940s which identified the skiing potential of the area. Because of growing evidence that a ski resort could be developed, the Forest Service grew to believe that it should be. There was never an opportunity to test the hypothesis that "what could be done, should be done."

Once the commitment to the mission became fixed with the Forest Service, anything that might call the decision into question was looked upon as an obstacle to be overcome. Justification studies were not done. Ecological studies were not done. Only snow and weather studies and development planning studies were done. These aided the project; the others might impede it. No judicious weighing process ever took place. By the 1940s the Forest Service had become committed, and it then became just a question of how to find the capital and the developer that would bring the project into being.

In the tradition of mission-oriented agencies, the Service as a whole simply maintained that all problems would be adequately dealt with. The Service, which nationally conceives of its main mission as cutting and selling timber, is accustomed to thinking that the landscape can be treated roughly without worry. It was sure some sort of mitigating provision could be put in the plan to deal with each objection that might arise. Since 1965 the Forest Service has engaged in patch-up planning to respond to various objections. A high-rise auto garage will be built instead of using up lots of space for parking. To avert the problem of traffic in a narrow valley, the garage has been moved steadily down valley. To respond to the problem of separating paying visitors from campers, provisions for camping have been phased out. In response to suggestions that the area will be too crowded, the initial traffic estimates have been cut back. Significantly, most of these problems were not anticipated by the planners. These were the planners' responses to public criticism. The planning process has proceeded in a single-minded, defensive fashion.

Not only has the Forest Service assumed that ecological problems could be overcome, it has also assumed that there were no real ecological limits on what it might do in the valley. The concept of any natural carrying capacity for the basin is absent from its planning. No studies have been done on the relationship of foot traffic to ground vegetation. No studies have been done on summer range needs of migratory deer herds, although the suggestion has been made that deer numbers should be reduced to minimize problems of conflict. This means "kill the deer" to solve the problem. No data have been gathered to suggest at what levels of human use, and in response to which patterns, the natural flora and fauna begin to decline, and whether there are any sharp dropoffs in the decline. In a wildlife refuge and an area that qualifies for national park status, the Service is flying blind. Its only studies are on how to replant bulldozed ski runs.

What is worse, historically the Service has assumed that the bigger the project is, the better. What was conceived of as a project costing a couple of hundred thousand dollars in the 1940s jumped to a $3 million mini-

Looking east across the Mineral King Game Refuge to the Great Western Divide in the distance. This view from near the Refuge's boundary with Sequoia National Park shows (right foreground) the lower portion of the glacially formed hanging valley below White Chief Peak. The Mineral King Valley is just behind and down from the foreground ridge on the left. The area would be extensively developed for skiing and other activities. (Allen Malmquist, Sierra Club)

mum project in February 1965, and to a $35 million project in late 1965. Now with accompanying private developments proposed by other entrepreneurs, the prospects are that $50 million to $75 million may be spent along the East Fork of the Kaweah River. Through a series of quantum jumps in size, the character of the development has changed. It now threatens to be a monstrous blot on the landscape.

The Forest Service has placed no ceilings on size in the use permits. The idea of ceilings seems foreign to the Service; it thinks of minimum requirements only. The Forest Service will look at each new increment of development on a case-by-case basis, and it is bound to be caught facing a developer always claiming he needs the added development to make a profit. Once developers have their investment in an area, experience proves how nearly impossible it is to control their demands, particularly when no limits were set in the first place.

The only limit the Forest Service has promised is to restrict skiing numbers to what the lifts can handle. Subject to this vague standard, additions will be made as needed to the parking garage, and overnight accommodations will be added to respond to the market. As the late Walt Disney himself said: ". . . our efforts now and in the future will be dedicated to making Mineral King grow to meet the ever increasing public need. I guess you might say that it won't ever be finished."

Action against the Development

The Sierra Club asked for hearings immediately after the Forest Service issued its second prospectus early in 1965. The club was turned down by the Forest Service, which held that it was engaged in a routine matter of issuing a special use permit. Thousands of these are issued for everything from a cabin permit to a right-of-way for a telephone line. However, these vary drastically in terms of impact and permanency. At the scale to which this project was now multiplied, it probably represents the largest and most permanent private installation ever contemplated on public land. The Service, however, treated it as if it were no different from any other. In terms of importance, the issue exceeds many that Congress deals with every day. The character of the land involved and the sums at stake make this a national issue, but the Service claimed this was exclusively a matter within its administrative prerogatives.

Not since 1953, when the Tulare County Chamber of Commerce held an invitational meeting, has the public been provided with a forum to raise objections to the development. The Forest Service has been unwilling to hold even one public hearing; it had made up its mind and did not want to clutter up the record with objections.

The California authorities, charged with building the access road, were equally reluctant to open the matter to public scrutiny. The road was placed in the state highway system in July of 1965 by means of a rider on another bill. No legislative hearings were held. By clandestine maneuver, the then president of the Senate was able to settle this matter. At meetings of the State Highway Commission, the question was never "Should the road be built?" It was always "Where should it be built, and when, and at what cost?"

The closed nature of the planning process forced the critics to shift their tactics. After failing to get a hearing from the Regional Forester in the summer of 1965, they appealed to the chief of the Forest Service in Washington and then to the Secretary of Agriculture. These appeals succeeded in delaying the award of a preliminary permit from September of 1965 to December of that year. Basically, however, the Secretary ended up ratifying what the service proposed.

Beginning in December 1965, when the Forest Service issued a preliminary permit to Walt Disney Productions, a three-year-long contest of wills developed between the Secretary of Agriculture and the Secretary of the Interior. The Interior Secretary was reluctant to grant the necessary right-of-way permit through Sequoia National Park for the new access road, because the Park Service's own studies showed it would actually harm the park. The critics of the Disney project had maintained a similar position at a hearing on wilderness zoning within Sequoia National Park; at the same hearing, they expressed their fears about misuse of the adjoining Forest Service land, complaining, alas, to the wrong agency.

However, by 1967 heavy pressure stemming from the White House was brought upon the Secretary of the Interior to relent. In a complicated set of trade-offs engineered by the Bureau of the Budget to keep peace between the two contending departments, Interior was instructed to promise to grant the right-of-way when it was needed. The trade-offs involved land exchanges in connection with a Redwood National Park and a jurisdictional transfer involving land needed for the new North Cascades National Park. Secretary of Interior Stewart Udall, acting under instructions, then did agree to the access road, but he imposed conditions that it always be limited to no more than a two-lane road and that it be constructed in a way that would minimize damage to the park.

The context in which these trade-offs were negotiated again precluded public involvement. By their nature these were secret proceedings which only happened to be exposed as an outgrowth of the heated controversy over establishing a Redwood National Park. Without public knowledge, the President made a decision to prevent the Secretary of the Interior from doing his duty, as he saw it, of protecting the parks that were, by law, placed in his care.

The secret and closed nature of the process by which decisions were made in this case left the critics with no recourse but the courts. In addition to all the policy grounds for questioning the project, there were grave questions about its lawfulness. When all other avenues of recourse had been exhausted and the Forest Service made it clear it was about to proceed with issuance of a final permit, the Sierra Club filed suit to stop the project. The lawsuit was filed in June of 1969 against both the Secretary of Agriculture and various Forest Service officers and also against the Secretary of the Interior. It sought both a finding that the project was illegal and preliminary and permanent injunctions against any steps being taken to build the project. In July of 1969 the Federal District Court in San Francisco granted the preliminary injunction. Thereafter the government appealed the decision to the Court of Appeals for the Ninth Circuit, which overruled the district court. The Sierra Club has now carried the appeal to the Supreme Court.

In its allegations against the Secretary of Agriculture, the lawsuit charges that the project will sprawl over more than the 80 acres that Congress has set as the limit for resorts developed under lease on national forest land. In point of fact, the resort will cover some 300 acres. The Service intends to put this excess acreage on so-called year-to-year leases while putting only 80 acres on a 30-year permit as provided for by statute. The suit alleges this arrangement is "a clear and patent effort to circumvent the 80-acre limitation."

The suit also alleges that the proposal to put a giant resort in Mineral King conflicts with the designation that Congress bestowed on the area in 1926 as the Sequoia National Game Refuge. The statute permits only such uses of the refuge as are "consistent with the purposes for which said game refuge is established." How can a $35 million development in the heart of the refuge's best habitat possibly be considered a compatible use? The suit cites a state game authority who says "considerable habitat would be lost and wildlife would suffer from human encroachment," and a Forest Service wildlife authority who states: "The extent and nature of the proposed alteration of the basin is unacceptable to us—the damages extend beyond the effects on fish and wildlife, and these alone are critical."

The Forest Service's jurisdiction over the refuge is also challenged. The club's brief asserts that jurisdiction over wildlife refuges was transferred to the Secretary of the Interior by the 1939 reorganization act. The suit also charges that the Secretary of Agriculture and the Forest Service have acted arbitrarily and capriciously in failing to consider necessary factors in planning the development, such as its impact on ecology. It also charges them with failing to follow proper principles of administrative law, especially in refusing to hold public hearings on the proposal.

Because the Secretary of the Interior was on the verge of issuing a right-of-way permit for the access road through Sequoia National Park, he was also joined as a party. The suit points out that the organic act of the National Park Service and the act establishing Sequoia National Park make the Secretary of the Interior responsible for preserving the natural objects, wonders, and timber in the park from injury. If the Secretary were to approve such a road, which could injure the natural objects of the park and does not serve its purposes, the suit alleges he would be violating his duty under these various acts of Congress.

The suit also claims that construction of the road violates administrative procedures of the Department of the Interior. It asserts the road is not being constructed in conformance with established standards and policies for park roads, nor are requirements for hearings on routing and design being followed. The suit asserts finally that congressional approval is required before an electrical power transmission line, needed by the project, can be routed through the Park.

It may be some time before the litigation is settled.

Implications

The central premise of the Mineral King project is that Southern California's skiers need more ski areas. The argument is put forth that their skiers have to travel too far and that the existing areas are crowded. Some of the nearest areas that might be developed are in wilderness areas, and the opportunities for new developments are limited. The advocates of skiing development assert that it is only fair that more remote, large capacity sites like Mineral King should be developed to meet this need.

At first glance this argument sounds plausible, even appealing. But the thinking behind it is very much like the thinking other developers advance: "Our electrical consumption doubles every ten years, therefore we must build more power plants." "We must continue to build more freeways to relieve traffic jams."

We are beginning to find out that the new freeways will be crowded too, and that the new power will be used quickly. And as we are finding that anything we provide will find users, we are also finding that our environment gets a bit less tolerable with every new urban freeway and power plant.

Thus we realize that we lose something with every new development as well as gain something. In balancing the losses and gains, we must look at the overall trends in our environment. The quality of the environment is deteriorating because of a host of decisions to move ahead with more development, each of which can be rationalized as based on a proper individual balance between losses and gains. In sum total, however, these decisions push us further and further into an intolerable situation: an environment overloaded with development, with traffic, with smog, and damaged habitats.

To stop the trend toward progressive degradation, there is a growing conviction that we must not only curb population growth but must also question the assumptions of those who say we must keep meeting projections of demand. Meeting these projections may be no more than an exercise in building self-fulfilling prophecies. These new developments will be attractive enough to build new clienteles, which can be retrospectively offered as proof that demand was in fact being met. Of course, it is equally arguable that if another option were offered—one involving environmental protection—a clientele could be developed for it too. In this instance, it is entirely likely that inclusion of Mineral King in Sequoia National Park would attract a large body of devoted park visitors. Although one could argue over comparative numbers of visitors under both alternatives, it is quite possible that just as large, if not a larger, sum total of satisfactions would accrue annually to park visitors as to visitors to the resort.

Moreover, once you play the game of meeting projected "demands," the question always remains of what you do next once the area fills up. The logic of "building, building, building" to meet demand suggests that the process must keep on going indefinitely, or at least as long as population keeps growing and entrepreneurs promote new markets. However, it is patently evident in this case that it is physically impossible to keep on going indefinitely. There are just not that many additional sites for major development. Sooner or later, the process of building ski areas must come to an end. With that in mind, we must ask: "In what shape ultimately do we want to have our mountain landscape?" "What is the balance between development and preservation that will really best serve society?" If skiing demand in any event is going to be left unsatisfied at some later stage, why let it go on proliferating past the point of proper balance? Shouldn't we really draw the line to keep fragile landscapes, such as the Sierra, in the best shape we now know how to provide, and keep overpowering projects like Mineral King out?

Another disturbing aspect of the case for the Southern California skiers is the unspoken assumption that there is some obligation on the part of public authorities to provide expanding opportunities for every kind of sport, no matter what the inherent obstacles. Clearly no part of the southern United States is ideally situated to serve skiing. The region suffers from basic climatic disabilities, just as northern regions suffer from disadvantages in not having sufficient opportunities for warm-water ocean bathing.

Nevertheless some California planners seem to feel that engineers should be able to rescue Southern California from all its climatic disabilities, from lack of water, to air, to skiing. If the area is too far from enough water, their answer is re-engineer the state's northern rivers to run them southward. If the area lacks nearby skiing, the answer is to move higher into the Sierra with subsidized access and plans for high volume use to make skiing there financially feasible.

In both cases the resource is bent drastically to rescue a demanding population from the shortcomings of a particular area in which they have chosen to locate themselves. The question can just as well be asked for skiing as for water: "Why not locate the people in places where these opportunities can be afforded with far less violence to the landscape and its resources?" Certainly, people are not going to move out of Los Angeles in droves just because Mineral King is not built, but preventing its population from reaching farther and farther to abuse resources may gradually make Los Angeles a less and less attractive place to live. This is one way to bring the Los Angeles syndrome to an end.

Conclusion

The course of events that gave birth to the controversy over Mineral King can be instructive for all public resource agencies. The proposal's history shows how an agency can be trapped into assumptions that may not stand the test of time and changing needs; how a plan cannot be projected into larger and larger formats; how difficult it is to get ecological thinking injected into planning once a developmental goal is assumed; and how closing the public out of the planning process is only likely to lead to litigation and further delay.

Mineral King stands as an object lesson of how not to plan a public resource. The answers about what can ultimately be done in Mineral King must await further studies, but it is quite clear that not enough study was done to justify the project that has been proposed. As it is now proposed, Mineral King is a symbol of what cannot be permitted to overtake the Sierra or our other fragile natural regions.

In one sense, the Mineral King controversy appears to be a forerunner of things to come. But in another sense, it is an example of an old kind of thinking that has outlived its usefulness. Controversies such as at Mineral King will provide a contest between ecological thinking and the pressures of the entrepreneurial spirit.

One can hope that public agencies will not continue to have such a hard time deciding which side they are on.

12. THE EVERGLADES: WILDERNESS VERSUS RAMPANT LAND DEVELOPMENT IN SOUTH FLORIDA

John Harte and Robert H. Socolow

Imagine yourself in the midst of a flat, vast expanse of American wilderness. The shrieking sound of "kree-ah" "kree-ah" pierces the night air and you think of western prairies. But it is the limpkin's cry you hear, and the land surrounding you is under four feet of water—it is the sawgrass marsh community of Everglades National Park in Florida.

The Seminole Indians called the Everglades Pahayokee, or River of Grass. Everglades National Park is located at the mouth of this river, at the tip of the Florida peninsula. The park, the third largest in the country after Yellowstone and Mount McKinley, contains an abundance and variety of wildlife to be seen nowhere else in the United States. Perhaps most impressive are the anhingas, sometimes referred to as water turkeys or snake birds, and the large wading birds, including the roseate spoonbill, the great white heron, the wood ibis (actually a stork, the country's only stork), the white ibis, and the limpkin. So productive are the soils and the waterways in the park that these and over 300 other species of birds are supported here, in some cases in great density. Although the mammals, fish, and reptiles are somewhat more elusive than the large wading birds, they are no less exotic; such species as the alligator, the porpoise, the Virginia white-tailed deer, the manatee, or seacow, and even the rare panther, or mountain lion, find their niche in the Everglades ecosystem.

The plant communities too—such as the junglelike hardwood forests, the cypress swamps, the sawgrass marshes, and the mangrove swamps—are unlike those found anywhere else in the United States. In short, the park is teeming with the plant and animal life of a tropical ecological community.

The park exists today because of the foresight of many individuals who, over the past decades, have loved the Everglades and fought to save them from de-

181

structive abuse by man. The National Audubon Society, which played a large role in establishing the National Park in 1947, is once again deeply engaged in the fight to save the Everglades. Over the years, the nature of the threat to the park has undergone a sinister evolution. Once it was hunters slaughtering egrets for their plumes, and alligator poachers satisfying the careless whim of the fashionable for alligator hide. The poaching still goes on today, unfortunately, and threatens the very existence of not only the alligator but also other species whose life cycle, we shall see, intertwines with that of the alligator. The new threat to the Everglades arises from activity that is not deliberately malicious, but is potentially more devastating because the technological arsenal man now employs in bending nature to his convenience is so formidable that the entire park is in the process of being overwhelmed. Although the fight to save the park is often joined, and rightly so, around such specific manifestations of that activity as the Army Corps of Engineers' flood control project or the proposed new supersonic jetport, the root causes lie deeper in the unrelenting pressure for growth in South Florida. We shall describe here the probable consequences of this pressure for the Everglades and for man himself. We shall see that the well-being of man and the park, in quite direct and material ways, are critically linked. In part I we describe the Everglades ecosystem, emphasizing those features that render it susceptible to collapse. In part II, we describe those present and proposed activities of man in South Florida which threaten the park's survival.

I. The Everglades

The profusion of plant and animal life in South Florida is only the more spectacularly visible part of an intricate and balanced ecosystem. What defines the Everglades and forges its unique qualities is the geology, the hydrology, and the climate of South Florida and their interrelationships with the plant and animal life that flourish there. In order to comprehend the severity and extent of the threat to the park, it will be helpful first to understand more fully the Everglades ecosystem.

The park comprises 2035 square miles of the southern tip of Florida. However, because the park is an integral part of a larger geological and ecological unit, it is misleading to talk about the park in isolation, and we must begin with a description of the entire region of South Florida shown in Figures 1 and 2.

South Florida did not always lie above sea level. Over the past hundreds of thousands of years the sea level has fluctuated in rhythm with glacial activity. As the glacial ice mass advanced southward, more of the sea's waters were locked up in the form of ice, and the sea level dropped; as the glacier retreated, the level rose. As a consequence, certain regions of the earth, including South Florida, were periodically submerged. During those periods in which the sea covered the land, a limestone deposit was continuously being formed on the floor of the sea, thus raising the elevation of the landmass. Now in a period of apparent glacial retreat, the sea level is rising.[1]

In addition, over the past few thousand years, fresh waters flowing southward from central Florida have deposited on the limestone base the silt that they were bearing, further extending the above-sea-level mass. This natural process is still probably continuing today, although its effects are dwarfed, as we shall see, by the influence of man on the balance between dry land and wet land in South Florida.

As might be expected from its geological history, South Florida is extremely flat. Within the boundary of the park, the land is never more than 10 feet above sea level. Only a slight ridge along the east coast, averaging 20 feet above sea level, disturbs the monotonous topography, and upon it squats the urban sprawl of greater Miami.

The same declination of the land which brought the silted waters southward from central Florida still exists today and is of profound importance to the park's ecology. This gentle slope, dropping on the average one inch per mile between Lake Okeechobee and the park, supports a surface flow of fresh water down to the southern regions of the park, where the fresh surface water merges with the salt waters of the Gulf of Mexico and Florida Bay. So gradual is this slope that it takes a drop of water on the average three months to complete the journey from the Lake Okeechobee region to the coast.

Of course, many things can happen to that drop of water to divert it from reaching the park directly. Under natural conditions that drop of water might evaporate, be transpired by a plant, or seep underground into semiporous rock layers, called the aquifer, where it then flows through natural underground channels to the sea. With man's presence firmly established in South Florida, the water might also irrigate a farmer's field and absorb a little DDT, cool an industrial engine, flush a toilet, or quench a human thirst. We shall be concerned in part II of this essay with the magnitude and consequences of man's influence on water flow in South Florida.

Approximately 60 inches of rain a year fall on South Florida, which you might compare with the United States annual average of 30 inches. Very little of that water would remain on the land's surface were it not for the fact that something unusual has been occurring in much of South Florida over the past centuries. The passage of water downward from the surface into the aquifer is retarded in South Florida by a relatively water-impervious layer, called marl, which lies beneath the soil and is believed to be formed from the calcified remains of decayed algae and snails. Despite the fact that the algal mat grows abundantly along the floor of the Everglades marshes, marl formation takes place so slowly that little is actually known about the detailed process. In some areas the marl is as much as a foot or two thick.

We have seen, then, that South Florida can be thought of as a vast,

[1] Estimates are that the present rate of rise is 2½ inches per century.

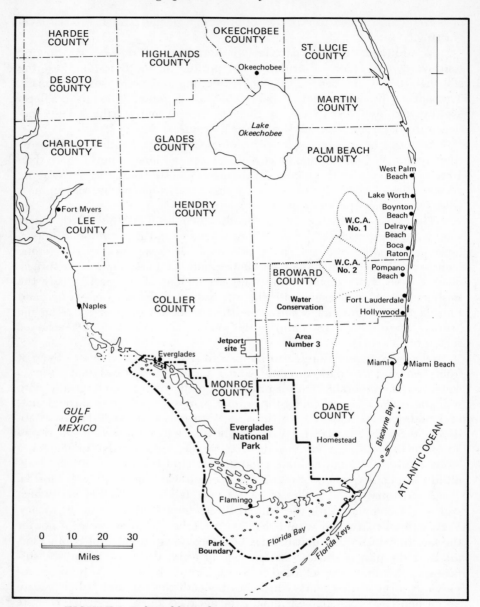

FIGURE 1 Political boundaries in South Florida.

shallow, slow-moving river, flowing from the headwaters in the Lake Okeechobee region south through the park and out into the Gulf; the region of
South Florida between the lake and the northern park boundary thus
serves as a huge watershed for the park.

However, this river differs in a number of respects from most rivers
with which the reader is familiar. In most places it resembles more a marsh

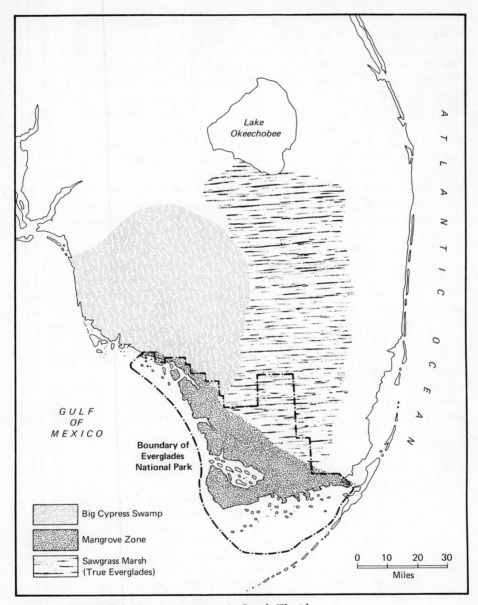

FIGURE 2 Vegetation zones in South Florida.

or a swamp than a river, for considerable vegetation protrudes above the water surface. Second, during a certain portion of the year much of the riverbed is dry; the area covered by water fluctuates enormously. This is because most of that 60 inches of rain, approximately 50 inches of it, falls between May and October, and thus there is a distinct dry season in the Everglades in the winter and early spring. Unusual adaptations of the wildlife to this cycle of wet and dry seasons, or hydroperiod, have developed, and we shall return to them shortly.

A variety of plant communities exist and compete in the park and in the inland regions north of the park; their characteristics are determined, to a great extent, by the hydrological and geological factors we have just discussed and, in addition, by the warm temperatures which average 68° in winter and 80° in summer. The dominant plant communities are illustrated in Figure 2.

In the eastern portion we find the true "Everglades," which is sawgrass marsh. Resembling somewhat a Kansas wheat field under several feet of water, the sawgrass marsh expanses are interrupted only by stretches of pine forest and tree islands or hammocks. Hammocks form either on naturally occurring higher ground where the marl is thicker and higher than in the surrounding marsh or on depressions in the marl in which decayed plant material accumulates around the roots of small plants and becomes peat, thus allowing the transition to larger plant forms.

Hammocks consist typically of large stands of tropical hardwood trees, such as the mahogany, a variety of palms, the coral bean, the gumbo-limbo, and a number of northern trees, including mulberries, oaks, and maples, which grow here at the southern limit of their range. Along the forest floor of the hammocks grow abundant ferns, orchids, ivies, and fruit-bearing shrubs; a hammock is truly a jungle of plants. The pine forests and hammocks provide for wildlife the high ground and protective cover needed for nesting.

Fires have undoubtedly influenced the pattern and growth of hammocks and sawgrass in the Everglades since Indians inhabited the area. An interesting and plausible hypothesis concerning the role of fire has been suggested by the ecologist Frank Egler:

> In the sawgrass country, the tangled herbaceous vegetation is ready to burn even before the soil is dry, while there may still be a few inches of water on the surface. Assuming that Indians were free and careless with fire, it follows that more often than not the fires would get started at the inception of the dry season in fall. In this manner, the fires would skim over the surface, not damaging the water-covered roots. The fires would smack against a dense hammock, and stop, pronto. The hammock itself may be under water; the foliage would be turgid and fire-resistant. Then, at the end of the dry season, when the peat soil and the hammock trees actually could burn, then there was no sawgrass debris on the surface with which a fire could get started. It is only by this hypothesis that I can logically account for the wall-like abruptness of the hammocks, existing quite paradoxically as dryseason-burnable islands in a sea of burned vegetation which sea, without the burning, would quickly be invaded by those same hammock trees.[2]

[2] Frank E. Egler, "Southeast Saline Everglades Vegetation, Florida, and Its Management," *Vegetatio*, Vol. III, p. 213 (1952).

The role of fire may be changing. According to Egler:

> The chief difference between Indian fires and whiteman fires: Indians burned with no conscience, as soon as things would burn. Whiteman, with a conscience, only delays burning, and when the vegetation does ignite, it creates a conflagration undreamed of to the Indian.[3]

Of course, even before man inhabited the Everglades, lightning fires must have had their impact on vegetation patterns in the region.

In order to convey some impression of the richness of life in the sawgrass marsh, we can compare the "net primary productivity per unit area" of various ecological communities. This quantity is defined as the number of dry grams of green plant matter produced per square meter per year, and in the sawgrass marsh averages about 2000. In comparison, in a temperate zone forest, in a cow pasture, and in a desert, the typical values are 1500, 500, and 30, respectively. For the entire earth, the average is 320. Only the oceans' estuaries, tropical forests, and farmland managed scientifically, compete with the marsh community in their net primary productivity per unit area.

Moving across South Florida to the west, in inland Collier County, and not within the boundaries of the park, one finds the Big Cypress Swamp. Here, several decades ago, grew the most magnificent cypresses in the country, often so big that three men could not reach around the trunk. Now few of these giant trees remain, thanks to an unregulated timber industry. Despite this loss, the cypress community is still a vital component in the ecosystem of which the park is a part, and, in its own right, it is a strange, lovely wilderness. Many of the large marsh birds which, during certain times of the year, reside in the park, feed and nest in the Big Cypress Swamp. In fact, a number of birds and mammals inhabiting the Big Cypress area for part of the year are on the Bureau of Sport Fisheries and Wildlife list of rare and endangered species in the United States. Among these are the wood ibis, the roseate spoonbill, the southern bald eagle, the Cape Sable seaside sparrow, the panther, and the alligator. By providing a congenial breeding and feeding environment, the Big Cypress Swamp serves to make their existence less precarious. Moreover, being upstream from the park, the cypress swamp is a vital part of the watershed from which the park derives its overland flow of water.

Moving southward, as you approach the Gulf Coast estuaries or brackish zones, where fresh water and salt water meet, the sawgrass and cypress communities give way to dense labyrinthine mangrove swamps. Mangrove trees grow in the semisaline waters of the estuaries as well as farther out in the shallow waters of the Gulf. The red mangrove, one of the species found in the park, propagates by dropping into the water seedlings which have already formed a simple root system while growing on the parent. These seedlings then float until they reach sufficiently shallow water to form a roothold in the muck. So numerous and intricately connected are the man-

[3] *Ibid.*

grove clusters that the result is a living maze of narrow waterways in which canoeists can become lost for days.

The estuaries are the nurseries of the sea; in them many of the oceans' fishes, crustacea, and other forms of sea life spawn and feed. In particular, the park's estuaries support the porpoise, the manatee, a large number of game fish (including the redfish, the barracuda, and various species of trout), and the large Tortugas pink shrimp. The shrimp breed in these estuaries and then are caught off the Dry Tortugas, islands west of Key West; they are of major commercial value in South Florida. These and many other forms of marine life, especially those lower down the food chain, are quite sensitive to the salinity of the water. If insufficient fresh water moves through the park and out into the estuaries, then oversalination occurs.

The seasonal variation in the rainfall, we noted, gives rise to a seasonal variation in the amount of fresh water that flows overland down through Collier, Dade, and Monroe counties and into the park. Whereas in the late summer the sawgrass community in the park may have been submerged under a depth of up to 3, 4, or even 5 feet of water, in the late winter (there is a time lag of approximately three months) the water level often drops to the point where much of the land is dry. How then do the animals survive the dry period? The alligator, it seems, is the animal that saves them.

All through the Everglades one finds depressions in the sawgrass marshes; these depressions measure anywhere from 10 to 100 feet in diameter and are typically several feet deeper than the surrounding marsh. Alligators actually scoop out these depressions with their tails, seasonally maintaining them against the leveling forces of the water. It is part of the mystery of the Everglades, however, that the origin of these depressions is unknown. In these depressions, known as alligator holes, scarce water collects and forms pools during the dry season; as a result, the aquatic and semiaquatic forms of life such as plankton, crustacea, fish, frogs, snakes, turtles, marsh birds, and, of course, the alligator have enough water to survive.

In fact, some species not only survive but with an apparent manlike obliviousness to their environment, choose to reproduce during the dry season and thus place the greatest demands on the food supply at this critical time. The wood ibis is a good example, for this bird can catch food efficiently only when its food supply is concentrated. Let us share J. J. Audubon's keen observation:

> This species feeds entirely on fish and aquatic reptiles, of which it destroys an enormous quantity, in fact more than it eats; for if they have been killing fish for half an hour and have gorged themselves, they suffer the rest to lie on the water untouched, when it becomes food for alligators, crows, and vultures, whenever these animals can lay hold of it. To procure its food, the wood ibis walks through shallow muddy lakes or bayous in numbers. As soon as they have discovered a place abounding in fish, they dance as it were all through it, until the water becomes

thick with the mud stirred from the bottom by their feet. The fishes, on rising to the surface, are instantly struck by the beaks of the ibises, and, on being deprived of life, they turn over and so remain. In the course of 10 or 15 minutes, hundreds of fishes, frogs, young alligators, and water snakes cover the surface, and the birds greedily swallow them until they are completely gorged, after which they walk to the nearest margins, place themselves in long rows, with their breasts all turned toward the sun, in the manner of pelicans and vultures, and thus remain for an hour or so.[4]

We can thus understand why the dry season in the Everglades is the propitious season for the wood ibis to nest; his groping method of food procurement can provide the large quantities of food needed in the nesting season only when the food supply is highly concentrated in the alligator holes.

One can think of the Everglades hydroperiod as a two-cycle engine: In the wet season there is a tremendous growth of purely aquatic life, and in the dry season there is a concentration of it. This cycle of production and concentration of aquatic life then affects the life cycle of the semiaquatic forms of life.

The success of these adaptations to the hydroperiod depends on adequate amounts of water flow into the park from the north. The rains cease in September, usually, but the surface flow continues on into the early winter and thus shortens the effective dry season from six months to more like two to four months. The best estimate is that 80 percent of the park water arrives in the form of rain falling directly on the park and only 20 percent flows into the park from the north, but nevertheless that 20 percent is crucial if the alligator holes are not to dry up. In addition, a certain minimal amount of moisture in the top layers of peat around the sawgrass roots is necessary to protect the peat from the periodic fires, mentioned above, which would otherwise do long-range damage to the viability of the sawgrass community.

The health of the park is as sensitive to variations in water *quality* as it is to the quantity and timing of surface water flow southward into the park, for the animals and plants in the park depend upon exceptionally pure water. Throughout most of the park, except, of course, in the brackish estuaries and the Gulf, the surface water is safe for human consumption. In fact, purer water is required by several species of the minute plankton at the beginning of the food chains than by man or the other higher animals at the top of the food chains.

Three prime sources of water pollution in the Everglades are agricultural fertilizers, urban and industrial sewage, and persistent chemical pesticides such as DDT. If runoff water carries nutrients from the farmers' fields or from sewage into the waterways of the Everglades, a bloom or rapid growth of vegetation may result. This phenomenon of increasing fecundity

[4] John James Audubon, *Ornithological Biography*, Vol. III, Edinburgh: Adam and Charles Black, 1835.

Three types of plant community in South Florida. *Top left:* A mangrove forest along the Gulf Coast. The clusters of mangroves dotted with egrets form a "rookery" or nesting area. *Below:* A cypress swamp. *Bottom:* A portion of the sawgrass marsh; the Park ranger is driving an airboat, the only rapid means of transportation through the sawgrass. In Everglades National Park, only park officials may drive such vehicles. (National Park Service; Allan Cruickshank, National Audubon Society; Florida State News Bureau)

The wood ibis and the alligator. (Allan Cruickshank, National Audubon Society)

of water is called eutrophication, and its effects may be seen in the Shark River Slough in the park, where there have been algal blooms in recent years. This has led to an unnaturally rapid filling in of the waterways and alligator holes and thus to a loss of aquatic habitat.

Nitrogen and inorganic phosphorus are the primary nutrients that are responsible for eutrophication. Average levels of concentration of these nutrients are measured as ratios of the weights of nitrate ions (NO_3) and of phosphate ions (PO_4) to the weight of the water sample in which they are found. In the Everglades, the concentrations are 1.5 parts per million (ppm) and 0.1 ppm, respectively. By way of comparison, secondary treatment applied to the waters used in the residential areas in South Florida results in treated water with concentrations of 20 to 30 ppm and 1 ppm, respectively. The treated sewage of the population centers in South Florida may present a potential hazard to the Everglades ecosystem if it is allowed to flow into the park's waters. The implication of this for future development in South Florida will be discussed later.

Persistent pesticides find their way to the Everglades because of their heavy use on citrus and vegetable farms and on lawns and home gardens in South Florida. Table 1 gives the present concentration of DDT, DDD, and DDE in a selection of plants and animals of the Everglades. In the Everglades and in Florida Bay the effects of this contamination on the bald eagle and the brown pelican populations have been especially severe in recent years. The essay by Professor Loucks in this book provides a detailed

TABLE 1. Concentration of DDT, DDD, and DDE
in the Everglades

Component	Sum of DDT + DDD + DDE in Parts per Billion
Fresh and estuarine water	0.02
Rainfall	0.08
Marsh soil	40.00
Algal mat	200.00
Small fish	500.00
Bald eagle	8,000.00
Brown pelican	8,000.00

SOURCE: Adapted from the National Academy of Sciences report, Washington, D.C., 1970.

description of the effects of DDT on wildlife and the process of biological magnification of DDT concentrations as the pesticides move up through the food chain.

As we have seen, the cycle of animal life in the park has evolved so as to be in rhythm with the hydroperiod. This adaptation has developed over millennia. The diversity of species and the complexity of the interrelationships among them reflect a viable natural community—one that has evolved into an equilibrium situation that is stable against naturally occurring variations in the environment, such as the hydroperiod. However, since this stability has been achieved by means of the rather specialized adaptations of many of the park's species to the hydroperiod, any severe alteration in the water cycle can threaten the survival of the park.

Thinking back over what we have described, the algae and the marl, the alligator and the ibis, the salinity of the estuaries and the shrimp, we see these adaptations, cycles, and intricate interrelationships of all forms of life as a source of wonder to man; they are also an intimation of the catastrophe that may occur if man tampers with the natural forces that have forged them.

II. Threats to Man and the Park

The future health of Everglades National Park will be seriously affected by any major new international jetport built in South Florida, especially if it is located near the park. In 1968 the Dade County Port Authority purchased 39 square miles of land in the Big Cypress Swamp for a new jetport; the southern boundary of the property lay only 7 miles north of Everglades National Park (see Figure 1). The National Audubon Society and other conservation organizations challenged the Port Authority at a series of hearings, and secured dramatic and favorable coverage of the jetport controversy in the national news magazines and on television. As a result of the efforts of these conservationists, the Department of Interior, the National Academy of Sciences, and a private group headed by former Interior Secretary Udall, investigated the jetport during the summer of 1969. All three produced reports that expressed alarm over the likely impact of the jetport on the National Park. All the while, the Port Authority was constructing a 10,500-foot runway at the site. During the autumn of 1969, the Nixon administration, asserting that the time had come for a new commitment to environmental quality in America, announced the decision that the development of a commercial jetport at the Big Cypress Swamp site was too severe a threat to Everglades National Park to be permitted to occur. The press wrote colorfully about the confrontation between the bird and the plane, and there were headlines like *Against all odds, the birds have won.*

In this section, we shall first review briefly some of the issues that surrounded the jetport controversy. Then we shall describe in somewhat greater detail the way in which Everglades National Park is endangered by development north of the park in the southeastern and southwestern parts of the state. The victory of the conservationists on the jetport issue was substantial, to be sure, but many more battles will have to be fought before the park has secured its vital supply lines, above all those which must bring it sufficient water of high quality.

Issues in the Jetport Controversy

The case for further airport facilities in South Florida has usually been considered a strong one. In 1968 Miami International Airport, the airport now handling the major commercial flights to South Florida, handled 445,000 operations (takeoffs or landings), which made it the eleventh busiest airport in the United States.[5]

The present airport cannot expand, for it is virtually surrounded by urban development; the city of Miami, spreading inland from the coast, has engulfed it like an ameba eating a food particle.

One-fourth of the operations at Miami International in 1968 were not commercial flights at all, but training flights for pilots and crews. The Nixon administration decision that had banned the development of a commercial jetport at the Big Cypress Swamp site explicitly authorized the site to be used for training operations for three years.

New commercial flights into Miami International are likely to replace the training flights as they are moved, and the Dade County Port Authority has claimed that additional facilities must be developed immediately to deal with the situation when Miami International is again saturated. It is easy for the Port Authority to find data that show that tourist and business

[5] The airports that ranked ahead of Miami International were only slightly busier, for O'Hare in Chicago, the busiest of them all, had 691,000 operations. Since there are 525,600 minutes in the year, all of the major airports handle approximately one operation every minute.

travel by air has increased steadily in recent years; it is then a simple matter to "predict on the basis of past trends" that the traffic will keep growing. If this is coupled with a sufficiently intense advertising campaign in the colder climates, it is possible that these predictions will come true.

Greatly expanded air traffic facilities represent a threat to the natural environment of South Florida quite generally, as well as to Everglades National Park. The most serious environmental problem of Miami International Airport—noise—will be diminished if some traffic is transferred to less populated areas, but there will always be some people under every new flight corridor. Indeed, the Big Cypress Swamp site, which was chosen in part because it was a full 40 miles from the dense population concentration of Miami, is only 5 miles from a quarter-mile-wide strip of land that is the reservation of the Miccosukee Indians. Moreover, the training flights now using the Big Cypress Swamp site pass over Everglades National Park, so that planes are intruding on the wilderness experience of park visitors. The concept of a wilderness experience was purposefully included in the act that established Everglades National Park, and this concept is being eroded.

Air pollution is another hazard of expanded air traffic in South Florida, no matter where an airport is built. Air pollution patterns from jetport operations are qualitatively different from the patterns resulting from ground-level sources. At increasing distances downwind from a highway, for example, the pollution levels drop rapidly because the dirty air undergoes vertical mixing with uncontaminated air above it; downwind from a jetport, however, the pollution levels will drop much more slowly, because the higher air will be dirty too, having received contaminants from descending and climbing aircraft. Because all planes take off and land in the same narrow east-west corridor, air pollution is confined to a strip roughly 2 miles wide and 20 to 40 miles long. Within this strip a jetport handling one million operations per year will produce levels of oxides of nitrogen estimated at twenty-millionths of a gram per cubic meter of air, over and above whatever oxides of nitrogen are produced in the strip by ground-level sources like automobiles. This level is comparable to present average nitrogen oxide levels due to all sources in Washington, D.C. (ten-millionths of a gram per cubic meter). Jetport air pollution has never been investigated extensively, in part because most major airports are near industrial cities which mask the effects, but it is a serious question in South Florida, where considerable effort has been spent on keeping air quality standards high.

Expanding the use of South Florida for *international* air travel poses a special set of problems, following from the necessity to maintain the highest health standards at a location where passengers disembark from trips all over the world. Insects capable of carrying many serious virus diseases are found within Everglades National Park, and to a lesser extent, in the water conservation areas. If an insect were to bite a passenger who was ill with a disease for which that insect was a vector (the technical word for "potential carrier"), the insect would become a carrier and the disease would be transmitted to the next person whom that insect attacked. There is something reckless, from a public health viewpoint, about locating an international airport in the vicinity of the park or of any other tropical wilderness where dangerous insect vectors are found.[6]

If an international jetport is built anywhere near the park, one can anticipate a major program of spraying with insecticides. World Health Organization standards require daily spraying with DDT within and even beyond the boundaries of any international jetport. Even if the standards are changed to allow the use of biodegradable pesticides, such a program would have serious ecological consequences for the park. These "pests," after all, are part of the Everglades food chain.

It is clear that the direct effects of a new jetport, which we have just outlined, will be deleterious for the human populations of South Florida as well as for the plant and animal populations of Everglades National Park. However, the most serious effect of a new jetport will be an indirect one. A new jetport will act as a stimulus to the further development of the South Florida economy, first of all in the immediate vicinity of whatever site is chosen, but also more generally throughout the region. When we now turn to the larger issues related to economic expansion in South Florida, we shall discover a similar confluence of interests: For the park to be protected, development must be restricted, and land use must be carefully planned; the same restrictions and plans are in the best interests of those who wish to see the optimal orderly development of South Florida.

Competing Demands for Water in Southeast Florida

North and east of the park, three of the most rapidly growing counties in America—Dade, Broward, and Palm Beach counties—are found. A narrow strip along the Atlantic coast includes the cities of Miami, Fort Lauderdale, and Palm Beach, which have been nearly fused together by additional settlement between them. Dade County, which includes Miami, more than doubled its population between 1950 (when it had just under half a million people) and 1968 (when it had more than 1.1 million people).

Land development in Southeast Florida is confined to a narrow coastal region by deliberate policy. Today, if you own land 20 miles inland from Miami, your land would lie under water for much of the year, and would be located within a Water Conservation Area administered by the Central and Southern Florida Flood Control District. The District, as a result of having purchased legal rights, known as water easements, has the exclusive right to determine what happens to the water in the Water Conservation Areas. To build a house on your land, unless you intend to put it

---

[6] The discussions of jetport air pollution and of insect vectors follow closely the discussion in Sections 4 and 5 of "Environmental Problems in South Florida," published by the National Academy of Sciences, Washington, D.C., 1970.

on stilts, you would have to dig a drainage ditch to get the water off your property, and that is just what the Flood Control District will not permit you to do.

Your land, with the water sitting on it, is worth a great deal to South Florida. The Army Corps of Engineers in the last 25 years has built a complex network of canals, levees, and sluice gates that lace the eastern half of southern Florida like an old-fashioned corset. As a result of these structures, which the Flood Control District administers, water can be moved out of Lake Okeechobee and onto your land to reduce the danger of flooding near the lake in the hurricane season. The District can also move water off your land in the growing season to supply a citrus grower with irrigation. The District can let the water simply stand on your land, in which case your land is functioning as a water reservoir for the urban coastal populations; the water on your land is then also helping to keep the salt water in the ocean from invading the coastal water supplies. Or the District can, in the appropriate or inappropriate season, move the water southward from your land, through one of several big sluice gates, and into Everglades National Park.[7]

From the point of view of the park, it is clear that it makes all the difference in the world which option the Flood Control District chooses. If the Flood Control District wants to, it can close the big sluice gates and no water at all will flow into the park from the Conservation Areas. This indeed is what happened for five years, from 1961 to 1965, with considerable damage to the park. In those years the rainfall was a few inches below normal, and water appeared to be in short supply; the cities and the agricultural interests got all the water they needed, but the park got nothing.

What will happen if the population of South Florida keeps growing? There will come a time, perhaps within this decade, when the choice again will be between a water shortage in Miami, a lower yield in the orange groves, and dried-up water holes in the Everglades. At first the choice will have to be made in the dry years only, and the Flood Control District will be able to take care of two out of the three demands. Later, the choice will have to be made in the wet years also, and the Flood Control District (if, as one would expect, it puts the urban needs first) will not be able to meet either the farmers' needs or the park's needs.[8] Eventually, there will not even be enough water for the urban population without greatly enlarging the region from which the urban needs are supplied. People scoff at such remarks, especially if they have seen the water standing on the land in the Everglades. But the fact is that the Flood Control District has to go further away every year to obtain water for the cities.

Obviously, a water supply can be expanded if water is used more than once, irrigating fields with the waste water from the cities, cycling bathtub

---

[7] The park has been used as a dumping ground for excess water in times of unusually heavy rainfall, when there has been flood danger to the north.

[8] The farmland may disappear anyhow, because property values are rising to the point where other uses of the land seem financially more attractive.

water into air conditioners, and so forth.[9] However, multiple use only makes sense if each successive use makes less severe demands on water quality, or if water is treated between uses, because in each use the water quality deteriorates. What this means is that without special water treatment plants the park cannot effectively share water with any other user, because, as explained in part I, the park requires water of an even higher quality than that required for public water supplies. In a sense, the water that irrigates a farm in the rapidly expanding agricultural area around Homestead and then drains into the park a few miles south *is* being used twice, but the second use, as the water enters the park laden with nitrates, phosphates, and pesticides, is as a poison! A serious threat to the park, perhaps no less significant than the reduction in the quantity of water flowing into the park, is the debasement of the quality of that water.

Through its appropriations to the Army Corps of Engineers, the federal government has some leverage on how the Flood Control District apportions its water. In legislation signed into law in June 1970, concerned members of Congress were able to include quantitative guidelines relating to the obligations of the Flood Control District to the national park:

> Delivery of water from the central and southern Florida project to the Everglades National Park shall be not less than 315,000 acre feet annually, prorated according to the monthly schedule set forth in the National Park Service letter of Oct. 20, 1967 to the Office of the Chief of Engineers, or 16.5 per centum of total deliveries from the project for all purposes including the Park, whichever is less. (Public Law 91282, Section 2)

Even minimal legislation of this kind can be enforced only if coastal development is restrained.[10] Otherwise, even the best intentions will not suffice to give the park its water. Implicit in the development plans for Florida's east coast are life-and-death decisions for Everglades National Park.

Potential Chaos in Southwest Florida

As we have just seen, the storage of water in the inland water conservation areas in the eastern half of the peninsula provides a reservoir of water for the coastal populations and also keeps the ocean's salt water from contaminating the coastal water supplies. We, the authors of this essay, happen to have studied the issue of future land development in the relatively unpopulated *western* half of the peninsula when we participated in a study of the Everglades during the summer of 1969, held under the auspices of the National Academy of Sciences and the National Academy of Engineering. The National Academy study group asked itself the ques-

---

[9] A second way of expanding water supplies, desalination, is briefly discussed in part IV of the essay on Water in this book.

[10] The reader will find a quantitative discussion of the water supply and the demands for water in South Florida in part II of the essay on Water.

tion: Are the same water conservation practices going to be necessary in the western half of the peninsula as in the east? After exploring this question for a month, we became convinced that the Big Cypress Swamp, roughly coextensive with the eastern half of Collier County and the part of Monroe County north of the Park boundary (See Figures 1 and 2), has a hydrological function in southwestern Florida which is quite analogous to that of the water conservation areas in southeastern Florida.

Why should we have become interested in this problem? The jetport controversy originally stimulated our investigation, because rapid land development is being forecast for Southwest Florida, and the jetport would have encouraged this development.[11] Land development and extensive drainage of the Big Cypress Swamp, we realized, would have a disastrous impact on the water supply to the park. Moreover, we suspected that it would also have a disastrous impact on the water supply available to the residents of Southwest Florida.

In order to start thinking sensibly about water management, we had to know how the water moves underground in the porous rock formations known as aquifers. Fortunately, the U.S. Geological Survey has described the aquifers in southeastern Florida and in southwestern Florida in two recent reports.[12] The information about the southeastern aquifers is quite complete; the information about the southwestern aquifers is more scanty, but it is adequate for a crude assessment of the problem.

The peninsula of southern Florida can be considered to be approximately symmetrical in that the eastern half resembles the western half. It turns out that this symmetry not only applies to the configuration of the land above sea level, but also applies to the underground aquifers. South Florida has two major aquifers: one east of the middle line, one west of it.[13] Both aquifers are made of limestone, and both have the shape of a wedge, a few feet thick near the middle of the peninsula (the two aquifers almost join one another, but not quite), 100 feet thick at the respective

[11] The plans for the jetport had already stimulated land development even before the jetport began its first operations: land was sold in the Big Cypress Swamp at around $150 per acre in 1961, but at about $450 per acre in 1968, after the jetport site had been selected. These prices refer to comparable land, located in the interior of the Big Cypress Swamp in areas inaccessible by road at the present time.

[12] Melvin C. Schroeder, Howard Klein, and Nevin D. Hoy, "Biscayne Aquifer of Dade and Broward Counties, Florida," Report of Investigations No. 17, Florida Geological Survey, Tallahassee, Florida, 1958; H. J. McCoy, "Ground-Water Resources of Collier County, Florida," Report of Investigations No. 31, Florida Geological Survey, Tallahassee, Florida, 1962.

[13] The eastern aquifer, in the Geological Survey reports, is called the Biscayne aquifer, and the western one is called the "shallow" aquifer. These two are the only aquifers which receive rainwater directly by seepage. Below them, there are artesian aquifers (aquifers with water under pressure, sealed off from the atmosphere by impermeable rock). Unless a desalination program is implemented the artesian aquifers will have low utility for man, because they contain about one thousand parts by weight of chloride ions per million parts of water, four times the maximum chloride content of potable water.

coasts. Because one continuous aquifer runs from mid-peninsula to the coast, tampering with the aquifer inland will inevitably have consequences at the coast.

The western aquifer, however, is of lower quality than the eastern aquifer in several respects. (1) The water in the western aquifer is more highly mineralized than the water in the eastern aquifer. There are several locations on the west coast where the water drawn up from wells has been found to be barely drinkable. (2) Rainwater does not seep down to the western aquifer as easily as it seeps down to the eastern aquifer (as hydrologists would say it, the recharge rate is lower). (3) Less water can be stored in a cubic foot of the western aquifer than in a cubic foot of the eastern aquifer (as hydrologists would say it, the storage capacity per unit volume is less). (4) Water pressure is not transmitted as easily through the western aquifer (the average transmissibility is less).

These differences tend to make even more severe the adverse consequences that will result if the inland water is drained off the land in the west. If less rainwater reaches the aquifer, the salinity of the water supply will increase. Since the capacity of the aquifer is smaller, the same amount of drainage means a larger percentage change in the water stored. And the pressure due to the higher water inland is more essential to drive the water through the aquifer when the transmissibility is lower.

To drain the land means to channel enough of the rainwater away from the land so that the upper surface of the standing water, the water table, will not rise above ground at any season of the year. This can be accomplished by digging canals to guide the water off the land, but because the land is so flat, pumps may be required as well. And, in this flat land, there may be nowhere else to drive the water except into the sea.

Once the water table is lowered, the volume of water stored is reduced. Nothing could be simpler. In principle, the process is reversible: close off the canals, and within a few years the water will again cover the land. In practice, drainage is one of the least reversible steps that men can undertake, for on drained land men will build houses and farms, and who then will say, let the water return?

The lowering of the water table inland will decrease the volume of the potential water supply for the west coast residents. In addition, if the water table is lowered, salt water from the sea will invade a larger portion of the aquifer under the land, so that, unless a well near the coast is very shallow, the water it will bring up will be contaminated.[14] Thus, where fresh water is at a premium, drainage of the natural inland reservoirs lowers the limit on the largest population which the area can sustain. The reason one has not heard this argument in most of the United States is that population densities are rarely as large, and watersheds as small, flat, and nearly surrounded by sea water as in the narrow southern Florida peninsula. One similar region is Long Island, New York.

As drainage begins in the west, the first evidence is appearing of the

---

[14] Part III of the essay on Water discusses salt intrusion in more detail.

contamination of coastal water supplies by seawater. A similar problem had arisen years before in the eastern part of the peninsula: inland drainage for the purpose of establishing agricultural land led to the intrusion of salt water into the coastal water supplies near the drainage outlets, the Miami River and the Tamiami Canal. As a by-product of having established the water conservation areas and having closed off these canals, the east coast aquifer has been flushed out again, but the process took many years.

One of the problems with salt-water intrusion is that it is difficult to establish whose drainage program has contaminated whose water supply, and hence the law is almost helpless to deal with the individual case. What is required, instead, is an overall plan. Two possible plans, at two extremes in terms of population distribution, are (1) the Big Cypress Swamp can be drained and settled, with the total population in fact limited by the available water supply, or (2) the Big Cypress Swamp can be left in its natural state, much of it flooded much of the year, with the population confined to a coastal strip.

The first option, draining the Big Cypress Swamp, would deprive the park of roughly 100 billion gallons of surface water during a year of average rainfall. This amount of water is about 30 to 40 percent of the total amount of surface water which the Park Service estimates that the park requires from the north. The remaining surface water is expected to come from the Flood Control District.

The development that would follow on the drained land would add problems of pesticide and nutrient contamination. The park, in all probability, would be devastated. The Big Cypress Swamp, itself a marvelous wilderness teeming with animal life, would vanish. And, in addition, the diminished volume of water stored in the inland aquifer and the intrusion of seawater into the coastal aquifer would seriously impair the water supplies of the coastal residents. Small wonder, then, that the National Academy of Sciences report argued for the *second* option:

> Our most important specific conclusion is that maintenance of a large portion of the Big Cypress Swamp as a natural water-conservation area would serve several useful purposes simultaneously, with respect to preservation of the Everglades Park and to an orderly development of Southwest Florida, as well as preservation of the Big Cypress wilderness itself.[15]

Because the western aquifer is less plentiful than the eastern aquifer, even with the second plan populations will be water-limited at levels substantially below those obtained on the east coast. A large west coast population will also have to contend with the shallowness of the Gulf of Mexico, and its slow flushing rate; it will be a struggle to keep the Gulf healthy, if it is overused for sewage. Moreover, the extraordinary beauty of the west coast today will only be preserved if aesthetic considerations play a major

[15] "Environmental Problems in South Florida," *op. cit.*, p. 8.

role in determining whatever coastal development does occur. But with the Big Cypress Swamp preserved as a natural water conservation area, at least the problem of salt-water intrusion should not arise.

Options for the Future

To preserve the Big Cypress Swamp requires money, for the land is now privately held, and if it is to be permanently set aside it must be purchased by the state or federal agency that will administer the land. We are talking about at least half a million acres at $100 to $200 per acre. Such sums could be threshed out of the many times larger profits which those who are involved in the development of South Florida will harvest in the next decade.[16] If the Big Cypress Swamp is not purchased and set aside, federal land reclamation (that is, drainage) projects costing hundreds of millions of dollars will probably be undertaken—at the taxpayer's expense. The immediate necessity is to buy time, so that drainage does not begin precipitously.

Fortunately, some time has been won by a decision of the Collier County government not to permit any changes in the zoning regulations in the eastern half of the county for a two-year period ending in October 1971. Since all of that land is currently zoned for agriculture, this effectively prohibits the formation of large-scale drainage projects, which only developers of major industrial or residential properties are prepared to undertake. In principle these zoning regulations could be extended indefinitely. In fact, when we talked with Collier County officials in January 1970, we were disturbed to find that these county zoning ordinances were regarded with distaste and with embarrassment. We heard several times, as an accompaniment to the phrase "You can't stop progress," the phrase "A man has a right to develop his land." Here was the ethic of rugged individualism, being transplanted to a situation where the individual landowner is a land speculator living somewhere remote from Florida (in some cases, Brazil or Japan), who, by checking the appropriate box in a printed inquiry, will permit some major land developer to go in, drain the land, and then sell the land for him. A man has a right not to have his property confiscated without compensation, to be sure, but does he really have a right to *develop* it?

We heard these phrases from a man of considerable authority, who also confided in us: "I'd personally just as soon see our county stay just the way it is right now, but you'll never catch me saying that in public." The odds are that in South Florida the god of unrestricted economic growth

---

[16] Former Interior Secretary Udall, in a recent report, has suggested that if a new jetport is built in South Florida, revenues could be raised from taxes on its activities ("Beyond the Impasse: The Dade Jetport and the Environment of South Florida," December 1969, obtainable from the Overview Corporation, New York). If such a jetport were to become as busy as the Dade County Port Authority predicts, handling 50 million passengers annually by 1980, one could tax the passengers a dollar apiece and buy the land in two years.

Anhinga and chicks. (Allan Cruickshank, National Audubon Society)

will continue to be obeyed, not challenged, until it is much too late. The Everglades Park will be ruined, and one can then drain it and pave it over as well. With the estuaries destroyed, there will be no fishing because there will be no place for the fish to breed. With no aquatic life to worry about, thermal pollution of Biscayne Bay, Florida Bay, and the Gulf becomes less troublesome, and so one could desalinate immense quantities of seawater and service maybe 20 million people living on quarter-acre plots across the whole of South Florida. The air would probably not be as bad as in Los Angeles, the traffic might not be as bad as in New York. Man would have subdued the River of Grass and eradicated that most offending of all the useless offspring of nature, the swamp.

Or, it is just barely possible, there will be another outcome, and our grandchildren will be able to see an anhinga.

III. THE EQUILIBRIUM SOCIETY

An idea is slowly developing which may have a revolutionary impact upon mankind. There is a growing realization that the resources of the earth are finite and that the capacity of the air, water, soil, and even our own psyches, to absorb the abuse associated with ever-increasing growth is limited.

If you accept this idea, then you will find it difficult to deny the conclusion that population, energy production, and consumption of material goods must eventually be limited. Man must achieve a state of equilibrium with nature.

Of course, this may all be foolish speculation; the psychic state of man may be so shell-shocked from centuries of war with nature that he is incapable of making the transition to peace. Or perhaps, as many suggest, future technologies will allow mankind to avoid facing the implications of a finite earth. We know something about these future technologies, and they leave us a little scared. We suspect that there must always be side effects that overwhelm the good intentions. We think of how miracle pesticides, designed to increase food production, engender immunities in the pest species and ravage the innocent, or of how a new jetport can trigger a chain of events that lead to the destruction of a National Park and endanger the water supplies of thousands of people. Time and again technology backfires as tolerances of the environment are exceeded. There is every indication that problems like these will keep arising; they are, quite literally, growing pains.

Are there, perhaps, severe constraints upon man's activities imposed by such "facts of life" as the laws of thermodynamics and the delicacy of the fabric of air, water, and soil? The case studies appear to support the view that where amelioration has occurred, it has generally resulted from man's restraint and not from his ability to utilize new technologies.

These essays on the Equilibrium Society deal with some of the implications of a finite earth. More than in the case studies, the emphasis here is on long-range solutions. How can a course be charted to the equilibrium society? What will the equilibrium society be like?

Alice and Lincoln Day deal with the problem of getting the population to stop growing. Underlying their analysis is the premise that adults bear children in large part to satisfy their own needs for security, and therefore a society that wishes to lower its birthrate can do so by being more supportive to the individual, especially the individual woman. The social modifications the Days propose for *getting to* the equilibrium society are likely to be woven into the equilibrium society as well.

The reader may want to call on his own experience, and perhaps do some investigations of his own, to discern what factors have influenced decisions about family size in those people he knows well. He may also want to think about a question the Days pose obliquely: Would a nation in which practically *all* families are two-child families be as interesting as one in which the *average* family is a two-child family, but many large families persist and large numbers of other couples have no children at all?

Professor Daly recognizes that a stationary population does not guarantee an equilibrium society if the fixed population goes on consuming more and more. He asserts that economic life will have to adjust to the constraints of a finite earth. If one looks at a small enough subsystem, as most economists do, growth there looks unlimited. But the constraints do not come from economics; they come from biology, geology, and, perhaps, psychology. Traditionally these constraints have been ignored. Professor Daly shows what happens when economic principles are applied to the *whole* system, and he also draws some provocative parallels between economies and ecosystems.

Professor Falk reminds us of the most sober lesson of all, which is that no nation can possibly solve its ecological problems in isolation from the other nations of the world, because ecological problems do not respect national boundaries. He points out, however, that global aspects of the environmental crisis will spur on the establishment of new international organizations. For example, the question of how the use of the oceans is to be regulated cannot remain unanswered for more than a few years; and in the process of answering it, nations may evolve new levels of cooperation. That such levels of cooperation could be first steps toward the transfer of real power to a world government may be more than just fanciful thinking. The environmental challenge is a planetary challenge, and the problems of planning and enforcement must ultimately be solved by all nations together.

What these three essays do not deal with, but what deserves serious thought, is where the sparkle will come from in the equilibrium society. A stationary-state economy in which innovation is stifled would be a disaster. Which of the forms of growth that contribute to the vigor of our present society would the reader most like to see encouraged?

We would hope to see scientific research kept high on the list. There are mysteries within mysteries in the living cell and within the atomic nucleus, and there are mysteries enveloping mysteries in the space beyond our solar system. Perhaps science, and art, and now undreamt-of forms of leisure will preserve the sense of innovation and open-endedness which seems essential for the human spirit.

Discussion of equilibrium is fine for the wealthy of the world, but where per capita consumption is woefully low, talk of equilibrium can seem like a cruel joke. Although the talent and wealth of the prosperous can be shared with the poor, this effort has not met with much success in times of growth. To improve on this unimpressive record and to begin to solve deep questions about the distribution of wealth will be an ever-present challenge to the equilibrium society, and another source of vigor.

13. TOWARD AN EQUILIBRIUM POPULATION

Alice Taylor Day and Lincoln H. Day

Today's rapid multiplication of human numbers is a remarkable departure from any previous pattern of world population increase. Though archaeological findings of human remains and artifacts strongly suggest that in the course of history populations of individual societies have risen and fallen markedly, human population on a world scale seems to have persisted for centuries in a state approaching numerical equilibrium. This equilibrium was essentially one of high mortality counterbalancing a typically high natality—with major losses of life occurring at the younger ages. Although present-day mortality levels in Western Europe produce a loss of about 2½ percent within the first four years of life, mortality levels in these same countries as recently as the beginning of the present century resulted in death rates eight to nine times as high.[1]

A new demographic equilibrium is needed if the earth is to remain a habitat fit for human life. Because of the high levels of recent population increase, this equilibrium will necessarily be based on much larger numbers. Whether it is based, as well, on the low mortality of the present rather than the high mortality of the past depends in large part on when and how men set about to achieve a reduction in the rate of their numerical growth.

Sources of Demographic Equilibrium

There is substantial evidence of efforts on the part of primitive peoples to reduce the pressures caused by rapid additions to their numbers. Early in this century, for example, the New Zealand anthropologist, Raymond Firth, found the inhabitants of the Polynesian island of Tikopia practicing birth control by means of *coitus interruptus*, and supporting this practice with a clearly enunciated idea of the relation of man and environment.

"Families by Tikopian custom are made corresponding to orchards in the woods," a Tikopian ex-

[1] Estimated from data in Nathan Keyfitz and Wilhelm Flieger, *World Population: An Analysis of Vital Data*, Chicago: University of Chicago Press, 1968.

plained to Firth. "If children are produced in plenty, then they go and steal because their orchards are few. So families in our land are not made large in truth; they are made small." [2]

That man has for centuries limited childbearing for more personal reasons, as well, is evidenced by the use of induced abortion among such peoples as the ancient Japanese and Egyptians, and by the Old Testament story of Onan who "spilled his seed upon the ground"—a case of withdrawal with contraceptive intent. Abortion and *coitus interruptus* have been widely practiced by generations of Europeans and Japanese, and are, even today, among the most commonly used methods of birth control. [3]

But the means of intentional control over one's offspring have extended beyond the natality side of the demographic ledger to infanticide —which, unlike birth control, permits selection by sex and physical condition. Malthus cites many examples of infanticide among the ancients and also among his contemporaries in both Europe and Asia. [4] Sometimes open and admitted, the practice seems to have been, at least among Europeans, more frequently covert and denied. Yet what, other than a deliberate move to dispose of excess offspring, was the widespread European practice, extending past the mid-nineteenth century, of placing unwanted children in foundling hospitals in which mortality rates seem generally to have exceeded 70 percent within a year's time? [5]

However, the historical importance of these more direct attempts to limit human numbers must not be exaggerated. Their effect on population size was hardly as great as that of more indirect social controls over natality: lengthy postponement of marriage in response to the requirements of apprenticeship or military service, or to the custom of having to be in a position to support a wife and family; the requirement of permanent celibacy for certain groups in the population (clergy in Europe and widows in India, for example); and periodic separation of the sexes.

On the whole, however, controls over natality, whether direct or indirect—even when supplemented by resort to infanticide—contributed less to maintaining the equilibrium of the human population than did the decimating effects of forces beyond man's control. Over most of human history, population has been primarily limited by man's incapacity to bend the environment to meet the requirements of increasing numbers of people. The techniques for producing and distributing food, for combating disease, and for providing clothing, shelter, and adequate sanitation were, until re-

[2] Raymond Firth, *We the Tikopia: A Sociological Study of Kinship in Primitive Polynesia*, 2nd ed., London: Allen and Unwin, Ltd., 1957, p. 491.

[3] D. V. Glass, "Family Limitation in Europe: A Survey of Recent Studies," in Clyde V. Kiser (ed.), *Research in Family Planning*, Princeton: Princeton University Press, 1962, pp. 231–261.

[4] T. R. Malthus, *An Essay on Population*, 7th ed., 1834 (little changed from the 2nd ed., 1803), London and Toronto: J. M. Dent & Sons, 1914, esp. Vol. I, pp. 130–135, 140, 146, 171, 181–185.

[5] William Langer, "Europe's Initial Population Explosion," *American Historical Review*, Vol. 69, No. 1, pp. 8–10 (October 1968).

cently, too rudimentary to sustain life for rapidly increasing numbers.

Today, these limits no longer obtain to their earlier degree. Over the past 100 years, science and technology have enabled man to increase the productivity of the land and to survive previously fatal diseases. As a consequence, world population has soared. To reach the first half billion people on this planet took mankind anywhere from 1000 to 10,000 centuries from his early beginnings to the middle of the seventeenth century. But a scant 150 years later, by 1800, that number had doubled to one billion. A second doubling occurred by 1930. Since then, the rate of population growth has increased further. As a result, the time it takes to double the world's population has decreased from 70 years at the 1930 rate of increase to somewhat less than 35 years at the 1970 rate of increase. By the end of the twentieth century, if present trends continue, the world's population will be almost 7 billion—twice what it is at present.

Although human beings have an immense capacity to accommodate themselves to the deterioration of their environment, eventually either ecological or social constraints, or both, will bring an end to population growth. The earth has a limited capacity to adjust to the intrusions of human and technological expansion, and men have a limited capacity to cope with the complexity of human relations in a mass, urbanized society. Man is no exception to the rule that numerical expansion of any species is temporary.

The Present Demographic Situation in the United States

What has just been said in a global context about rapid population expansion, environmental limits, and the need to restore numerical equilibrium applies with equal force to the United States. The United States entered the twentieth century with a population of 76 million. Today, that population numbers 205 million: representing an increase of 170 percent within but seven decades. (See Figure 1.) The implications of this population increase have been obscured by two factors of unusual significance in the United States. One is the widespread practice of birth control that has enabled the great majority of American couples to keep their family sizes within the national norm of two to four children. The other is the rising level of material comfort that a majority of Americans have enjoyed simultaneously with the rapid increase in their numbers. Thus, compared with most of the world's people, Americans may well feel that they need have no concern about their demographic conditions.

The situation does not warrant complacency, however. It is true that the birthrate in the United States steadily declined after 1957 to an all-time low of 17.4 births per 1000 population in 1968—less than half the birthrate for the world as a whole—and that it has stayed at about that level for the past two years. Yet even with this low birthrate, there is an addition, each year, of some 1.5 million Americans—enough to fill a city the size of Minneapolis–St. Paul and its surrounding suburbs.

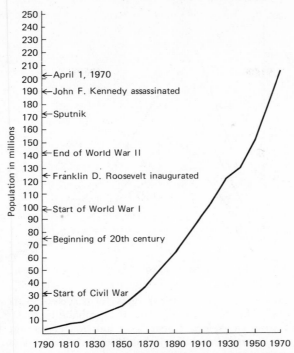

FIGURE 1 The growth in the population of the United States from 1790 to 1970.

It should be noted that it is the "crude" birthrate that is being referred to here. Though the crude birthrate—in combination with the crude deathrate—determines the growth rate of a population, the crude birthrate is a poor indicator of actual reproduction levels because it can be so greatly affected by differences in age structure and by changes in the timing of childbirth. In fact, the all-time low crude birthrate of 1968 masks a gross reproduction rate (a quantity closely reflecting the number of children born to each woman) still some 20 percent higher than that characterizing the depression years of the 1930s.° The number of children born to women at each age level (the age-specific birthrate) is the direct target of population control programs; age-specific birthrates for 1935, 1957, and 1967 are shown in Figure 2.

The increase in population stretches our resources to meet the demands for housing, roads, automobiles, airports, shops, and recreational facilities that most Americans have come to regard as necessities. Paradoxically, the very capacity through science and technology to postpone a downturn in material standards constitutes a real danger to the quality of the American environment. For our wealth and technology cover us with a veneer of material well-being that masks the high ecological and social

° See Appendix 2 on the Mathematics of Demography for a precise definition of the gross reproduction rate. [J.H. and R.H.S.]

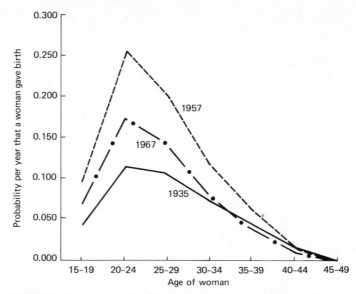

FIGURE 2 Age-specific birth rates for selected years in the United States. The figure shows directly the probability that a woman had a child during the selected year, as a function of the age of the mother. The data average over the probabilities for women whose ages fall in a five-year age grouping. The data are from U.S. Bureau of the Census, *Statistical Abstract of the United States.*

costs of additional numbers. "How can things be getting worse," Americans may well ask, "when our gross national product keeps going up?" Yet, each year, Americans spend an ever higher proportion of this high gross national product to escape from the consequences of congestion and to provide for ever larger numbers.

The high rate of population increase, particularly since World War II, has another hidden cost for American society. The fact that the average annual rate of population increase has been 40 percent greater over the past 2½ decades than in the preceding 45 years of this century has left a population rather younger than that found in most countries of low mortality: a population with, in fact, a considerable potential for further increase, even in the event of a sizable drop in average family size. By 1980 American women of childbearing age (15 to 44) will be 40 percent more numerous than they were in 1969. Those age 20 to 29 (the peak years for childbearing in the United States) will be some 57 percent more numerous. This is not conjecture; these women have already been born.

What this age distribution means so far as the task of halting population increase is concerned can be seen in some recent calculations concerning the number of years the United States population would continue to increase once natality dropped to a point where women bore exactly the number of children needed to replace their generation (that is, at present levels of mortality, an average of 2.13 children per woman).[6] Eventually, this level of natality would lead to a population growth rate equal to zero. But population is not like water issuing from a tap: you cannot simply turn it off when the desired level has been reached. With the current age distribution of the United States, assuming natality at exact replacement levels, and assuming, also, no change in the ages at which women bear their children, it would still take between 65 and 70 years before population stopped increasing. For example, had natality declined to exact replacement level in 1965 (actually, it was 50 percent higher than replacement level that year), and remained at that level indefinitely and with no change in the timing of childbirth, people would continue to be added to the population of the United States until the year 2035, when the total population would number about 272 million. Were this pattern of natality to be achieved in 1980 instead, population (under these assumptions) would continue increasing until it ultimately reached about 305 million in the year 2045.[7]

Demographic Characteristics and the Optimum Human Environment

There are three demographic conditions that would augment man's chances of realizing the optimum conditions for human life: low mortality, an unchanging age and sex distribution, and an average growth rate equal to zero. Low mortality is an end in itself. Despite sporadic resort to infanticide and other such frontal methods of limiting excess numbers, all peoples have sought to minimize losses from death to members of their own group. Reducing the risk of death, particularly to infants and small children, is also a factor lessening the likelihood of a couple's deciding to have additional children as "insurance" against possible loss. Studies of childbearing behavior among couples who have experienced the death of a child—even indirectly—suggest that there is a definite tendency for such couples to compensate by having at least one "extra" child.[8]

An unchanging age and sex distribution (what demographers term a "stable" distribution) would be one in which there would always be the

[6] If women have 2.13 children on the average, almost exactly 1.00 daughters per woman will survive to have children of their own. This follows from two facts: 105 boys are born for every 100 girls, and 4 percent of all girl babies die before reaching childbearing age. Thus with 2.13 babies a woman will have 1.09 boys and 1.04 girls, and 1.00 girls will live to have daughters themselves.

[7] Tomas Frejka, "Reflections on the Demographic Conditions Needed to Establish a U.S. Stationary Population Growth," *Population Studies*, Vol. 22, No. 3, p. 389 (November 1968).

[8] David M. Heer and Dean O. Smith, "Mortality Level, Desired Family Size, and Population Increase," *Demography*, Vol. 5, pp. 104–121 (1968); and Shafick S. Hassan, "Religion versus Child Mortality as a Cause of Differential Fertility," paper presented at the annual meeting of the Population Association of America, 1967.

same proportion of persons within any particular age and sex category. The same proportion would be, say, male and age 28 in one year as in the following year, the year after that, and so on. The undesirability of a changing age distribution can be understood by recalling the problems Americans have faced in adjusting to the post-World War II "baby boom," where the birth rate jumped 30 percent in only two years, from 20.4 to 26.6 per thousand population between 1945 and 1947 (corresponding to a jump from 2.7 to 3.7 million annual births). In contrast, a stable age distribution means, for example, that a constant fraction of the population is of school age and a constant fraction is of working age. Thus a society is relieved of the burden of planning for large yearly fluctuations in the fraction of the population at different age levels.

However, a stable distribution does not imply a zero growth rate (or vice versa). Yet if the average growth rate continues to exceed zero (regardless of the degree of stability in the age and sex distribution), society will eventually break down. Some suggestion of how this breakdown might occur can already be gleaned from current experience of transportation and communication tieups, delays attending travel to work and to recreational facilities, and the rising costs of adequate waste disposal, housing, schooling, health care, and other municipal services in rapidly expanding metropolitan areas.

What demographers call a "stationary" population is a population with a stable age distribution *and* a zero growth rate. With a stationary population (as with any population with a stable age distribution) the age and sex distribution is completely determined by the male and female death rates at each age level, that is, by the "age- and sex-specific death rates." It is therefore possible to calculate what age and sex distribution would eventuate from the continuous operation of any set of age- and sex-specific death rates. In Figure 3 the result of such a calculation for a stationary population based on current United States death rates is shown superimposed upon the actual current population of the United States in order to show what changes a transition to a stationary population would entail.°

Strategies for Population Control

What can be done to bring about the changes necessary to attain a zero rate of population increase in the United States and elsewhere? The last few years have seen a wide variety of proposals that range all the way from the bizarre, such as encouraging homosexual relations, to the coercive, such as enforced sterilization after the birth of a given number of offspring. The ideological and political overtones associated with population policy make it a highly controversial issue. It is hardly surprising, therefore, that there are differences in judgment about the best way to proceed,

---

° The characteristics of a stationary population are further discussed in Appendix 2 on the Mathematics of Demography. [J.H. and R.H.S.]

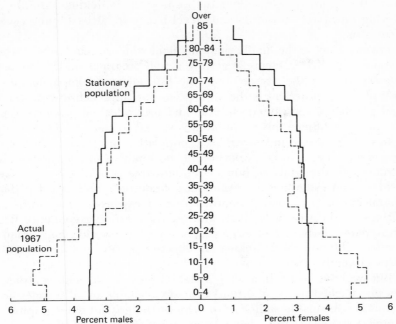

FIGURE 3 The 1967 distribution of population in the United States (dotted line) and the stationary population which is determined by the 1967 death rates (solid line). The fact that 105 male children are born for every 100 female children has been built into the computation of the stationary population. Comparing the stationary population with the actual population, we see that the former has a larger number of elderly people and fewer children, but nearly as many people in the age range of the labor force; one replaces young dependents by retired people in the transition to equilibrium.

corresponding to differences in personal values, in theories about the determinants of human behavior, and in images of the future.

One line of approach, exemplified by the family planning movement, emphasizes the improvement and extension of birth control techniques and services. Until very recently this approach had a near monopoly on action programs to influence the rate of human reproduction. Initiated by Margaret Sanger, some 50 years ago, the goal of the "Planned Parenthood" movement has been to bring the childbearing of individual couples entirely within the realm of personal choice, that is, to make every child a "wanted" child. Historically, the movement has been concerned especially with those individuals for whom "unwanted" children represent a financial or emotional burden. The means promoted to achieve this end has been contraception, although as a backstop for contraceptive failure and as a means of securing to each woman the right to control her own reproductive capacities, abortion has become increasingly accepted. Nonetheless, whatever the means, the primary concern has been with individual childbearing. Lowering birthrates on the national level has been, at best, only a secondary concern.

Professionals in the family planning field who are familiar with the political and cultural problems of getting action programs under way remain convinced that at the present time the full, energetic implementation of family planning programs is the most promising approach to both the individual problem of unwanted children and the collective problem of population increase.[9] Since this assumption is shared by most of those in a position to allocate funds and expertise to population programs, it has resulted in the channeling of a large portion of the funds available for population programs into the study of human reproductive physiology as the key to improving and extending the techniques of contraception. It is felt that the development of a superior contraceptive—inexpensive, easy-to-use, and effective—would greatly facilitate the task of extending voluntary natality control, particularly to that large proportion of the world's population who are poor, illiterate, and lacking in effective means to implement their choices of family size.

But will it work? If a safe, reliable, and convenient contraceptive (even one backed up by abortion) is made available to all couples around the world, will this be sufficient to achieve the goal of demographic equilibrium? Obviously, this depends on what fraction of the population growth of a country can be assigned to "unwanted" babies, a question that is very hard to answer. One study in the United States, conducted by Charles Westoff and Larry Bumpass, worked with 1965 interview data from a large national sample of wives under age 45. Westoff and Bumpass concluded that 22 percent of the births to this group were "unwanted" by at least one parent before conception. The incidence of these unwanted births ranged from a low of 17 percent among the "nonpoor" to a high of 42 percent among the "poor." Applying their findings to the country as a whole, Westoff and Bumpass conclude that the elimination of unwanted births during the period 1960–1965 would have reduced the rate of natural increase (that is, the excess of births over deaths) in the United States by 35 to 45 percent.[10]

The relation of family size and population increase to income bracket in the United States is shown in Figure 4. As a National Academy of Sciences study in 1965 pointed out, "The larger numbers of children in low-income families have relatively little effect on the average family size, on the total number of children born, or on the United States birth rate." [11] Hence a successful population policy must be addressed to all income brackets and not just to the poor. Despite considerable attention accorded popula-

[9] Bernard Berelson, "Beyond Family Planning," *Science*, pp. 533–543 (February 7, 1969).

[10] Larry Bumpass and Charles F. Westoff, "The Perfect Contraceptive Population," *Science*, Vol. 169, p. 1177 (1970).

[11] National Academy of Sciences—National Research Council, *The Growth of U.S. Population*, Washington, D.C., 1965.

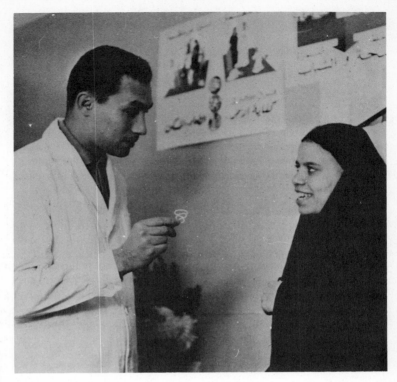

Above: A clinic gynecologist explains a simple intra-uterine coil device for preventing conception to a young woman in Cairo. *Below:* Margaret Sanger (1883–1966), one of the first leaders of the family planning movement, in Brooklyn in 1916. She is standing in front of her first family planning clinic that she was later jailed for establishing. (UPI; Planned Parenthood)

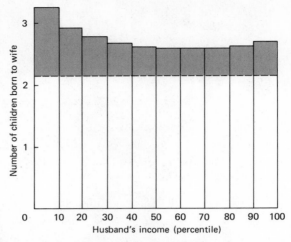

FIGURE 4 The average number of children born to women who were between the ages of 35 and 39 in 1960, as a function of the annual income of their husbands at that date. Only women who had husbands at that date are included. The husbands' incomes are grouped according to the decile of the population in which they fall. The horizontal dotted line shows the "replacement level" family size.

tion problems by the media of mass communication over the last few years, the most recent national opinion polls show little evidence that young people intend to have much smaller families than the average of three born to their parents. Yet, in a society with low levels of mortality and high proportions marrying and having children, the difference between a typical family of two children and a typical family of three children, over many consecutive years, is the difference between a population that is not growing and a population that is doubling roughly every half century.

The evidence in most other countries is equally disturbing. Despite the rapid spread of public acceptance of family planning in the last decade, birthrates in many underdeveloped countries are actually rising, not falling—chiefly because the proportions in the reproductive ages are rising. A decline in natality has occurred in coincidence with highly organized family planning programs in Taiwan, South Korea, Ceylon, and Singapore —countries that were already relatively far along the road to modernization. These countries had more popular education, higher per capita incomes, more people living in cities, and lower birth and death rates to begin with than did those countries, such as India, in which the advent of family planning programs has brought no overall decline in natality. Declining natality could have come about in these countries, says the demographer Kingsley Davis, as much in response to general social and economic change accompanying modernization, as in response to the greater accessibility of effective contraceptives.[12]

[12] Kingsley Davis, "Population Policy: Will Current Programs Succeed?", *Science*, pp. 730–739 (November 10, 1967).

A recent study of resistance to "family planning" in Indian villages illustrates how individual desires for children can lead to a high rate of population increase, independent of the availability of contraceptive devices.[13] The lack of close communication between husband and wife, the status accruing to the possessor of many grandsons, the lowly status of the childless young bride in the house of her mother-in-law, and the fear of loneliness and destitution in the event of early widowhood, all perpetuate an interest in large families. Moreover, the couples studied had a realistic sense of the risks of infant mortality. They tolerated the burdens of having four to six children early in marriage in order to attain the support of one or two surviving adult sons later in life. It was only when they had borne the number of children they felt was sufficient to achieve their personal goals that the availability of birth control became relevant. But, by then, of course, they had already contributed substantially to population increase.

So far as is known, there is hardly a nation in the world (with the possible exception of Hungary, Rumania and Japan) where the fulfillment of individual aspirations about childbearing would produce birthrates low enough to halt population increase within the next generation. The prevention of "unwanted" births (the actual result of a successful family planning program) could still leave a rapid rate of population increase.

How, then, can a society deal with the problem of too many *wanted* babies? Economic incentives are frequently suggested, and Senator Packwood (Republican of Oregon) has proposed a bill to eliminate tax exemptions for any children in excess of two per family (born after 1972). India has already resorted to a similar measure: income tax deductions for dependent children are now given only for the first and second child, and in certain states, educational concessions and benefits are granted only to those children whose parents restrict the size of their families.[14] The state of Maharashtra, moreover, is beginning to exert pressure for family limitation by withholding maternity leaves, educational grants, and housing privileges from couples with more than three children. However, in India, such measures are unlikely to have a significant demographic impact because they are limited to but a small fraction of the population, namely urbanites of some means and education.

In the absence of equal access to birth control (including abortion), economic measures applied across the board would penalize the low-income groups in the population and increase the already substantial burdens faced by these families in providing for their children. Moreover, even the short-run consequences of economic measures are uncertain; their impact on a family's decision about how many children to have would depend not only on the economic position of the family but also on the needs that children filled in that family. Applying economic penalties directly could simply serve to place more families on the welfare rolls. Positive inducements to limit family size, such as making scholarships or educational

[13] Thomas Poffenberger, "Motivational Aspects of Resistance to Family Planning in an Indian Village," *Demography*, Vol. 5, pp. 757–767 (1968).

[14] Bernard Berelson, *op. cit.*, p. 537.

grants available to families with no more than two children, could be tried without running as much risk of negative consequences for the children of large families. But even the distribution of privileges on the basis of family size would indirectly penalize the child of the large family for the behavior of his parents.

Another approach begins with the conjecture that certain economic and social trends in the industrialized countries, by creating burdens for families, are already having a depressing effect on the birthrate. This has led Kingsley Davis to enumerate ways to implement a "realistic government policy" for lowering the birthrate. His "catalogue of horrors" (as he terms it) reads very much like an actual description of pressures that are beginning to mount in the United States today:

> Squeeze consumers through taxation and inflation, make housing very scarce by limiting construction, force wives and mothers to work outside the home to offset the inadequacy of male wages, yet provide few childcare facilities; encourage migration to the city by paying low wages in the country and providing few rural jobs; increase congestion in cities by starving the transit system; increase personal insecurity by encouraging conditions that produce unemployment and haphazard political arrest.[15]

We share Davis's view that these are "horrors," for they are inconsistent with the basic intent of halting population increase, which is to promote those conditions of life most hospitable to human development and the satisfaction of human interests. Whatever the immediate demographic value of, for example, intentionally "increasing congestion in cities" or "increasing personal insecurity by encouraging conditions that produce unemployment and haphazard political arrest," to foster such hardship for the purpose of lowering the rate of human increase would obviously be to yield to the very degradation that one seeks to forestall by preventing overpopulation.

But philosophical considerations aside, the long-run demographic efficacy of such measures is, itself, uncertain. A policy of indirect repression (for example, limiting the supply of housing or squeezing consumers) might have quite different consequences for the birthrates of successive generations. Fluctuations in family size seem to occur more in response to abrupt *changes* in social conditions—such as those accompanying war, depression, epidemics, political unrest—than to any particular attributes of the social situation; for what one generation perceives as an onerous constraint can become but a fact of life to the next. There is even the possibility that in response to certain types of controls natality might actually increase as individuals adjusted to a harsh social setting by withdrawing more than ever into the nuclear family for support.

Even more drastic methods have been proposed to deal with the problem of too many "wanted" children. Some biologists have raised the possi-

[15] Kingsley Davis, *op. cit.*, p. 739.

bility of adding "temporary sterilants" to the water supply to hold births at a predetermined "optimum" level, and Kenneth Boulding, an economist, has suggested (apparently only to stimulate discussion) issuing marketable licenses to have children, in just those numbers which would ensure an average of two children per couple. Melvin Ketchel, a physiologist, estimates that, although control agents that could be used to regulate the natality of whole populations without depending on the voluntary action of individual couples have not yet been developed, they could be made available for field testing after but 5 to 15 years of intensive research.[16]

Such involuntary controls would appear to be, by their very nature, rather difficult to apply. Just how, for example, does one actually implement a policy of involuntary sterilization following the birth of a third child; or how, as in many nonindustrialized countries where few eat factory-produced foods or obtain their water from a municipal supply, does one achieve much reduction in natality from a program of introducing sterility-producing compounds into food and water? And could one be sure that whatever natality-repressing substance was developed would not add substantially to the biological harm already being caused by such by-products of our technological way of life as carbon monoxide, DDT, radiation, drugs, and food preservatives? If there is one thing the layman has learned from the new interest in ecology, it is to exercise caution about the long-run effects of all alien matter introduced into the biosystem.

It is, of course, conceivable that in such a country as the United States, where the means to control births are widely available and widely used, a condition of overpopulation could be perceived as so threatening to human well-being that the public would accept—as a necessary public health measure—a program of, say, society-wide involuntary sterilization after birth of the nth child. However, were such a widespread, negative perception of the situation to eventuate, the climate of social opinion against excess childbearing would probably be so pervasive that sufficient numbers of individuals would already be curtailing their childbearing to a degree that would obviate the need to resort to compulsory measures.

Population Control through Social Change

If the more extreme types of involuntary controls are rejected as either politically infeasible, unworkable, or inconsistent with other social goals, and if—under present social conditions—leaving the choice of family size entirely to individuals is inadequate, then what alternatives are we left with? What we should like to propose is a population policy with four long-term goals: (1) further separation of sex from reproduction; (2) changes in norms about ideal family size; (3) encouragement of those life styles that compete with early marriage and large families; and (4) enhancement of the security of the individual per se—apart, that is, from the security he might derive from his status as member of a family.

[16] Cited in Bernard Berelson, *op. cit.*, p. 533.

If you try to escape from crowded subways, you may find that others have had the same idea. (*N. Y. Daily News*, Richard Corkery; Zimbel from Monkmeyer Press)

There are many specific proposals to foster these kinds of social change, but all are linked to levels of natality through their influence on only five elements of demographic behavior: (1) proportion of population marrying, (2) age at marriage, (3) distribution of family size, (4) intervals between successive births, and (5) ages at commencement and at completion of childbearing. Other things being equal, a lower overall natality would result to the extent that social arrangements reinforced the following kinds of choices about marriage and childbearing: the decision of fewer people to marry; of more couples to marry at a later age; of more couples than at present to remain childless or have but one child, and of fewer to have three or more children; of more people to space their children further apart; and of more women to start their childbearing later and to complete it earlier.

Further Separation of Sex from Reproduction

This is fundamental to reducing natality because of its bearing on the prevention of unwanted conceptions and births, whether within or outside of marriage. If individuals are to remain childless or restrict their family size further, these ends will be facilitated if there is ready access to a variety of means to control birth. All means, so long as they are effective and do not endanger the well-being of the persons involved, should be available; and they should be available to all who need them, regardless of age, sex, or marital status—and, for minors, regardless of parental consent. The emphasis in family planning programs has been on providing female contraceptives that need some medical assistance for their use. Though effective and long-lasting, these techniques meet only one type of need. It would also be helpful if all male and female contraceptives, except those a trained person must administer, were widely dispensed in public places, including grocery stores. This has been done for some time in Sweden and was recently begun in India. Sterilization should also be made more freely available and more socially acceptable as one of the safest and surest methods of birth control known today. So also should abortion be made more readily available: to protect the health and well-being of the woman, to lessen class distinctions in access to this important sociomedical practice, and to provide an effective, medically sound method to avoid an unwanted birth in the case of accidental conception.

Changing Norms about Ideal Family Size

Improvement and extension of the means to control birth are vital to a lowering of levels of natality, but it will be insufficient unless supported by norms that place a high value on small family size. As with the other projects described in this book, achieving a genuine cessation of population increase will require above all else changes in certain attitudes that have been characteristic of human cultures over nearly all of man's history. Considerable change in this direction has already occurred in the industrialized countries, but in the nonindustrialized countries norms of childbearing for the great majority are probably little different from what they were in the earliest agricultural societies. To urge family limitation upon the sub-sistence peoples of Asia, Africa, and Latin America is to ask that they con-trol their childbearing largely in the absence of those conditions of life that made the smaller-sized family seem desirable to Europeans. In addition to planning and administering more effective means for the delivery of birth control services, there is the need in these countries for greater efforts to reduce infant mortality, increase public health services, raise the levels of agricultural production and industrial employment, and improve educa-tional opportunities for men and women; in short, to alter the context in which norms about childbearing are shaped.

So far as the United States is concerned, a greater predisposition to-ward having small families will be closely related to changes in other values—those glorifying growth, bigness, and high consumption, for example—that place individual rights before the requirements of commu-nity well-being, and that view the environment as invulnerable and capa-ble of sustaining whatever demands are made upon it.

Increasing the public's awareness of population dynamics and their consequences might also contribute to a revision of norms about childbear-ing, if it is done with sensitivity. But to scare people with projections of large numbers could be counterproductive. It could lead to defeatism about the efficacy of taking action to curb population growth, or it could foster a callousness toward social needs that would lead to further reliance on private means to personal satisfactions: increased consumption, for ex-ample, or an attempt to find security through larger families. In an effec-tive presentation, facts about population should be related to their conse-quences for a particular age category, racial group, business, community, nation. No one should leave school without a basic understanding of popu-lation dynamics in a finite world, and of the part individual decision mak-ing plays in the shaping of demographic conditions.

Once the relation between individual childbearing and population in-crease is widely understood, a new ethic of childbearing can come into existence: A parent is no longer judged exclusively by the time, energy, and money that he expends on his offspring. Instead, the number of off-spring becomes an additional matter of public concern. The parent be-comes conscious of the demands that each of his children will make on global natural resources. The unfavorable stereotypes associated with the one-child family and the no-child family disappear. And the couple with one or two children is praised for not having more.

Greater Variety of Life-Styles

Exercise of a greater range of individual options with respect to whether or not to marry, age at marriage, and the timing of childbearing is everywhere associated with lower natality. Extending this range of options could lead to lower natality without resort to a rigid standardization of family size. Childbearing could be the province of a small proportion of the population who would be characterized by large families; or of a large proportion whose family sizes would be either concentrated at the average number per couple necessary for a zero growth rate or spread over a wider range, but still with an average at this level. In a low mortality country like the United States, an average natality at replacement level would mean that about one out of four women would have to remain childless if those who became mothers had an average of three children each; about 45 percent would have to remain childless if mothers averaged four children each; and 65 percent if mothers averaged six children.

Better housing arrangements for unmarried persons, and a system of taxation that does not penalize them for their unmarried state, could be instituted to allay some of the pressures individuals experience toward marriage and, with marriage, childbearing. The provision of housing for married students—regardless of age—that is now a feature of some universities in the United States would appear to be a movement in a pronatalist direction. So also, perhaps, is the current trend to do away with all "monosexual" educational institutions; for whatever may be its advantages, there is no denying that coeducation puts men and women together in a relatively unchaperoned situation at a highly nubile time of life. Still, it is possible that the pressures toward marriage stemming from this closer association may be more than offset by a greater emotional intimacy and sexual access—backed up by more readily available birth control—of a sort that would lessen the pressure to achieve these satisfactions by early entry into marriage.[17]

Certainly there is a need to alter the status of women in many industrialized countries hardly less than in the non-industrialized. The opportunity for employment, particularly for employment outside the home, seems to affect quite considerably a woman's decisions about marriage and children.[18] If they are to refrain—voluntarily—from having large families, women need to be able to think of themselves as something in addition to wives and mothers. This requires a number of changes in women's roles in society, especially the provision of suitable jobs—suitable with respect to scheduling, remuneration, and opportunities for advancement no less than with respect to skills and interests. The society must provide opportunities for training and for retraining at ages ordinarily considered too advanced to permit a favorable economic return. And innovations are required with respect to childcare facilities and patterns of leaves of absence. In the United States, real options for women to combine interesting work with a

[17] See, for example, Harold T. Christensen and George R. Carpenter, "Timing Patterns in the Development of Intimacy," *Marriage and Family Living*, Vol. 24, No. 1 (February 1962); and Harold T. Christensen, "Cultural Relativism and Premarital Sex Norms," *American Sociological Review*, Vol. 25, No. 1 (February 1960).

[18] See, for example, A. J. Jaffe and Koya Azumi, "The Birth Rate and Cottage Industries in Underdeveloped Countries," *Economic Development and Cultural Change*, Vol. 9, No. 1, pp. 52–63 (October 1960); and Jeanne Clare Ridley, "Number of Children Expected in Relation to Non-Familial Activities of the Wife," *Milbank Memorial Fund Quarterly*, Vol. 37, pp. 277–296 (1959).

small family are increasingly restricted by the dispersed residential pattern of our suburbs and the dependence on the automobile as the exclusive means of transportation. For many suburban women, at least, the length of the trip to work, in addition to the expense of maintaining an extra car and hiring household help, outweighs whatever advantages might accrue to obtaining employment outside the home.

Enhancing the Security of the Individual

Encouraging a greater variety of life-styles could be supplemented by a number of measures to enhance the security of the individual per se, that is, apart from his status as a member of a family. The provision of financial security in old age, for example, has been proposed to reduce the tendency to bear children (particularly sons, in the nonindustrialized countries) for the purpose of ensuring economic support in old age. Along the same lines would be provision of a guaranteed income and comprehensive health care on an individual, rather than family or household basis. Similarly, such changes as improving public transportation to community facilities, increasing the access of urban populations to outdoor recreation, and providing more opportunities for people of all ages to engage in socially useful activities could all conceivably increase the individual's scope for satisfying his interests and, at the same time, lower natality by reducing his dependence on marriage and childbearing.[19]

Some have proposed establishment of a permanent national service corps, whose ostensible purpose would be to alleviate pressing domestic needs—from the tutoring of educationally disadvantaged children to the beautification of public places and the protection of the environment. The demographic consequences of such opportunities for service would be to reduce the temptation to resort to childbearing—either early or often—for want of other meaningful options in life.

Another way to lessen desires for childbearing would be to counter tendencies toward privatism in child-rearing. For some few, this might be achieved through facilitating adoption. For others, a variety of opportunities could be provided for increasing the frequency of rewarding contacts with children other than one's own (for example, through special education and school programs, music groups, outings, and environmental improvement projects). There is also the need to devote more attention to strengthening the man-made environment in ways conducive to social cohesion, participation, and emergence of the feeling of belongingness. There is, for example, a widespread need in many of today's cities for more parks and playgrounds, more plazas and pedestrian ways, more places for people to meet that are safe and quiet, and for more attention to aesthetic values. Underlying all these specific approaches is the need for a heightened

---

[19] Alice Taylor Day, "Population Control and Personal Freedom: Are They Compatible?", *The Humanist*, pp. 277–296 (November–December 1968).

awareness of the benefits that would accrue to all from a greater reliance on the community as an extension of the family.[20]

The proposal that these changes can affect childbearing is predicated on the assumption that children fulfill important personal and social needs, that it is, therefore, necessary to foster those arrangements that will enable individuals to meet their needs in ways other than through childbearing. Unless we provide conditions in which individuals limit the number of their children because it is in their own self-interest to do so, we will be compelled to resort to establishing quotas and to penalizing parents for bearing more than their share. The fall in the death rate that led to the precipitous rise in population required no particular change in outlook. Extension of life is a universal value. Thus, measures to improve health could be introduced on a mass basis with relatively little opposition, and with immediate effect. However, a further reduction in childbearing goes counter to many strongly held values, and for these reasons is not to be approached in the same brisk manner. We cannot avoid the uncomfortable observation that there is no simple way, and no single way, to achieve and maintain a zero rate of population increase.

[20] *Ibid.*

14. TOWARD A STATIONARY–STATE ECONOMY

Herman E. Daly*

> "... he looked upon us as a sort of animals to whose share, by what accident he could not conjecture, some small pittance of reason had fallen, whereof we made no other use than by its assistance to aggravate our natural corruptions, and to acquire new ones which nature had not given us; that we disarmed ourselves of the few abilities she had bestowed; had been very successful in multiplying our original wants, and seemed to spend our whole lives in vain endeavors to supply them by our own inventions ...."
>
> JONATHAN SWIFT

Any discussion of the relative merits of the stationary, no-growth economy, and its opposite, the economy in which wealth and population are growing, must recognize some important quantitative and qualitative differences between rich and poor countries, and between rich and poor classes within countries. To see why this is so consider the familiar ratio of gross national product (GNP) to total population (P). This ratio, per capita annual product (GNP/P), is the measure usually employed to distinguish rich from poor countries and which, in spite of its many shortcomings, does have the virtue of reflecting in one ratio the two fundamental life processes of production and reproduction. Let us ask two questions of both numerator and denominator for both rich and poor countries—namely, what is its quantitative rate of growth; and qualitatively, exactly what is it that is growing?

1. The rate of growth in the denominator, P, is much higher in poor countries than in rich countries. Although mortality is tending to equality at low levels throughout the world, fertility [1] in poor nations remains roughly *twice* that of rich nations. The average gross re-

* The author is grateful to John Harte and Robert H. Socolow for stimulating discussion and helpful comments.

[1] Fertility refers to actual reproduction as opposed to fecundity, which refers to reproductive potential or capacity. One measure of fertility is the Gross Reproduction Rate, defined in footnote 2.

226

production rate (GRR) [2] for rich countries is around 1.5, and that for poor countries is around 3.0 (that is, on the assumption that all survive to the end of reproductive life, each mother would be replaced by 1.5 daughters in rich countries and 3 in poor countries). Moreover, all poor countries have a GRR greater than 2.0, and all rich countries have a GRR less than 2.0, with practically no countries falling in the area of the 2.0 dividing point. No other social or economic index divides the world so clearly and consistently into "developed" and "underdeveloped" as does fertility.[3]

2. Qualitatively the incremental population in poor countries consists largely of hungry illiterates; in rich countries it consists largely of well-fed members of the middle class. The incremental person in poor countries contributes negligibly to production, but makes few demands on world resources—although from the point of view of his poor country these few demands of many new people can easily dissipate any surplus that might otherwise have been used to raise productivity.[4] The incremental person in the rich country contributes to his country's GNP, and to feed his high standard of living contributes greatly to depletion of the world's resources and pollution of its spaces.

3. The numerator, GNP, is growing at roughly the same rate in rich and poor countries, around 4 or 5 percent annually, with the poor countries probably growing slightly faster. Nevertheless, because of their more rapid population growth, the per capita income of poor countries is growing more slowly than that of rich countries. Consequently the gap between rich and poor is widening.[5]

4. The incremental GNP of rich and poor nations has an altogether different qualitative significance. This follows from the two most basic laws of economics: (a) the law of diminishing marginal utility, which really says nothing more than that people satisfy

[2] GRR is roughly the ratio of one generation to the preceding generation, assuming that all children born survive to the end of their reproductive life. It is usually defined in terms of females only. The length of a generation is the mean age of mothers at childbirth. GRR and other demographic concepts are discussed more precisely in Appendix 2 on the Mathematics of Demography.

[3] United Nations, *Population Bulletin of the United Nations, No. 7, 1963*, New York: UN, 1965.

[4] Goran Ohlin, *Population Control and Economic Development*, Paris: Development Centre of the Organization for Economic Cooperation and Development, 1967.

[5] According to Robert E. Baldwin: "In the 1957–58 to 1963–64 period, the less developed nations maintained a 4.7% annual growth rate in gross national product compared to a 4.4% rate in the developed economies. The gap in per capita income widened because population increased at only 1.3% annually in the developed countries compared to a 2.4% annual rate in the less developed economies." (*Economic Growth and Development*, New York: Wiley, 1966, p. 8).

their most pressing wants *first*—thus each additional dollar of income or unit of resource is used to satisfy a less pressing want than the previous dollar or unit; (b) the law of increasing marginal cost, which says that producers *first* use the best qualities of factors (most fertile land, most experienced worker, and so on) and the best combination of factors known to them. They use the less efficient (more costly) qualities and combinations only when they run out of the better ones, or when one factor, such as land, becomes fixed (nonaugmentable). Also, in a world of scarcity, as more resources are devoted to one use, fewer are available for other uses. The least important alternative uses are sacrificed first, so that as more of any good is produced, progressively more important alternatives must be sacrificed—that is, a progressively higher price (opportunity cost) must be paid. Applied to GNP the first law means that the marginal (incremental) benefits from equal increments of output are decreasing, and the second law means that the marginal cost of equal increments in output is increasing. At some point, perhaps already passed in the United States, an extra unit of GNP costs more than it is worth. Technological advance can put off this point, but not forever. Indeed it may bring it to pass sooner because more powerful technologies tend to provoke more powerful ecological backlashes and to be more disruptive of habits and emotions. To put things more concretely, growth in GNP in poor countries means more food, clothing, shelter, basic education, and security, whereas for the rich country it means more electric toothbrushes, yet another brand of cigarettes, more tension and insecurity, and more force-feeding through more advertising. In sum, extra GNP in a poor country, assuming it does not go mainly to the richest class of that country, represents satisfaction of relatively basic wants, whereas extra GNP in a rich country, assuming it does not go mainly to the poorest class of that country, represents satisfaction of relatively trivial wants.

For our purposes the upshot of these differences is that for the poor, growth in GNP is still a good thing, but for the rich it is probably a bad thing. Growth in population, however, is a bad thing for both: for the rich, population growth is bad because it makes growth in GNP (a bad thing) less avoidable; for the poor, population growth is bad because it makes growth in GNP, and especially in per capita GNP (a good thing), more difficult to attain. In what follows we shall be concerned exclusively with a rich, affluent-effluent economy such as that of the United States. Our purposes will be to define more clearly the concept of stationary state, to see why it is necessary, to consider its economic and social implications, and finally to comment on an emerging political economy of finite wants and nongrowth.

The Nature and Necessity of the Stationary State

The term *stationary state* is used here in its classical sense.[6] Over a century ago John Stuart Mill, the great synthesizer of classical economics, spoke of the stationary state in words that could hardly be more relevant today, and will serve as the starting point in our discussion.

> But in contemplating any progressive movement, not in its nature un-limited, the mind is not satisfied with merely tracing the laws of its movement; it cannot but ask the further question, to what goal? . . .
> It must always have been seen, more or less distinctly, by political economists, that the increase in wealth is not boundless: that at the end of what they term the progressive state lies the stationary state, that all progress in wealth is but a postponement of this, and that each step in advance is an approach to it . . . if we have not reached it long ago, it is because the goal itself flies before us [as a result of technical progress].
> I cannot . . . regard the stationary state of capital and wealth with the unaffected aversion so generally manifested towards it by political economists of the old school. I am inclined to believe that it would be, on the whole, a very considerable improvement on our present condition. I confess I am not charmed with the ideal of life held out by those who think that the normal state of human beings is that of struggling to get on; that the trampling, crushing, elbowing, and treading on each other's heels which form the existing type of social life, are the most desirable lot of human kind, or anything but the disagreeable symptoms of one of the phases of industrial progress. The northern and middle states of America are a specimen of this stage of civilization in very favorable circumstances; . . . and all that these advantages seem to have yet done for them (notwithstanding some incipient signs of a better tendency) is that the life of the whole of one sex is devoted to dollar-hunting, and of the other to breeding dollar-hunters.
> . . . those who do not accept the present very early stage of human improvement as its ultimate type may be excused for being comparatively indifferent to the kind of economical progress which excites the congratulations of ordinary politicians; the mere increase of production and accumulation. . . . I know not why it should be a matter of congratulation that persons who are already richer than anyone needs to be, should have doubled their means of consuming things which give little or no pleasure except as representative of wealth. . . . It is only in the backward countries of the world that increased production is still an im-

---

[6] The term *stationary state* has been burdened with two distinct meanings in economics. The classical meaning is that of an actual state of affairs toward which the real world is supposed to be evolving; that is, a teleological or eschatological concept. The neoclassical sense of the term is entirely mechanistic—an epistemologically useful fiction like an ideal gas or frictionless machine—and describes an economy in which tastes and techniques are constant. The latter sense is more current in economics today, but the former meaning is the relevant one in this discussion.

portant object: <u>in those most advanced, what is economically needed</u> is a <u>better distribution, of which one indispensable means is a stricter restraint on population.</u>

There is room in the world, no doubt, and even in old countries, for a great increase in population, supposing the arts of life to go on improving, and capital to increase. But even if innocuous, I confess I see very little reason for desiring it. The density of population necessary to enable mankind to obtain, in the greatest degree, all the advantages both of co-operation and of social intercourse, has, in all the most populous countries, been attained. A population may be too crowded, though all be amply supplied with food and raiment. It is not good for a man to be kept perforce at all times in the presence of his species. . . . Nor is there much satisfaction in contemplating the world with nothing left to the spontaneous activity of nature; with every rood of land brought into cultivation, which is capable of growing food for human beings; every flowery waste or natural pasture plowed up, all quadrupeds or birds which are not domesticated for man's use exterminated as his rivals for food, every hedgerow or superfluous tree rooted out, and scarcely a place left where a wild shrub or flower could grow without being eradicated as a weed in the name of improved agriculture. If the earth must lose that great portion of its pleasantness which it owes to things that the unlimited increase of wealth and population would extirpate from it, for the mere purpose of enabling it to support a larger, but not a happier or a better population, I sincerely hope, for the sake of posterity, that they will be content to be stationary, long before necessity compels them to it.

It is scarcely necessary to remark that a stationary condition of capital and population implies no stationary state of human improvement. There would be as much scope as ever for all kinds of mental culture, and moral and social progress; as much room for improving the Art of Living and much more likelihood of its being improved, when minds cease to be engrossed by the art of getting on. Even the industrial arts might be as earnestly and as successfully cultivated, with this sole difference, that instead of serving no purpose but the increase of wealth, industrial improvements would produce their legitimate effect, that of abridging labor.[7]

The direction in which political economy has evolved in the last hundred years is not along the path suggested in the quotation. In fact, most economists are hostile to the classical notion of stationary state and dismiss Mill's discussion as "strongly colored by his social views"[8] (as if the neoclassical theories were not so colored!), and "nothing so much as a prolegomenon to Galbraith's *Affluent Society*" (which also received a hostile reception from the economics professions). While giving full credit to Mill for his many other contributions to economics, most economists con-

[7] J. S. Mill, *Principles of Political Economy*, Vol. II, London: John W. Parker and Son, 1857, pp. 320–326, with omissions.

[8] All quotes in this paragraph are from Mark Blaug, *Economic Theory in Retrospect*, Homewood, Ill.: Richard D. Irwin, 1968, pp. 214–221. Blaug's views are, I think, representative of orthodox economists.

sider his discussion of the stationary state as something of a personal aberration. Also his "relentless insistence that every conceivable policy measure must be judged in terms of its effects on the birth rate" is dismissed as "hopelessly dated." The truth, however, is that Mill is even more relevant today than in his own time.

Enough of historical background and setting. Let us now analyze the stationary state with a view toward clarifying what Mill somewhat mistakenly thought "must have always been seen more or less distinctly by political economists," namely "that wealth and population are not boundless."

By "stationary state" is meant a constant stock of *physical* wealth (capital), and a constant stock of people (population).[9] Naturally these stocks do not remain constant by themselves. People die, and wealth is physically consumed—that is, worn out, depreciated. Therefore the stocks must be maintained by a rate of inflow (birth, production) equal to the rate of outflow (death, consumption). But this equality may obtain, and stocks remain constant, with a high rate of throughput (equal to both the rate of inflow and the rate of outflow), or with a low rate. Our definition of stationary state is not complete until we specify the rates of throughput by which the constant stocks are maintained. For a number of reasons we specify that the rate of throughput should be "as low as possible." For an equilibrium stock the average age at "death" of its members is the reciprocal of the rate of throughput. The faster the water flows through the tank, the less time an average drop spends in the tank. For the population a low rate of throughput (a low birthrate and an equally low death rate) means a high life expectancy, and is desirable for that reason alone—at least within limits. For the stock of wealth a low rate of thoughput (low production and equally low consumption) means greater life expectancy or durability of goods and less time sacrificed to production. This means more "leisure" or nonjob time to be divided into consumption time, personal and household maintenance time, culture time, and idleness.[10] This too seems socially desirable, at least within limits.

[9] By "stock" is meant a quantity measured at a point in time; for example, a population census or a balance sheet of assets and liabilities as of a certain date. By "flow" is meant a quantity measured across some actual or conceptual boundary over a period of time; for example, births and deaths per year or an income and loss statement for a given year.

The boundary lines separating the stock of wealth from the rest of the physical world may sometimes be fuzzy. But the main criterion is that physical wealth must in some way have been transformed by man to increase its usefulness over its previous state as primary matter or energy. For example, coal in the ground is primary matter and energy, coal in the inventory of firms and households is physical wealth, coal after use in the form of carbon dioxide and soot is waste matter. The heat produced by the coal is part usable and part unusable. Eventually all the heat becomes unusable or waste heat, but while it is usable it is a part of the physical stock of wealth. For some purposes one may wish to define proven reserves in mines as part of wealth, but that presents no problems.

[10] Staffan B. Linder, *The Harried Leisure Class*, New York: Columbia University Press, 1970.

To these reasons for the desirability of a low rate of throughput, we must add some reasons for the impracticability of high rates. Since matter and energy cannot be created, production inputs must be taken from the environment, which leads to depletion. Since matter and energy cannot be destroyed, an equal amount of matter and energy in the form of waste must be returned to the environment, leading to pollution. Hence lower rates of throughput lead to less depletion and pollution, higher rates to more. The limits regarding what rates of depletion and pollution are tolerable must be supplied by ecology. A definite limit to the size of maintenance flows of matter and energy is set by ecological thresholds which if exceeded cause a breakdown of the system. To keep flows below these limits we can operate on two variables: the *size* of the stocks and the *durability* of the stocks. As long as we are well below these thresholds, economic cost-benefit calculations regarding depletion and pollution can be relied on as a guide. But as these thresholds are approached, "marginal cost" and "marginal benefit" become meaningless, and Alfred Marshall's erroneous motto of "nature does not make jumps" and most of neoclassical marginalist economics become inapplicable. The "marginal" cost of one more step may be to fall over the precipice.

Of the two variables, size of stocks and durability of stocks, only the second requires further clarification. "Durability" means more than just how long a particular commodity lasts. It also includes the efficiency with which the after-use "corpse" of a commodity can be recycled as an input to be born again as the same or as a different commodity. Within certain limits, to be discussed below, durability of stocks ought to be maximized in order that depletion of resources be minimized.

One might suppose that the best use of resources would imitate the model that nature has furnished: a closed-loop system of material cycles powered by the sun (what A. J. Lotka called the "mill wheel of life" or the "world engine").[11] In such an "economy" durability is maximized, and the resources on earth could presumably last as long as the sun continues to radiate the energy to turn the closed material cycles.

Now man can set up an economy in imitation of nature in which all waste products are recycled. Instead of the sun, however, man chooses to use other sources of energy, because of the scale of his industrial activity. Even modern agriculture depends as much on geologic capital (to make fertilizers, machines, and pesticides) as on solar income. That capital (fossil fuels and fission materials), from which we now borrow, may not last more than a couple of centuries, but there is another possible energy source, controlled thermonuclear fusion, which may someday provide a practically inexhaustible supply of energy with little radioactive waste, thereby alleviating problems of resource depletion and radioactive contamination.

Nevertheless, the serious problem of waste heat remains. The second law of thermodynamics tells us that it is impossible to recycle energy, and

[11] A. J. Lotka, *Elements of Mathematical Biology*, New York: Dover Publications, 1946 (republication). See especially Chapter 24.

that eventually all energy will be converted into waste heat. Eventually all life will cease as entropy or chaos approaches its maximum. But even before this very long-run universal thermodynamic heat death occurs, the second law of thermodynamics implies that we will be plagued by thermal pollution, for whenever we use energy we must produce unusable waste heat. When a localized energy process causes a part of the environment to heat up, we call this thermal pollution, and it can have serious effects on ecosystems, since life processes and climatic phenomena are regulated by temperature.

We have already argued that, given the size of stocks, the throughput should be minimized, since it is really a cost. But the throughput is in two forms, matter and energy, and the ecological cost will vary, depending on how the throughput is apportioned between them. The amount of energy throughput will depend on the rate of material recycling. If we recycle none of our used material goods, then we must expend energy to replace those goods from raw materials and this energy expenditure is in many instances greater than the energy needed to recycle the product. For example, the estimated energy needed to produce a ton of steel plate from iron ore is 2700 kilowatt-hours, whereas merely 700 kilowatt-hours is needed to produce the same ton by recycling scrap steel.[12] However, this is not the whole story. The mere expenditure of energy is not sufficient to close material cycles, since energy must work through the agency of material implements. To recycle aluminum cans requires more trucks to collect the cans as well as more energy to run the trucks. More trucks require more steel, glass, rubber, and so forth, which require more iron ore and coal, which require still more trucks. This is the familiar web of interindustry interdependence reflected in an input-output table.[13] All of these extra intermediate activities required to recycle the aluminum cans involve some inevitable pollution as well. If we think of each industry as adding recycling to its production process, then this will generate a whole chain of direct and indirect demands on matter and energy resources which must be taken away from final demand uses and devoted to the intermediate activity of recycling. It will take more intermediate products and activities to support the same level of final output.

As we attempt to recycle more and more of our produced goods, we will reach the point of diminishing returns; the energy expenditure alone will give rise to a ruinous amount of waste heat or thermal pollution. On the other hand, if we recycle too small a fraction of our produced goods, then nonthermal pollution and resource depletion become a severe problem.

The introduction of material recycling permits a trade-off—that is, it

[12] Report of the Committee for Environmental Information, before Joint Congressional Committee on Atomic Energy, January 29, 1970. Quoted in *Environment* (March 1970).

[13] Herman E. Daly, "On Economics as a Life Science," *Journal of Political Economy*, July 1968.

Boy Scouts from the San Fernando Valley Council clean a California beach in a reclamation project encouraged by the Reynolds Metals Company which pays one-half cent for each all-aluminum can returned. (Reynolds Metals Company)

allows us to choose that combination of material and energy depletion and pollution which is least costly in the light of specific local conditions. "Cost" here means total ecological cost, not just pecuniary costs, and is extremely difficult to measure.

In addition to the trade-offs involved in minimizing the ecological cost of the throughput for a given stock, we must recognize that the "total stock" (consisting of wealth and people) is variable both in total size and in composition. Since there is a direct relationship between the size of the stock and the size of the throughput necessary to maintain the stock, we have a trade-off between size of total stock (viewed as benefit) and size of the flow of throughput (viewed as a cost)—that is, an increase in benefit implies an increase in cost. Furthermore, a given throughput can maintain a constant total stock consisting of a large substock of wealth and a small substock of people, or a large substock of people and a small substock of wealth. Here we have a trade-off in the form of an inverse relationship between two benefits. This latter trade-off between people and wealth is imposed by the constancy of the total stock and is limited by minimal subsistence per capita wealth at one extreme and by minimal technological requirements for labor to maintain the stock of wealth at the other extreme. Within these limits this trade-off essentially represents the choice of a standard of living. Economics and ecology can at best specify the terms of this trade-off. The actual choice depends on ethical judgments.

In sum, the stationary state of wealth and population is maintained by an inflow of low-entropy matter-energy (depletion) and an outflow of an equal quantity of high-entropy matter-energy (pollution). Stocks of wealth and people, like individual organisms, are open systems that feed on low entropy.[14] Many of these relationships are summarized in Figure 1.

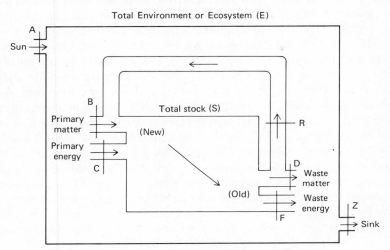

FIGURE 1 Rectangle (E) is the total ecosystem which contains the total stock (S) of wealth and people as one of its mutually dependent components. The ecosystem imports energy from outer space (sun) and exports waste heat to outer space (sink).

The stock contains matter in which a considerable amount of available energy is stored (mined coal, oil in oil tanks, water on high ground, living things, wood products, etc.) as well as matter in which virtually no available energy is stored. Matter and energy in the stock must be separately maintained. The stock is maintained in a steady state when $B = D$ and $C = F$. Throughput is defined only for the steady state, where throughput equals both input $(B + C)$ and output $(D + F)$.

From the second law of thermodynamics we know that energy cannot be recycled. Matter may be recycled (R), but only by using more energy (and matter) to do it. In the diagram energy moves only from left to right, whereas matter moves in both directions.

For a constant S the lower the rate of throughput the more durable or longer lived is the total stock. For a given throughput, the lower the rate of recycling (R), the more durable are the individual commodities. The optimum durability of an individual commodity is attained when the marginal production cost of increased durability equals the marginal recycling cost of not increasing durability further. "Cost" is total ecological cost and is extremely difficult to measure.

Both the size of the stock and the rate of throughput must not be so large relative to the total environment that they obstruct the natural ecological processes which form the biophysical foundations of wealth. Otherwise the total stock (S) and its associated throughput become a cancer which kills the total organism (E).

[14] Erwin Schroedinger, *What Is Life?* New York: Macmillan, 1945.

The classical economists thought that the stationary state would be made necessary by limits on the depletion side (the law of increasing costs or diminishing returns), but the main limits in fact seem to be occurring on the pollution side. In effect, pollution provides another foundation for the law of increasing costs, but has received little attention in this regard, since pollution costs are social, whereas depletion costs are usually private. On the input side the environment is partitioned into spheres of private ownership. Depletion of the environment coincides, to some degree, with depletion of the owner's wealth, and inspires at least a minimum of stewardship. On the output side, however, the waste absorption capacity of the environment is not subject to partitioning and private ownership. Air and water are used freely by all, and the result is a competetive, profligate exploitation—what biologist Garrett Hardin [15] calls the "commons effect," and welfare economists call "external diseconomies," and what I like to call the "invisible foot." Adam Smith's "invisible hand" leads private self-interest unwittingly to serve the common good. The "invisible foot" leads private self-interest to kick the common good to pieces. Private ownership and private use under a competetive market give rise to the invisible hand. Public ownership with unrestrained private use gives rise to the invisible foot. Public ownership with public restraints on use gives rise to the visible hand (and foot) of the planner. Depletion has been partially restrained by the invisible hand, while pollution has been encouraged by the invisible foot. It is therefore not surprising to find limits occurring mainly on the pollution side—which, of course, is not to deny depletion limits.

It is interesting that the first school of economists, the Physiocrats, emphasized the dependence of man on nature. Only the "natural" activity of agriculture was for them capable of producing a net product of value. Indeed, the word "physiocracy" meant "rule of nature." Something of the Physiocrats' basic vision, if not their specific theories, is badly needed in economics today.

Economic and Social Implications of the Stationary State

The economic and social implications of the stationary state are enormous and revolutionary. The physical flows of production and consumption must be *minimized, not maximized* subject to some minimum population and standard of living.[16] The central concept must be the stock of wealth, not as presently, the flow of income and consumption. Furthermore the stock must not grow. For several reasons the important issue of the station-

[15] Garrett Hardin, "The Tragedy of the Commons," *Science*, Vol. 162 pp. 1243–1248 (December 13, 1968).
[16] Kenneth E. Boulding, "The Economics of the Coming Spaceship Earth," in Henry Jarrett, ed., *Environmental Quality in a Growing Economy*, Baltimore: Johns Hopkins Press, 1966.

ary state will be distribution, not production. The problem of relative shares can no longer be avoided by appeals to growth. The argument that everyone should be happy as long as his absolute share of the wealth increases, regardless of his relative share, will no longer be available. Absolute and relative shares will move together, and the division of physical wealth will be a zero sum game. Also the arguments justifying inequality in wealth as necessary for savings, investment, and growth will lose their force. With production flows (which are really *costs* of maintaining the stock) kept low, the focus will be on the distribution of the stock of wealth, not on the distribution of the flow of income. Marginal productivity theories and "justifications" pertain only to flows, and therefore are not available to explain or "justify" the distribution of stock ownership. It is hard to see how ethical appeals to equal shares can be countered. Also, even though physical stocks remain constant, increased income in the form of leisure will result from continued technological improvements. How will it be distributed, if not according to some ethical norm of equality? The stationary state would make fewer demands on our environmental resources, but much greater demands on our moral resources. In the past a good case could be made that leaning too heavily on scarce moral resources, rather than relying on abundant self-interest, was the road to serfdom. But in an age of rockets, hydrogen bombs, cybernetics, and genetic control, there is simply no substitute for moral resources, and no alternative to relying on them, whether they prove sufficient or not.

On the question of maximizing versus minimizing the flow of production, there is an interesting analogy with ecological succession. Young ecosystems (early stages of succession) are characterized by a high production efficiency, and mature ecosystems (late stages of succession) are characterized by a high maintenance efficiency. For a given B (biomass stock) young ecosystems tend to maximize P (production flow), giving a high production efficiency P/B, whereas mature ecosystems tend to minimize P for a given B, thus attaining a high maintenance efficiency, B/P. According to ecologist Eugene P. Odum young ecosystems seem to emphasize production, growth, and quantity, whereas mature ecosystems emphasize protection, stability, and quality.[17] For the young, the flow of production is the quantitative source of growth, and is maximized. For the mature, the flow of production is the maintenance cost of protecting the stability and quality of the stock, and is minimized. If we conceive of the human economy as an ecosystem moving from an earlier to a later stage of succession (from the "cowboy economy" to the "spaceman economy" as Boulding puts it), then we would expect, by analogy, that production, growth, and quantity would be replaced by protective maintenance, stability, and quality, respectively, as the major social goals. The cardinal virtues of the past become the cardinal sins of the present.

With constant physical stocks economic growth must be in nonphysi-

[17] Eugene P. Odum, "The Strategy of Ecosystem Development," *Science* (April 18, 1969).

cal goods: services and leisure.[18] Taking the benefits of technological progress in the form of increased leisure is a reversal of the historical practice of taking the benefits mainly in the form of goods, and has extensive social implications. In the past, economic development has increased the physical output of a day's work while the number of hours in a day has, of course, remained constant, with the result that the opportunity cost of a unit of time in terms of goods has risen. Time is worth more goods, a good is worth less time. As time becomes more expensive in terms of goods, fewer activities are "worth the time." We become goods-rich and time-poor. Consequently we crowd more activities and more consumption into the same period of time in order to raise the return on nonwork time so as to bring it into equality with the higher returns on work time, thereby maximizing the total returns to total time. This gives rise to what Staffan Linder has called the "harried leisure class." [19] Not only do we use work time more efficiently, but also personal consumption time, and we even try to be "efficient in our sleep" by attempting subconscious learning. Time-intensive activities (friendships, care of the aged and children, meditation, and reflection) are sacrificed in favor of commodity-intensive activities (consumption). At some point people will feel rich enough to afford more time-intensive activities, even at the higher price. But advertising, by constantly extolling the value of material-intensive commodities, postpones this point. From an ecological view, of course, this is exactly the reverse of what is called for. What is needed is a low relative price of time in terms of material commodities. Then time-intensive activities will be substituted for material-intensive activities. To become less materialistic in our habits we must raise the relative price of matter. Keeping physical stocks constant and using technology to increase leisure time will do just that. Thus a policy of nonmaterial growth, or leisure-only growth, in addition to being necessary for keeping physical stocks constant, has the further beneficial effect of encouraging a more generous expenditure of time and a more careful use of physical goods. A higher relative price of material-intensive goods may at first glance be thought to encourage their production. But material goods require material inputs, so that costs as well as revenues would increase, thus eliminating profit incentives to expand.

In the 1930s Bertrand Russell proposed a policy of leisure growth rather than commodity growth, and viewed the unemployment question in terms of the distribution of leisure. The following words are from his delightful essay "In Praise of Idleness."

> Suppose that, at a given moment, a certain number of people are engaged in the manufacture of pins. They make as many pins as the world

[18] Services are included in GNP and are not in themselves physical outputs. However, increasing service outputs often require increases in physical inputs to the service sector, so that there is an indirect physical component. Leisure is not counted in GNP, and more physical inputs are not necessarily required as the amount of leisure is increased.

[19] Staffan B. Linder, *op. cit.*

needs, working (say) eight hours a day. Someone makes an invention by which the same number of men can make twice as many pins as before. But the world does not need twice as many pins. Pins are already so cheap that hardly any more will be bought at a lower price. In a sensible world, everybody concerned in the manufacture of pins would take to working four hours instead of eight, and everything else would go on as before. But in the actual world this would be thought demoralizing. The men still work eight hours, there are too many pins, some employers go bankrupt, and half the men previously concerned in making pins are thrown out of work. There is, in the end, just as much leisure as on the other plan, but half the men are totally idle while half are still overworked. In this way it is insured that the unavoidable leisure shall cause misery all round instead of being a universal source of happiness. Can anything more insane be imagined? [20]

In addition to this strategy of leisure-only growth, and the resulting reinforcement of an increased price of material-intensity relative to time-intensity, we can internalize some pollution costs by charging effluent taxes. Economic efficiency requires only that a price be placed on environmental amenities—it does not tell us who should pay the price. The producer may claim that the use of the environment to absorb waste products is a right that all organisms and firms must of necessity enjoy, and whoever wants air and water to be cleaner than it is at any given time should pay for it. Consumers may argue that the use of the environment as a source of clean inputs of air and water takes precedence over its use as a sink, and that whoever makes the environment dirtier than it otherwise would be should be the one to pay. Again the issue becomes basically one of distribution—not what the price should be but who should pay it. The fact that the price takes the form of a tax automatically decides who will receive it—the government. But this raises more distribution issues; and the "solutions" to these problems are ethical, not technical.

Another possibility of nonmaterial growth is to redistribute wealth from the low utility uses of the rich to the high utility uses of the poor, thereby increasing total "social utility." Joan Robinson has noted that this egalitarian implication of the law of diminishing marginal utility was "sterilized . . . mainly by slipping from utility to physical output as the object to be maximized." [21] As we move back from physical output to nonphysical utility, the egalitarian implications become "unsterilized."

Economic growth has kept at bay two closely related problems. First, growth is necessary to maintain full employment. Only if it is possible for nearly everyone to have a job can the income-through-jobs ethic of distribution remain workable. Second, growth takes the edge off of distributional conflicts—if everyone's absolute share of income is increasing there

[20] Bertrand Russell, *In Praise of Idleness and Other Essays*, London: Allen and Unwin, Ltd., 1935, pp. 16–17.

[21] Joan Robinson, *Economic Philosophy*, London: C. A. Watts and Co., Ltd., 1962, p. 55.

is a tendency not to fight over relative shares, especially since such fights may interfere with growth and even lead to a lower absolute share for all. But these problems cannot be kept at bay forever, because growth cannot continue indefinitely.

Growth, by allowing full employment, permits the old principles of distribution (income-through-jobs link) to continue in effect. But with no growth in physical stocks, and a policy of using technological progress to increase leisure, full employment and income-through-jobs are no longer workable mechanisms for distribution. Furthermore, we add a new dimension to the distribution problem—how to distribute leisure. The point is that distribution issues must be squarely faced and not left to work themselves out as the by-product of full-employment policies aimed at promoting growth.

A stationary population, with low birth and death rates, would imply a greater percentage of old people than in the present growing population, though hardly a geriatric society as some youth worshippers claim.° Since old people do not work, this further accentuates the distribution problem. However the percentage of children will diminish, so, in effect there will be mainly a change in the direction that payments are transferred. More of the earnings of working adults will be transferred to the old, and less to children.

What institutions will provide the control necessary to keep the stocks of wealth and people constant, with the minimum sacrifice of individual freedom? This, I submit, is the question we should be struggling with. It would be far too simpleminded to blurt out "socialism" as the answer, since socialist states are as badly afflicted with growthmania as capitalist states. The Marxist eschatology of the classless society is based on the premise of complete abundance; consequently, economic growth is exceedingly important in socialist theory and practice. Also, population growth, for the orthodox Marxist, cannot present problems under socialist institutions. This latter tenet has weakened a bit in recent years, but the first continues in full force. However, it is equally simpleminded to believe that the present big capital, big labor, big government, big military type of private profit capitalism is capable of the required foresight and restraint, and that the addition of a few effluent and severance taxes here and there will solve the problem. The issues are much deeper, and inevitably impinge on the distribution of income and wealth.

All economic systems are subsystems within the big biophysical system of ecological interdependence. The ecosystem provides a set of physical constraints to which all economic systems must conform. The facility with which an economic system can adapt to these constraints is a major, if neglected, criterion for comparing economic systems. This neglect is understandable because in the past, ecological constraints showed no likelihood of becoming effective. But population growth, growth in the physical stock

° See Figure 3 on page 213. [J.H. and R.H.S.]

of wealth, and growth in the power of technology all combine to make ecological constraints effective. Perhaps this common set of constraints will be one more factor favoring convergence of economic systems.

Why do people produce junk and cajole other people into buying it? Not out of any innate love for junk or hatred of the environment, but simply in order to earn an income. If, with the prevailing distribution of wealth, income, and power, production governed by the profit motive results in the output of great amounts of noxious junk, then something is wrong with the distribution of wealth and power, the profit motive, or both. We need some principle of income distribution independent of and supplementary to the income-through-jobs link.[22] Perhaps a start in this direction was made by Oskar Lange in his *On the Economic Theory of Socialism* [23] in which he attempted to combine some socialist principles of distribution with the allocative efficiency advantages of the market system. However, at least as much remains to be done here as remains to be done in designing institutions for stabilizing population. But before much progress can be made on these issues, we must recognize their necessity and blow the whistle on growthmania.

An Emerging Political Economy of Finite Wants and Nongrowth

Although the ideas expressed by Mill have been totally dominated by growthmania, there is a growing number of economists who have frankly expressed their disenchantment with the growth ideology. Arguments similar to those in the first section, stressing ecological limits to wealth and population, have been made by Boulding and Spengler (both past presidents of the American Economic Association).[24] Recently E. J. Mishan, Tibor Scitovsky, and Staffan Linder have made penetrating antigrowth arguments.[25] There is also much in Galbraith that is antigrowth—at least against growth of commodities the want for which must be manufactured along with the product.[26]

In spite of these beginnings, most economists are still hung-up on the assumption of infinite wants, or the postulate of nonsatiety as the mathematical economists call it. Any single want can be satisfied, but all wants in the aggregate cannot be. Wants are infinite in number if not in intensity,

[22] Robert Theobald, *Free Men and Free Markets*, Garden City, N.Y.: Doubleday, 1965.

[23] Oskar Lange, *On the Economic Theory of Socialism*, ed. Benjamin E. Lippincott, New York: McGraw-Hill, 1964.

[24] Kenneth E. Boulding, *op. cit;* J. J. Spengler, Public Address, Yale Forestry School, Summer 1969.

[25] E. J. Mishan, *The Costs of Economic Growth*, New York: Praeger, 1967; Tibor Scitovsky, "What Price Economic Growth," *Papers on Welfare and Growth*, Stanford, Calif.: Stanford University Press, 1964; Staffan B. Linder, *op. cit.*

[26] J. K. Galbraith, *The Affluent Society*, Boston: Houghton Mifflin, 1958.

and the satisfaction of some wants stimulates other wants. If wants are infinite, growth is always justified—or so it would seem.

Even while accepting the foregoing hypothesis, one could still object to growthmania on the grounds that, given the completely inadequate definition of GNP, "growth" simply means the satisfaction of ever more trivial wants, while simultaneously creating ever more powerful externalities which destroy ever more important environmental amenities. To defend ourselves against these externalities we produce even more, and instead of subtracting the purely defensive expenditures we add them! For example, the medical bills paid for treatment of cigarette-induced cancer and pollution-induced emphysema are added to GNP when in a welfare sense they should clearly be subtracted. This should be labeled "swelling," not "growth." Also the satisfaction of wants created by brainwashing and "hogwashing" the public over the mass media represent mostly swelling. A policy of maximizing GNP is practically equivalent to a policy of maximizing depletion and pollution.

One may hesitate to say "maximizing" pollution on the grounds that the production inflow into the stock can be greater than the consumption outflow as long as the stock increases as it does in our growing economy. To the extent that wealth becomes more durable, the production of waste can be kept low by expanding the stock. But is this in fact what happens? In the present system if one wants to maximize production one must have a market for it. Increasing the durability of goods reduces the replacement demand. The faster things wear out, the greater can be the flow of production and income. To the extent that consumer apathy and weakening competition permit, there is every incentive to minimize durability. Planned obsolescence and programmed self-destruction and other waste-making practices, so well discussed by Vance Packard, are the logical result of maximizing a marketed physical flow.[27] If we must maximize something it should be the stock of wealth, not the flow—but with full awareness of the ecological limits that constrain this maximization.

But why this perverse emphasis on flows, this "flow fetishism" of standard economic theory? Again I believe the underlying issue is distribution. There is no theoretical explanation, much less justification, for the distribution of the stock of wealth. It is a historical datum. But the distribution of the flow of income is at least partly explained by marginal productivity theory, which at times is even misinterpreted as a justification. Everyone gets a part of the flow, call it wages, interest, rent, or profit—and it all looks rather fair. But not everyone owns a piece of the stock, and that does not seem quite so fair. Looking only at the flow helps to avoid disturbing thoughts.

Even the common-sense argument for infinite wants—that the rich seem to enjoy their high consumption—cannot be generalized without committing the fallacy of composition. If all earned the same high income,

[27] Vance Packard, *The Waste Makers*, New York: Pocket Books, 1963.

a consumption limit occurs sooner than if only a minority had high incomes. The reason is that a large part of the consumption by plutocrats is consumption of personal services rendered by the poor, which would not be available if all were rich. Plutocrats can easily spend large sums on consumption, since all the maintenance work of the household can be done by others. By hiring the poor to maintain and even purchase commodities for them, the rich devote their limited consumption time only to the most pleasurable aspects of consumption. The rich only ride their horses; they do not clean, comb, saddle, and feed them, nor do they clean out the stable. If all did their own maintenance work, consumption would perforce be less. Time sets a limit to consumption.

The big difficulty with the infinite wants assumption, however, is that pointed out by Keynes, who in spite of the use made of his theories in support of growth, was certainly no advocate of unlimited growth, as seen in the following quotation:

> Now it is true that the needs of human beings may seem to be insatiable. But they fall into two classes—those needs which are absolute in the sense that we feel them whatever the situation of our fellow human beings may be, and those which are relative in the sense that we feel them only if their satisfaction lifts us above, makes us feel superior to, our fellows. Needs of the second class, those which satisfy the desire for superiority, may indeed be insatiable; for the higher the general level, the higher still are they. But this is not so true of the absolute needs—a point may soon be reached, much sooner perhaps than we are all of us aware of, when those needs are satisfied in the sense that we prefer to devote our further energies to non-economic purposes.[28]

For Keynes, real absolute needs are those that can be satisfied, and do not require inequality and invidious comparison for their very existence; relative wants are the wants of vanity and are insatiable. Lumping the two categories together and speaking of "infinite wants" in general can only muddy the waters. The same distinction is implicit in the quotation from Mill who spoke disparagingly of "consuming things which give little or no pleasure except as representative of wealth."

Some two and a half millennia before Keynes, the Prophet Isaiah, in a discourse on idolatry, developed the theme more fully.

> [Man] cuts down cedars; or he chooses a holm tree or an oak and lets it grow strong among the trees of the forest; he plants a cedar and the rain nourishes it. Then it becomes fuel for a man; and he takes a part of it and warms himself, he kindles a fire and bakes bread; also he makes a god and worships it, he makes a graven image and falls down before it. Half of it he burns in the fire; over the half he eats flesh, he roasts meat and is satisfied; also he warms himself and says, "Aha, I am warm, I have seen the fire!" And the rest of it he makes into a god, his idol; and

---

[28] J. M. Keynes, "Economic Possibilities for our Grandchildren," *Essays in Persuasion*, New York: Norton, 1963 (originally 1931).

he falls down to it and worships it; he prays to it and says, "Deliver me, for thou art my god!"

They know not, nor do they discern; for he has shut their eyes so that they cannot see, and their minds so that they cannot understand. No one considers, nor is there knowledge or discernment to say, "Half of it I burned in the fire, I also baked bread on its coals, I roasted flesh and have eaten; and shall I make the residue of it an abomination? Shall I fall down before a block of wood?" He feeds on ashes, a deluded mind has led him astray, and he cannot deliver himself or say, "Is there not a lie in my right hand?" [Isa. 44:14–20]

The first half of the tree burned for warmth and food, the finite absolute wants of Keynes, the bottom portion of GNP devoted to basic wants —these are all approximately synonymous. The second or surplus half of the tree used to make an idol, Keynes' infinite relative wants or wants of vanity, the top or surplus (growing) portion of GNP used to satisfy marginal wants—are also synonymous. Furthermore the surplus half of the tree used to make an idol, an abomination, is symbolic of the use made of the economic surplus throughout history of enslaving and coercing other men by gaining control over the economic surplus and obliging men to "fall down before a block of wood." The controllers of the surplus may be a priesthood that controls physical idols made from the surplus, and used to extract more surplus in the form of offerings and tribute. Or they may be feudal lords who through the power given by possession of the land extract a surplus in the form of rent and the *corvée;* or capitalists (state or private) who use the surplus in the form of capital to gain more surplus in the form of interest and quasi-rents. If growth must cease, the surplus becomes less important, and so do those who control it. If the surplus is not to lead to growth, then it must be consumed, and ethical demands for equal participation in the consumption of the surplus could not be countered by arguments that inequality is necessary for accumulation. Accumulation in excess of depreciation, and the privileges attached thereto, would not exist.

We no longer speak of "worshiping idols." Instead of "idols" we have an abomination called "GNP," large parts of which, however, bear such revealing names as Apollo, Poseidon, and Zeus. Instead of "worshiping" the idol, we "maximize" it. The idol has become rather more abstract and conceptual and rather less concrete and material, while the mode of adoration has become technical rather than personal. But fundamentally idolatry remains idolatry, and we cry out to the growing surplus, "Deliver me, for thou art my god!" Instead we should pause and ask with Isaiah, "Is there not a lie in my right hand?"

15. ADAPTING WORLD ORDER TO THE GLOBAL ECOSYSTEM

Richard A. Falk [*]

The Peace of Westphalia in 1648 brought an end to the Thirty Years' War and marked the beginning of the modern system of world order which accepts, as its basis, the autonomy of sovereign states. To this day, the quality of world order reflects the interactions between national governments. Threats and warfare have been the most salient of these interactions, constituting both the principal energy of change and the main instrument of order in world affairs.

World Wars I and II brought about a determined effort to mitigate the effects of a global politics based on war by building up central international institutions of peace and security. The League of Nations and the United Nations were the main products of these efforts, but these organizations have acted primarily as *instruments* of sovereign cooperation rather than as *substitutes* for the dominance of sovereign states. The military power of states continues to reside at national levels, and the driving forces in world affairs continue to be associated with intense competition for disputed territory, economic control and political influence. From 1914 to 1970 is not a long period of time in which to transform governing structures, public attitudes, and the political consciousness of elite groups. The great historical question, accentuated by the persisting danger of large-scale nuclear war, is whether the political life of mankind can be reorganized before rather than after some kind of catastrophic breakdown.

National governments are no longer able to solve the most serious problems facing the welfare and security of their own populations. We have grown accustomed to this basic reality in relation to nuclear weaponry. We now know that immense budgetary expen-

[*] A different form of this essay will appear as a paper in the 1970 Proceedings of the American Society of International Law. The essay also reflects an approach that is more fully developed —with supporting evidence and concrete proposals—in my book *This Endangered Planet*, Random House, 1971. I wish to thank Claudia Cords for giving me so much help in preparing this essay and John Harte and Robert H. Socolow for their diligence and imagination in clarifying and otherwise improving my text.

245

ditures on military hardware by the United States have led, not to greater security, but to offsetting measures by principal adversary governments, and that the two strongest states in the world would be likely to lose at least 100 million people as well as their major cities in the first 24 hours of an all-out attack by either side. There is no defense system that either of these states can now construct that is likely to work if the other state makes an all-out effort to penetrate it. Therefore, peace and security depend on the will and wisdom of the elite of a foreign government, and not upon the military prowess and preparedness of a defensive state. Such a reality makes us aware of the thin line, precariously maintained by fallible human beings and complex electronic equipment, which separates security from catastrophe in our world.

In the last few years this fundamental international condition has been further complicated by a new set of dangers arising from mounting pressures on the global environment. These pressures stem from the cumulative interplay of population growth and technological development. At this stage we do not yet have the facts and figures to enable a full appreciation of the scale of danger, nor do we know enough about the tolerance limits of oceans, river systems, and the atmosphere to identify with any precision the danger points and, especially, to specify thresholds of irreversibility. Ecologists have started to warn us that many environmental systems do not deteriorate gradually but, rather, are able to maintain the basic integrity of their character virtually until the point of collapse. It is this deception of man by nature that has contributed to the ecological collapse of large inland water systems such as Lake Erie: warnings about deterioration were discounted for years because of fairly favorable quality reports until the time of abrupt collapse, at which point the processes of decay could no longer be feasibly arrested. We do not know the extent to which the increased pollution of the oceans and atmosphere is generating a process of decay that will soon cross thresholds of irreversibility, nor do we even have effective means at present to collect such information. We do know that there has been an immense buildup of harmful pollutants in the oceans, most especially of oil, lead, mercury, and DDT, and that major disruptive impacts on marine ecosystems are likely to occur at some point in the future. Similarly, we know that these pressures are likely to continue and to grow worse as more and more people organize to live at higher standards of living, and as industrial societies develop more and more sophisticated technologies to facilitate their further mastery over nature.

The frequency and the severity of environmental crises having international implications is clearly increasing. The *Torrey Canyon* breakup is an example of the inevitable outcome of a rapidly expanding volume of oceanic transport of oil by tankers. The record shows that the increasing number of tanker accidents each year corresponds roughly to the increasing volume of activity. The increasing rate at which man exploits the planet is leading to other problems and challenges that may be less specific but are potentially even more disastrous. The possibility that radioactive and other highly toxic wastes may have a serious impact on life in the oceans, the fear of global weather modification arising inadvertently from the buildup of CO_2 and particulate matter in the atmosphere, and the chances of earthquakes resulting from underground nuclear explosions are but a few examples. At this stage we have not even drawn up a complete or accurate agenda of problems, but it is evident that the political fragmentation of mankind into separately administered states handicaps the efforts to solve any of them. The basic ecological premise posits the wholeness and interconnectedness of things.[1] It already seems clear that the basis of life on earth is imperiled by the absence of any central mechanisms of effective guidance and control on an international level.

The Present System of Sovereign States: Permissive Exploitation

Unlike domestic society, the activities of men and nations in the international realm are virtually free from consistent patterns of regulation. In the present system, land on the continents and the airspace above it are viewed as the private property of individuals or nations, while the oceans and the airspace over oceans are viewed as belonging to everyone, to be shared and used for mutual benefit. People are not revising their attitudes toward property, in spite of developments in modern technology which render obsolete the premises underlying the present system. It is helpful to isolate the three basic premises to which these remarks apply.

The Premise of Excess Capacity

In earlier decades, the volume of human demands being made on the environment was small in relation to the capacity of the environment to sustain life. Although local shortages of food, water, and land have long existed, and rivalries among social groups have generated wars, the dynamics of conflict have seemed to be fully consistent with the indefinite continuation of life on earth. Even the forecasts about the end of the world which are found in several major religious traditions do not reveal any awareness of the finiteness of the earth as an island in space sustaining a limited quantity of life and vulnerable to ecological disaster.

The situation today has been greatly moderated by the development of modern public health and by dramatic advances in agriculture. As a result, the world can accommodate a far larger number of people in the sense of keeping them alive, but in the course of so doing, especially in industrialized societies, great pressures on the environment have built up.

---

[1] On ecological perspectives, see David W. Ehrenfeld, *Biological Conservation*, New York: Holt, Rinehart and Winston, 1970; Eugene Odum, *Ecology*, New York: Holt, Rinehart and Winston, 1969; Paul Shepard and David McKinley (eds.), *The Subversive Science; Essays Toward an Ecology of Man*, Boston: Houghton Mifflin, 1969.

The limited capacity of our air and water to disperse our wastes, the limited yields of minerals which can be extracted from rocks close to the surface of the earth, and the vanishing of species are among the factors which have recently made us aware of global scarcity and have undermined our earlier confidence in global abundance. In spite of the fact that a laissez-faire system of organization is only effective in the absence of scarcity, international society continues to be almost completely unregulated.

Regulation is currently limited to special situations where overuse creates "conservation" issues, as with whale hunting. The experience with whale hunting is a dismal one, but is worth recounting because it illustrates the limited prospects for effective enforcement when sovereign states are not held accountable for failure to comply.

In 1946 those nations involved in whaling negotiated and signed a convention which established the principle of quotas for annual whale catches. Whale catches were to be limited in such a way that whaling interests could sustain profitable yields over long periods of time, a typical management goal whenever a renewable resource (timber is another example) is involved. However, the International Whaling Commission, which was to administer the convention, possessed neither the power nor the independence from the national whaling industries that was necessary to carry out the conservation program. At first, the quotas were set at levels beyond what could be caught, and hence were meaningless. Then, when in spite of reduced catches whaling companies still said that there was no scientific evidence that whales were endangered, the International Whaling Commission formed a committee of scientists, authorities in population dynamics, to study the problem in detail. In 1963 they filed a report to the International Whaling Commission substantiating the dangers, but the Commission refused to act on it. A report filed in 1964 with additional information met a similar fate. At this point the International Whaling Commission almost broke up, since by the convention it was obliged to act in accordance with any scientific evidence presented to it. Finally, in 1965 a special meeting was called, and quotas which began to take account of the scientific recommendations were agreed upon. Since then, the quotas have slowly come more into line with the recommendations, although with no enforcement machinery and no international observers it is difficult to judge to what extent even these limits have been heeded. Meanwhile, the blue whale, the largest animal that has ever lived on earth, is headed for extinction. Probably fewer than 300 blue whales remain at this time.

At present Russia and Japan, the only major whaling nations, are responsible for about 85 percent of the activity. The United States annually has bought about 20 percent of all whale products, mostly from Japan. Recently the United States decided to act unilaterally. On December 5, 1969, President Nixon signed into law the Endangered Species Act; on June 2, 1970, the Department of the Interior published a list of the species which were to be protected; all species of giant whales were included. Because of this new law, no whale or whale products in any form can be brought into the United States. But enlightened, though belated, actions by single governments cannot disguise the fact that this experiment in international cooperation has so far been a failure.

The Premise of Local Impact

A system of sovereign and independent states is appropriate only for an era when the consequences of a nation's actions are confined within its own boundaries. Even today most events continue to be of local significance in this sense, and can be regulated by local governing bodies. Special situations sometimes arise, as when state A allows its industrial corporations to dispose of raw wastes upstream, and the downstream users in state B become victims of pollution; if causation is clear, the states generally deal with one another directly. Measures to curb water pollution and to provide for navigation on the Danube and Rhine rivers illustrate the capacity of the present system to evolve generally satisfactory cooperative procedures. The treaty method has acted as a flexible instrument of adjustment where the actions in one state have been causing damage to another.

But when the effects are more diffuse and represent the cumulative outcome of numerous, separate, small instances, each of which may seem trivial, even benign, as with industrial processes underlying ocean and air pollution, then the present system shows almost no capacity for successful response. Nuclear testing in the atmosphere is a spectacular example of the widening scope and lengthening duration of events located specifically in space and time (that is, at the test site at the time of the test). Similarly, the balance of gases in the global atmosphere can be altered, with implications for the global climate, by the combustion of fossil fuels in a single country, or even, perhaps, by the operation of a single nation's fleet of supersonic aircraft.*

The Premise of Compatible Use

In the past, when resources were plentiful and actions were localized, the use of one geographical arena rarely restricted the use of any other, nor did different uses of the same arena often overlap in dangerous ways. Exploitation of the land and the rivers did not have serious repercussions upon the use of the oceans, and vice versa. In addition, the principal uses of the oceans for navigation, fishing, and naval operations were generally mutually compatible. Certain specific conflicts might occur—for instance, by overfishing in a particular area—but these could usually be either resolved by specific agreement or allowed to result in the temporary deterioration of a particular resource. The international law of the oceans accommodated basic needs by finding compromises between national sovereignty and community control. Coastal nations, for example, were granted a belt of special authority over offshore waters in recognition of special security,

---

* See the essay on Energy in this book for further discussion of these points.
[J. H. and R. H. S.]

economic, and health interests—a procedure that worked well so long as the territorial needs of these nations were limited to within a few miles of the shore. On the "high seas" nations have agreed not to interfere with each other's activities, and these agreements have worked largely because the separate activities were mutually compatible.

In recent years, however, these arrangements have come under increasing pressure, and territorial sovereignty has been expanded at the expense of the areas to which a community concept once applied. This has happened for several reasons. First, the technology of war has increased the distance from the shore at which a country can be threatened militarily. Second, the technology of fishing and mining has made it possible for the most advanced countries to operate at great distances from their homeland, and more successfully than coastal states relying on more primitive techniques. As a result poorer states have claimed protective custody over vast stretches of ocean water; Chile, Ecuador, and Peru, for instance, have claimed exclusive sovereign control over waters 200 miles from their shores. Third, the value of mineral resources on the continental shelf has led states to claim this wealth for their own nationals.

Beyond this, evidence is emerging that there are incompatibilities of use even in the areas of the ocean not claimed by any state. Reliance on persistent pesticides for agricultural development on land causes damage to marine ecology in the middle of the ocean in a variety of ways, still not fully understood.° Similarly, attempts to mine the oceans' mineral resources may endanger fishing industries. And finally the use of the ocean for the disposal of lethal nerve gas may cause harm to the entire community of ocean users and to the ocean itself.

In summary, current methods of regulating activity on land and in the oceans are geared to lower levels of demand for goods and to less sophisticated technologies of exploitation than now exist. The existing system of competition among sovereign states is not only obsolete, however; it is also counterproductive.

The logic of competition induces maximum self-assertion, contrary to the collective good.[2] Countries agree to cooperate today only when the issue is trivial or merely facilitative (for example, fixing the conditions of international postal service or governing the exchange of diplomats). But countries cannot be compelled to cooperate, and the objective of a majority of the nations of the world is often thwarted when cooperation is not universal. Thus China and France can continue to contaminate the globe with radioactive fallout, in disregard of the Limited Test Ban Treaty, and

---

° A discussion of some of the mechanisms by which this damage occurs is found in the essay by Professor Loucks in this book. [J. H. and R. H. S.]

[2] This phenomenon has been called "The Tragedy of the Commons." For an analysis of this issue, see Garrett Hardin, "The Tragedy of the Commons," *Science*, Vol. 162, pp. 1243–1248 (December 13, 1968); see also Beryl Crowe, "The Tragedy of the Commons Revisited," *Science*, Vol. 166, pp. 1103–1107 (November 29, 1969).

Japanese whalers can ignore the international regime set up to assure whale conservation. South Africa can continue with racial oppression, Brazil can encourage population growth, and the United States can dump as much nerve gas into the oceans as it wishes.

The sale of arms provides a good example of how competition between nations impedes the development of international controls. Most national governments are eager to build up a favorable balance of trade and regard it as highly desirable to earn large quantities of foreign exchange. Arms sales to foreign countries are very tempting, especially when the decision to forego the sale is likely merely to shift the transaction to another less scrupulous country, with the result that the earnings and, quite possibly, political leverage are lost. Sales can be restricted only if *all* principal suppliers curb their activities effectively and simultaneously. Given the degree of divisiveness in international society, it should not be surprising that it is difficult to make such a cooperative system work reliably. In fact, the realization of difficulty induces a sense of futility. Why try if trying is likely to penalize the more civic-minded national governments? The experience with international sanctions also illustrates the limits of cooperation in much the same way. The high degree of verbal consensus in support of economic sanctions against the Smith regime in Rhodesia has not achieved its goals, principally because South Africa and a few other countries do not want these sanctions to work.

The competition between nations which is partially responsible for the environmental crisis is also responsible for distracting attention from it. Can one imagine a discussion of environmental quality between an Arab and an Israeli leader? These and other nations are preoccupied by international rivalries, and devote their resources and energies mainly to promoting national security.

Habits of competition also impel the poor nations to pursue what the rich nations possess. Rich and poor alike specify self-interest in terms of more wealth, more power, more growth, and sometimes even more people. The possibility of a global environmental crisis is on the verge of providing the rich nations with new excuses to neglect the demands of the poor nations, however. There is serious reason to question whether the globe could remain habitable if the entire world population were to live the way Americans live at the present time, with all the inefficiency and waste now present in American society. The rich nations may perceive this as a valid reason for trying to persuade the poor nations to modify their plans for industrialization. But the poor nations are likely to repudiate this advice and urge the rich nations to control *their* consumption. Moreover, the poor countries are sure to point out that the gap between standards of living in the rich and poor countries is still widening.

The present inequalities and rivalries among nations thus complicate any attempt to replace competitive patterns of behavior by cooperative ones. The current negotiations over the terms of the international inspection of nuclear power installations in countries not possessing nuclear weapons is a case in point. In accordance with the Nuclear Non-Proliferation Treaty, national nuclear programs are to be inspected by the International Atomic Energy Agency (IAEA), after a state voluntarily offers to have its program subject to external control. When IAEA inspection finally begins in 1972, inspectors are likely to find their role limited to that of auditors. Italy, Japan, and West Germany want inspection confined to the input and output of the nuclear plants, for fear that inspectors who obtain knowledge of techniques used inside the plants might give useful information to competitors.[3]

About the only way in which present inequalities could have beneficial implications for the global environment is by opening up opportunities for ecological trade-offs. For example, industrial countries might establish preferential trade relations with those poorer countries that agree to use safe substitutes for DDT. In the great majority of instances, however, the habits of competition among nations will retard the search for methods of global environmental control.

There are some significant exceptions to this bleak picture that illustrate the potentiality for international cooperation within the present system of sovereign states. A treaty signed on December 1, 1959, has so far effectively ruled out a competitive struggle for sovereign control over the Antarctic and has allowed the 12 interested nations to pursue their separate courses of scientific investigation and discovery in a spirit of harmony and cooperation. The World Health Organization (WHO) works on problems common to all countries, including control and eradication of diseases and the protection of health through public health services. Its concern with environmental health problems, research, and education is the concern of all countries, and thus WHO can work harmoniously, effectively, and without fanfare. On a binational level, joint commissions that have been in existence for a long time are suddenly being reinvigorated in response to the recent upsurge of interest in environmental regulation. For instance, representatives of the governments of the United States and Canada held a meeting in June 1970 to discuss common problems of pollution and eutrophication in the Great Lakes. They agreed to adopt and enforce certain common measures, including regulations requiring removal of at least 80 percent of phosphates from sewage and industrial waste disposal. Also, the Soviet Union and the United States have had some success in reaching agreements that prevent the arms race from being fully carried over to outer space or the ocean floor and have stimulated wider agreements to halt the further spread of nuclear weapons, such as the Limited Test Ban Treaty and the Non-Proliferation Treaty. The current proposals before the United Nations which would prohibit weapons from being stationed on the ocean floor are further encouraging signs.

[3] See the *New York Times*, July 27, 1970, p. 5. For a detailed analysis of the IAEA's functions and problems, see Lawrence Scheinman, "Nuclear Safeguards, the Peaceful Atom, and the IAEA," *International Conciliation*, No. 572, pp. 5–64 (March 1969).

Prospects for an Adequate World-Order System

At the present stage of international development, there is no general or common appreciation of the threats to the environment or what to do about them. The United Nations is scheduling an initial world conference in 1972, but it is unlikely to provide more than an exchange of views and suspicions, and possibly the formulation of highly abstract declaratory standards. There is not as yet any appropriate sense of the magnitude of the task and of its urgency, and there is no understanding of the extent to which the protection of the environment will require an organizational and attitudinal revolution on a global level. Such a lack of awareness persists in government circles, even in a country such as the United States which has grown alarmed about environmental issues in recent years.

Yet, eventually, change must come. Separate, unequal, competing sovereign states will have to be replaced as prime centers of international decision making in the environmental field, and world political institutions to which responsibility will pass must be given the authority to monitor the environmental condition of the planet, to react quickly to disasters, to ration scarce resources, to zone international waters according to permissible uses, to set pollution standards, and to enforce its rules and regulations. It is likely that global institutions will be expected to resolve conflicts, apportion resources, and secure human justice in a wider range of situations than would be suggested by a narrow interpretation of conservation and environmental management.

Suppose a limited form of world government were to come into existence, with power to regulate all uses of the ocean. How should it use this power? It certainly would change the present system of registering ships, by which a small country like Liberia can establish exceptionally permissive licensing procedures because it is more concerned with revenues from tanker registries than with the protection of the oceans against unseaworthy vessels. The *Torrey Canyon,* a substandard oil tanker registered in Liberia, might not have been permitted to operate in a more ecologically conscious world. But what about the extraction of natural resources from the ocean? Should our hypothetical world organization forbid all mining by the enterprises of individual nations and do the mining itself? Should it instead issue licenses? Should it collect royalties or taxes, and if so, how should the revenues be used? Most important, should it regulate the *distribution* of the minerals extracted from the oceans, and if so, according to what principle? There are a host of alternative economic and political models of operation that could be incorporated into such a world organization.

The distinct ecological consequences of various modes of operation may be difficult to discern. For example, an ocean resource policy that puts a heavy penalty on ocean extraction may drive individual nations to exploit their domestic reserves beyond wise limits. A country might decide not to extract oil from shale by underground nuclear explosions if deep

Above: A United Nations Emergency Force of Yugoslav troops on patrol near El Arish, in the Sinai Peninsula, in 1957. *Below:* At the McMurdo Station in Antarctica, three U.S. Navy icebreakers team up to move a mass of ice which closed the shipping channel. McMurdo Station is the site of much United States research on that continent. (United Nations; UPI)

ocean oil were readily available. It might decide to station fewer power plants on its rivers and lakes if deep ocean sites were permitted. Should a world organization encourage or discourage exploitation of the ocean resources? If it discourages their exploitation, is there any way in which it can simultaneously affect resource use within the territorial limits of the sovereign states?

The last question suggests that we reexamine our hypothetical world organization. Does it make sense to separate the problem of ocean management from land management? Politically, this separation could be a practical way to gain experience with the kinds of jurisdictional problems that arise from the interplay of modern technology and ecology. A more extensive form of world order could emerge from success in undertaking such a modest first step. Ecological realities, however, could make such a scheme unworkable. Improper land-use practices can lead to the deterioration of coastal estuaries, which in turn can disrupt the life cycles of much of ocean life. Thus it may make better ecological sense for a world organization to acquire authority over land use and ocean use in a coordinated fashion. To maintain political equilibrium during the coming period of transition will tax the skills of the world's diplomats and international lawyers to an unprecedented degree.

It is clear that the establishment of a world government in and of itself does not guarantee a saner approach to the environment than what we now have. Indeed, we can speculate that if we had had an obtuse world government for the last few decades, things might have become a lot worse. Suppose 20 years ago, before the environmental crisis was widely recognized, worldwide pressures to remove the threat of nuclear war had brought about the replacement of the present political system with a world government. That government would probably have sought to overcome some of the inequalities among nations by stimulating capital transfers and by building huge power and agro-industrial projects to take advantage of the aggregate capacities of river basin systems and other large natural networks. Several projects of this kind have been retarded, thus far, by national boundaries and sovereign competition (even among friendly neighbors such as the United States and Canada or Mexico). Such political obstacles to man's intrusion upon nature would almost certainly have been cast aside, if world government had been instituted in a period prior to the emergence of some sensitivity about environmental quality.

Even the present period may be too soon from this point of view, for every potent world ideology today continues to maintain a strong commitment to achieving a maximum economic growth through the rapid expansion of the industrial sector, and this commitment accelerates environmental decay. However, the decay, in turn, is producing a shock of recognition in the richest and most industrialized countries; a greater awareness of the tolerances of the planet is beginning to emerge. Therefore it is fortuitous that environmental awareness, which strengthens the antiwar pressures for a more cooperatively conceived organization of life on earth, may at the same time reduce some of the hazards of concentration of authority. The sequence of pressures may lead to a more rational eventual solution. Yet it remains questionable whether there is enough time available to make adjustments in the world-order system. These adjustments cannot take place until *after* significant movements for world-order reform emerge in the main cultural, ethnic, political, and economic centers of world activity.

The future of human society depends on making the case persuasive that the present pattern of relations governing man-in-society and man-in-nature endangers the whole species and the entire planet, and that positive alternatives exist and can be brought into being. The mere depiction of disaster is likely to discourage action unless it is coupled with a program for positive action. Fear in isolation induces immobility, not conversion and passionate action. We need to develop plausible alternatives to global disaster, and to establish a humane set of substitutes for the automatic checks of war, disease, and famine—which never were compatible with the dignity of man and society. The need is for new visions of world order based on the conditions of dynamic equilibrium between man and nature.

In the meantime stopgap measures and educational activities are useful kinds of initiatives. It will be desirable to take action to avoid specific disasters, for instance, radioactive leakage from nuclear power plants or the further pollution of international rivers and lakes. It may also be possible to engage in joint ventures to ensure that only beneficial appropriation of minerals from the oceans takes place. Institutions through which nations cooperate are rapidly expanding in number, variety, and role to meet the needs of an increasingly interdependent world, and they should provide considerable experience and a cadre of experts with careers and values built around a more cooperative approach to international relations. Only a shared sense of the problems facing mankind will make it possible to work toward a shared solution. We possess the technology of communication, information-dispersal, and transportation that will facilitate centralized management. Indeed the efficiency of these new technologies poses a new set of threats to human welfare that will need to be taken into account.

Minor adjustments within the existing international order can do no more than gain time for the initiation of drastic changes in the world-order system. The need for drastic change suggests the likelihood of struggle between those who operate and benefit from the present political system and those who support the creation of an increasingly powerful world government. Good education, as always, should pursue a strategy of subversion, weakening confidence in existing arrangements, and even converting the old elite to the new vision; but it seems likely that the defenders of the status quo will condemn and suppress those who work visibly and effectively toward a new world system based on an ecological vision of wholeness. Prospects for change may not be at all serious until a countermovement emerges, perhaps one that identifies environmentalists as what they are, or should be: subverters of the existing order, apostles of a new order aiming to do away with the war system, repudiators of the ideology of national sovereignty and of the mindless exploitation of the natural resources of the earth.

The shape of this new order cannot now be blueprinted. It must be an expression of a collaborative process among the peoples of the world. It is certainly not a matter merely of extrapolating existing tendencies and designing a world state that draws inspiration from the nation-state form. At most we can make the case for the inadequacy of the present system of world order, combine it with a demonstration that the technological means exist to support a new equilibrium, and advance an argument for the realization of certain dominant values. The exact structures of order, processes of transition, and shifts in wealth-producing capabilities will depend on the way in which world-order reeducation and interregional bargaining proceeds in various parts of the world. How the new world order evolves will depend as well on how ecological deterioration manifests itself in the years ahead. The search for new solutions will surely grow more intense as the evidence mounts that we are faced with a crisis of survival.

IV. CONSTRAINTS ON GROWTH: THE SCIENTIFIC ARGUMENT

Never before in the history of this planet has a living species caused modifications of the natural surroundings on a scale comparable to the scale of natural processes. To make this statement precise, we must be quantitative. We must learn the magnitudes of the parameters that characterize the planet: the amount of water that flows from land to sea each year, the amount of energy that the earth receives from the sun, the amount of radiation in the rocks and air. Then we can compare such quantitites with quantitites that characterize the activities of man: the amount of water and the amount of energy we use, and the amount of radiation our medical care and our nuclear power plants contribute.

We believe that learning to think in terms of such comparisons is one of the best ways to begin to understand the sensitivity of the planet to man's activities. Accordingly, we have prepared three supplementary essays, on water, energy, and radiation, describing both our natural environment and the scale of man's alterations. We have included the necessary science within the essays, introducing the reader to the vocabulary of each field, including the customary units.

Both the Cayuga Lake and the Everglades case studies identify threats to our water resources. The essay on Water attempts to place these and other symptoms of water resource afflictions in a more general context. It also gives a quantitative discussion of how salt water can contaminate coastal water supplies and of how fresh water can be extracted from the sea.

The reader who has been following discussions of energy production in the popular media may well be confused in several areas, for in recent years a number of myths have propagated with a life of their own. We have been particularly careful in the essay on Energy to address ourselves to both the myths and the facts related to oxygen depletion and climate modification.

Nuclear physics has a reputation for inscrutability. This arises in part from an unfamiliar vocabulary, some-

259

thing which is inevitable when one describes an unfamiliar domain of experience. Partly, the reputation arises from the lack of confidence that characterizes the communication between expert and citizen in this field. The reputation is due, also, to the sinister overtones of some of the subject matter, which make some people want not to understand. Our intention in the essay on Radiation is to cut through some of the mystery. The new vocabulary cannot be avoided, but a little effort on the reader's part ought to reward him with a useful understanding of this increasingly important subject.

16. WATER

John Harte and Robert H. Socolow

Your share of the water on the earth is 100,000,000,000 gallons. Your share of the fresh water on earth, not counting the water in the two polar icecaps, is 600,-000,000 gallons. Your share of daily rainfall is 80,000 gallons, but 75 percent of this rain falls on the sea, so 20,000 gallons is your share of the average daily rain on the land. Your share of the daily rain on the land which does not evaporate but instead flows into the sea (the so-called runoff water) is 7500 gallons.[1]

How much of this water can man actually use? Obviously, the question is not well defined until we specify what kind of use. Let us try to establish some of the general concepts that help make precise any discussion of water scarcity.

I. A General View of Water Use

The oceans can be used by man for transportation and exploited by man for natural resources. This will almost never interfere with evaporation from the ocean, which is the source of 37 percent of the rainfall onto the land.[2] (The other 63 percent evaporates from the land and rains back on the land.) This 37 percent, 9.8×10^{15} gallons for the whole world each year, moves in a cycle, evaporating from the sea, carried by winds to the land, raining onto the land, and running downhill in rivers and underground aquifers back to the sea. This entire cycle is powered by the sun and represents the world's most colossal desalination program.

Some uses of fresh water leave the water unsuitable for further use without costly treatment. These are the uses of fresh water that must be closely regulated lest contamination render unusable the permanent storage areas of fresh water such as lakes and underground rock formations. Examples of such uses are waste disposal, industrial chemical processing, and irrigation on farms that are subject to modern agricultural chemicals (insecticides, herbicides, and fertilizers).

[1] The volume of stored water in the world by location is given in Table 1. Water flow rates in the world and in the United States are given in Table 2.

[2] Extensive oil spills on the ocean surface *can* increase the fraction of the sunlight reflected by the ocean, reducing evaporation.

TABLE 1. Distribution of the Earth's Water Resources

Location	Water Volume (gallons)	Percentage of Total Water
Oceans	348,700 × 10^{15}	97.2
Icecaps and glaciers	7700 × 10^{15}	2.15
Atmosphere	34.1 × 10^{15}	0.01
Fresh-water lakes	33 × 10^{15}	0.009
Saline lakes and inland seas	27.5 × 10^{15}	0.008
Average in stream channels	0.3 × 10^{15}	0.0001
Soil moisture	17.6 × 10^{15}	0.005
Groundwater within depth of half a mile	1100 × 10^{15}	0.31
Groundwater— deep lying	1100 × 10^{15}	0.31

SOURCE: Adapted from Brian J. Skinner, *Earth Resources,* Foundation of Earth Science Series. Englewood Cliffs, N.J.: Prentice-Hall, 1969.

Other uses of water, such as most forms of recreation, leave the water very nearly as it was. Intermediate between these two situations is the use of water for cooling a power plant: the heated water may be lethal to fish, or may speed the eutrophication of a lake, but after flowing a little while it will cool off and may be attractive to the managers of a second power plant (perhaps downstream on the same river) or may become municipal water.

The great natural fresh-water storage areas, above and under the ground, are part of the legacy we would like to pass on to our descendants altered as little as possible. These reservoirs lose water to the oceans and the atmosphere, and they are replenished by rainfall. We can interrupt this flow of water in order to use it, but if we go to the reservoir and remove year after year more water than falls as rain, the reservoirs will dry up. In this sense the rainfall, not the size of the reservoirs, determines the maximum amount of water that we can use. Only if we take care to return the used water to the reservoir as clean as when it was borrowed can we exceed the limit set by rainfall.

Averaged over the whole world, rainfall on land is about equally likely to be transpired by plant life, to evaporate directly from the land, or

TABLE 2. Water Flow Rates for the World and the United States

Water Flow	Gallons per Year
World rainfall on the sea	85.5×10^{15}
World rainfall on the land	26.1×10^{15}
World evaporation from the sea	95.3×10^{15}
World evapotranspiration from the land	16.4×10^{15}
World runoff	9.8×10^{15}
United States rainfall	1.55×10^{15}
United States evapotranspiration	1.08×10^{15}
United States runoff	0.47×10^{15}

SOURCE: Adapted from Brian J. Skinner, *Earth Resources*, Foundation of Earth Science Series. Englewood Cliffs, N.J.: Prentice-Hall, 1969.

to run into the sea. Man is very limited in his ability to alter this distribution; he cannot alter the transpiration rate without affecting the amount of plant life nor the evaporation rate without affecting rainfall. In this sense, the rate at which runoff water flows into the sea provides a kind of upper limit on the rate at which water can be used in ways which prevent its reuse.

In Comment 1: *Water for cooling and runoff water—an exercise,* the assumptions behind a frequently made comparison between runoff water and water used for cooling are made explicit.

Comment 1: Water for cooling and runoff water—an exercise

In considering water uses that do not prevent reuse, such as power plant cooling, the concept of runoff is not particularly relevant. Nevertheless, it is fashionable to compare annual runoff with the annual cooling requirements of power plants. The reader may find it interesting to investigate the statement quoted by Professor Eipper in his essay on page 115: "By the year 2000 . . . the total cooling water requirements of the United States' steam electric industry are expected to approximate one-third of the country's entire yearly supply of runoff water." This conclusion can be derived [3] from that fact that current (1969) United States electricity production is 1.3×10^{12} kilowatt-hours per year and from the following six assumptions:

[3] Appendix 3, on units, may be helpful.

1. The annual growth rate for electrical production, 7 percent, will continue.

2. Modern electrical power plants will continue to operate at 40 percent efficiency.

3. Water that cools power plants will be heated 20° F (11° C) on the average.

4. Half of the heat will be rejected to runoff water, the other half bring rejected to the atmosphere through cooling ponds and cooling towers and to the oceans.

5. Runoff water will never be used twice on its way to the sea.

6. The United States runoff, which is now 30 percent of rainfall (average annual rainfall is 30 inches), will not be modified.

The reader should also consider how such assumptions could be modified in practice and whether some other estimate might be more realistic.

Water Management

Water scarcity is a problem in many parts of the world and has caused much attention to be paid to technologies that either permit the reuse of water or reduce evaporation. Waste water treatment is conventionally described in terms of three levels of technology. *Primary treatment* separates sewage by gravity, allowing the removal of the scum that floats and the sludge that settles. *Secondary treatment* uses biological organisms to break down organic matter in sewage before it is discharged; the intent is to lessen the burden on the biological organisms present in whatever body of water receives the sewage. *Tertiary treatment* refers to any chemical method designed to remove materials not separated out or broken down by primary or secondary treatment, including phosphates, nitrates, pesticides, and dissolved metals. Chemical methods exist to remove virtually any contaminant from water, but extensive tertiary treatment is rarely implemented because of its high cost.

The task of purifying water, to be sure, involves technical challenges that go beyond finding the optimum chemical treatment process: cities where both municipal sewage and storm water flow through the same system of pipes have to contend with enormous variations in the volumes of water requiring processing. And agricultural water cannot be purified unless it is first collected—a formidable task.

If we do not purify the water we use after we are done with it, it is possible that natural biological processes will purify it for us. Biological processes remove many contaminants from solution and break down others. When man seeks to "use" a body of water for sewage disposal, however, it is easy for him to overwhelm the natural purification system. Nor can nature deal effectively with *all* contaminants: an example of a contaminant that cannot be broken down is radioactive waste—short of using a cyclotron there is no way to deactivate a radioactive nucleus before it decays. On the contrary, biological systems are capable of concentrating many radioactive isotopes, thereby increasing their danger to man.

Water as menace and opportunity. *Above:* Children around a polluted water hole that may be a source for water for cooking and drinking. *Below:* The dedication in New York City in 1967 of a special cap on a fire hydrant that gives a spray suitable for playing on hot summer days. (UNATIONS; New York Police Athletic League)

Nature also purifies water by evaporation. Some contaminants, among them DDT, remain bonded to the water as it moves into the atmosphere, but many contaminants are left behind. Sewage disposal by spreading wastes on a field is an example of water purification by evaporation; if the sewage is properly selected, the field will be fertilized at the same time, and the disposal represents a kind of recycling.

It is often dangerous to allow contaminated water to evaporate, however. For example, consider what has happened to some of the land in West Pakistan that has been heavily irrigated. Irrigated water comes in laden with salts, and because the water evaporates from the land instead of flowing out again, the salts remain behind. As a result, the soil quality has deteriorated.

New technologies for retarding the evaporation of water have been proposed; they are designed, of course, to increase the amount of fresh water available to man at a single site. If such technologies are ever applied on a large scale, they will not only affect rainfall but also air and water temperature. A large amount of energy is required to evaporate the annual rainfall, and is ultimately dissipated in the form of heat. (See Comment 2: *Evaporation, rain, and energy.*) This heat is distributed over vast regions of the earth's surface and the lower atmosphere. If evaporation is retarded, some solar energy must find another outlet: local heating of lakes is likely to result. A lake, just like a person, keeps cool by evaporation, and the effects on life in the lake are likely to be harmful.[4]

In Egypt an *increased* amount of evaporation (and a reduced runoff) has been accepted in return for electric power and a greater capacity to control the Nile. Aswan Dam, now being completed, will cause a lake 300 miles long to form behind it. Evaporation from the lake will greatly exceed the earlier evaporation from the river. There is no chance that the increased evaporation will result in rainfall that gives back to Egypt all of the water that evaporates, for the low rainfall in Egypt (about 4 inches a year) is determined by broad geographical features. The evaporated water will fall somewhere else. Thus Aswan Dam will permanently reduce the amount of water available for use in Egypt.

Comment 2: Evaporation, rain, and energy

Considerable energy flows with the hydrocycle. Consider, for example, the energy needed to evaporate the 1.1×10^{17} gallons of water that fall as rain each year in the world. It takes 540 calories to evaporate a cubic centimeter of water, a quantity nearly independent of the temperature at which this is done. Using Appendix 1 to convert gallons to cubic centimeters, you can show that 2.2×10^{23} calories per year are required to evaporate the annual rainfall. This is far greater than the rate at which energy is utilized for photosynthesis and represents an energy flux equal to roughly one-third of the solar flux on earth. (See essay on Energy.)

When the vaporized water condenses, as it does before a rainstorm,

---

[4] Cayuga Lake is an example of a lake that would be adversely affected by warming. See Professor Eipper's essay in this book.

the 540 calories per cubic centimeter of water are released to the atmosphere. The result is that the moist air is warmed by this heat of vaporization and rises. This explains why you can tell that a rainstorm is coming when the leaves on the trees are drawn skyward in an updraft.

Of course, a rainstorm is a complicated process. Some of the energy released to the atmosphere upon condensation is radiated to outer space, and some may power the winds that drive the rain from sea to shore.

II. Water Limits in Southeast Florida

The amount of water available for use in southeast Florida must be estimated on the basis of the watershed for that region. This watershed extends from the northern reaches of the Kissimmee River near Orlando to the southern tip of the peninsula, and eastward from a line running almost down the middle of the peninsula all the way to the Atlantic Coast. (See Figure 1.) This watershed occupies about 15,000 square miles, is roughly coextensive with the Central and Southern Florida Flood Control District and Everglades National Park, and serves about 3 million residents, approximately half of whom live in metropolitan Miami. We consider this particular watershed in some detail in order to stress that water use is only sensibly discussed at the regional level and to give the reader a more quantitative appreciation for some of the issues raised in the essay about Everglades National Park earlier in this book.

The average annual rainfall in South Florida is 60 inches, which is double the national average. The annual runoff to the sea, however, is only 20 percent of the rainfall, compared to an average figure of 30 percent nationally. As a result, the number of inches of rain which become runoff water is 12 inches, not much above the national average of 9 inches. The amount of water which evaporates or is transpired by plants in South Florida, 48 inches, is so large for two reasons: (1) the climate is warm and (2) the land is flat, so that there is a large amount of surface water.[5]

Sixty inches of water falling on 15,000 square miles in a year corresponds to a volume of 1.6×10^{13} gallons of water. Divided among 3 million people, this comes to 14,000 gallons per person per day. The runoff water (20 percent of this amount) is 2800 gallons per person per day—less than half of the average per capita daily runoff for the United States as a whole, which is about 6500 gallons. The per capita daily runoff in South Florida scarcely exceeds the national average for per capita daily *use*, which is about 2000 gallons per person per day. (The 2000 gallons are accounted for approximately as follows: 150 gallons per person per day for domestic use, 1000 gallons for irrigation, and the rest for industrial purposes including cooling. All of these figures are rising, since more people are taking care of lawns and more people have swimming pools, to cite

---

[5] Transpiration would be even larger than it is were it not that many plants have adapted to the need to retain water by developing long narrow leaves which retard transpiration.

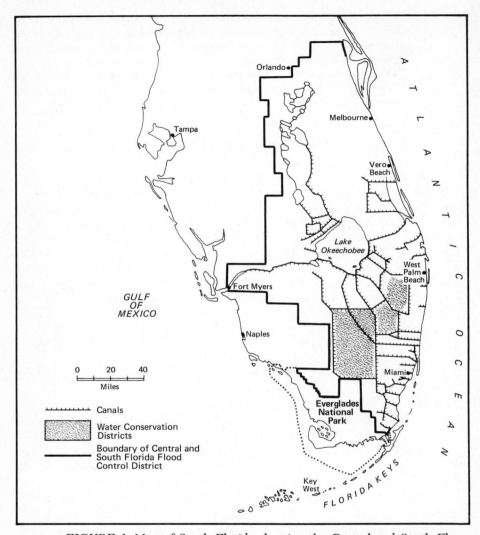

FIGURE 1 Map of South Florida showing the Central and South Florida Flood Control District. (Adapted from "Environmental Problems in South Florida." Report of the National Academy of Sciences and National Academy of Engineering, Washington, D.C., 1970)

two obvious examples.) The comparison of runoff water with water use is certainly crude, as we have remarked earlier, but it gives the correct impression that water is in short supply in South Florida.

Actual per capita water use in South Florida is difficult to estimate. Much of the water used for irrigation is not metered; a lot of it does not become runoff water either. Industrial water use is less than the national average; the factory that made the automobile of the Miami resident was cooled by water in Michigan, not Florida. Domestic water use is probably about 150 gallons, not much different from the national average.

The needs of Everglades National Park for water compete with those of agriculture, industry, and municipalities. The northern boundary of the park cuts right across the watershed region that we are discussing, and the park has customarily depended on water flowing south across that boundary from what are now the water conservation areas. New legislation, quoted in Essay 12, page 197, requires the Flood Control District to supply at least 315,000 acre-feet (100 billion gallons) of water per year to the park when water is plentiful. This corresponds to the household needs (at 150 gallons per person per day) of 1.9 million people, or the total water needs (at 2000 gallons per person per day) of 140,000 people. In 1968, a year of abundant rainfall, the park received 290 billion gallons from the Flood Control District which is about three times the new legal minimum and approximately the amount the park would have received from that area under natural conditions. The new legislation is intended to prevent a repetition of what happened during the years of low rainfall in the early 1960s, when the Flood Control District shut the sluice gates at the northern border of the park and cut the flow to the park nearly to zero.

The Central and Southern Florida Flood Control District has built so many dikes, canals, and pumping stations that the natural water flow patterns have been completely altered. This construction work has given the Flood Control District the capacity to bring water from the less populated regions north of Lake Okeechobee to the more populated regions south of the lake. However, population is growing rapidly all along the east coast, and inland near Orlando, and it is clear that there will be competing urban demands for the District's water, superimposed upon the already existing competition with the needs of agriculture for irrigation and the needs of the park. We can imagine waterworks extended northward up the peninsula almost indefinitely, the cost of water and the loss due to evaporation rising with each extra mile the water is pumped. But eventually the process must cease.

III. The Intrusion of Salt Water into Coastal Aquifers

Another water management problem in Florida is the problem of salt water intrusion into coastal aquifers. This problem is faced by any coastal settlement that derives its water from wells, and is especially serious when the land near the coast is flat. The problem is easily explained.

To begin with, ask yourself whether a well dug near the ocean shore will encounter salt water or fresh water. The answer turns out to depend on how deep the well is and how high the water table is where the well is being dug. A simple model will tell us where the salt water leaves off and the fresh water begins.

Water permeates any porous subsurface rock and moves within it. This is just as true for the sand and limestone beneath the ocean as for the soil and limestone beneath the land. Water in either case keeps on seeping downward until some kind of impermeable rock is encountered; on the two coasts of Florida, this is roughly 100 feet below sea level. In the permeable rock under the ocean, far from the coast, we find salt water, and in the permeable rock under the land far from the coast we find fresh water that has earlier fallen as rain. But near the coast there must be a combat zone.

If salt and fresh water had the same density, the combat zone would lie directly beneath the shoreline. But, in fact, salt water is 2½ percent heavier than fresh water (that is, its density is 1.025 grams per cubic centimeter). Accordingly, beneath a column of salt water, the pressure is 2.5 percent greater than beneath a column of fresh water of the same height. Thus, right at the coast, where the water table must be at sea level, if one started with a column of fresh water beneath the shoreline, it would literally be pushed inland by the sea. Hence the combat zone is under the land.

The only thing that finally stops the salt water is the fact that away from the coast, the water table has a chance to rise above sea level, and thereby the fresh water has a chance to build up extra pressure. Below any point on land, the boundary between the fresh water and the salt water occurs where the fresh water depth (measured from the water table) is just 2½ percent greater than the depth below sea level. At that depth, the pressures of the salt-water column and the fresh-water column are equal. The fresh-water column is 1.025 (or 41/40) times as high as the salt-water column.

Thus, near the coast one will find 40 times as much fresh water below sea level as above sea level in the aquifer. For example, if the water table is 2 feet above sea level, the boundary between salt and fresh water will be 80 feet below sea level.

Accordingly, the boundary surface that marks the farthest advance inland of the ocean waters reflects the shape of the water table like an elongating image. Figure 2 should make this clear. Of course, in practice, the boundary between salt and fresh water is not sharp, but the transition zone

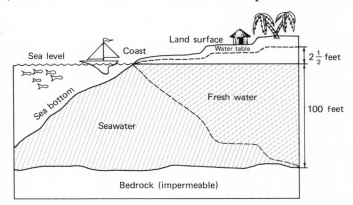

FIGURE 2 Schematic representation of the extent of salt water intrusion in a coastal aquifer. The boundary region between fresh and saltwater mimics the shape of the water table above it. If the water table is lowered a given distance, the boundary region rises 40 times that distance.

of intermediate salinity is actually quite well defined, and this crude model is reasonably accurate.

If the water table is lowered anywhere between the coast and that inland region where fresh water is found all the way down to the impermeable rock, there will evidently be increased salt intrusion. Inland drainage where the land is as flat as in South Florida will lower the water table over large distances, not just near the drainage area.

There are ways to prevent salt intrusion, but they are costly, complicated, and unpredictable. It is possible, for example, to force fresh water *down* wells into the aquifers to drive the salt water out. One can admire the prowess of hydrologists, which has permitted Holland and Israel, for example, to live at the outermost limits of what the local water situation will permit. Still, these same examples serve to reinforce an obvious lesson: that water quantity eventually poses limits on regional development.

IV. Desalination

New technologies for extracting fresh water from seawater have been suggested as a possible means of breaking once and for all the bonds of water scarcity. Obviously, much of our previous discussion of water limits, which carried connotations of imminent restrictions on some of man's activities, would be quite misleading if desalination were a possible large-scale solution.

There are three main ways to remove the salt from seawater: (1) One can boil the seawater and condense the vapor in a different place. The vapor will have no salt in it and the salt will be left behind. (2) One can freeze the water. The water that freezes first will be lower in salt concentration than the remainder; this relatively salt-free ice can be separated out and then allowed to melt. (3) We can use certain membranes that water molecules pass through more easily than the sodium, chloride, and other ions that are "salt" in solution. The second and third methods lead to a step-by-step reduction in salinity; they have proved difficult to operate on a large scale. The first method, which is called *distillation*, is the one that has been most often exploited commercially (for example, on board ships and more recently in large plants in locations like Key West, Florida, which are remote from fresh water supplies), and it will be discussed here.

It takes 540 calories to vaporize 1 gram of water—an amount of energy that is quite insensitive to the temperature at which vaporization occurs and to the concentration of dissolved salts (provided the solution is dilute enough that the great majority of the molecules are water molecules). In a completely inefficient distillation process, these 540 calories per gram would not be recovered. The most obvious way to improve the efficiency of the process is to have the steam condense back to water on the pipes that are bringing the salt water into the area where the boiling is taking place. In condensation 540 calories are given off, and this energy can serve to heat the incoming salt water. It is by use of such a multistage process that one of today's most efficient distillation plants, on the island of Aruba in the West Indies, has brought down the energy cost per gram from 540 calories to 40 calories.[6]

It is illuminating to consider how much energy would be required to supply the water needs of a typical American. Taking 40 calories per gram as the energy required for desalination and 2000 gallons per day as an estimate of per capita water use for all purposes (including irrigation and industrial cooling), we find that 3×10^8 calories per day would be required. Just to supply this individual with his domestic water (150 gallons per day) would require 2×10^7 calories per day.

Let us compare this energy expenditure with the energy required to supply our typical American with all his electricity. Assuming that he consumes electric power at a rate of 0.75 kilowatt, and assuming 40 percent efficiency in energy production at the power plant, we find that 4×10^7 calories per day are required, only one-seventh of what would be needed for his total water needs, and twice what would be needed for his domestic water.

In other words, if any coastal city that had a water shortage were to get its water, just for domestic use, from the sea, this would raise its energy consumption for electricity by 50 percent.[7] With all of the problems that the present scale of energy consumption present, this is hardly a welcome prospect. Like most energy-intensive "panaceas," desalination looks much less attractive as a large-scale enterprise when it is viewed in a larger environmental context, which evaluates the injurious side effects of energy production (resource depletion, air pollution, thermal pollution, and others).

No matter what improvements in technology occur, there is an absolute minimum amount of energy required to reduce the concentration of salt in water by any given amount, working with water of any given temperature. The existence of such a minimum energy is a consequence of the laws of thermodynamics. Salt dissolved in water is a highly dispersed arrangement of salt and water; salt separated from water is a more ordered —or lower entropy—arrangement. It takes energy, which must be supplied from an external source, to go from a less ordered to a more ordered arrangement, that is, to decrease entropy.

At 20° C the minimum energy to separate the first gram of pure water from a very much larger but fixed volume of seawater is 0.66 calories. (This result is derived in Comment 3: *The minimum energy required to desalinate seawater.*) As one removes fresh water from a fixed volume of seawater, the seawater becomes steadily saltier, and the energy required to remove each successive gram of fresh water steadily increases. In practice, one does not continue desalination past the point where half of the initial seawater has been extracted as fresh water; at that point, the minimum en-

[6] Barnett F. Dodge, "Fresh Water from Saline Waters. An Engineering Research Problem," *American Scientist*, Vol. 48, p. 476 (1960).

[7] To supply the water to an inland city from the sea would cost even more energy because of transportation requirements.

ergy cost is about 0.9 calories per gram of fresh water. These limits are completely general; that is, they apply whatever desalination method is used.

At the absolute theoretical limit (that is, 0.66 calories per gram), we still find an energy cost for the 2000 gallons of water per person per day of 5×10^6 calories per day.

It is amusing to discuss absolute limits like these, but in commerical installations limits from physics are almost always dwarfed by limitations imposed by engineering and economics. There is always a trade-off in any industrial process between fixed costs of construction and operating costs. Energy consumption is an operating cost, of course, and to approach thermodynamic minima for energy consumption requres using enormous installations and greatly increasing fixed costs. The reason for this is that the minimum energy requirements dictated by thermodynamics are calculated on the basis of reversible changes, which means, in practice, small temperature differences between parts of the apparatus and slow movement of parts; to accomplish the same rate of output under such constraints requires much larger installation than if temperature gradients and velocities of moving parts are increased.

Comment 3: The minimum energy required to desalinate seawater

Suppose the desalination of seawater takes place by the method idealized in Figure 3. A membrane divides a vessel into two compartments, and the membrane, by being mounted on a piston, can move in such a way as to change the relative sizes of the two compartments. The membrane permits water molecules to pass from one compartment to the other but blocks the movement of the ions into which salt breaks up in solution. (A membrane with this property is called semipermeable.) If seawater is placed in one compartment and fresh water in the other and the piston is locked in place, then water molecules will pass freely back and forth through the membrane until the pressures attributable to the water molecules on each side of the membrane are equal. There is an additional pressure on one side of the membrane due to the salt ions striking the membrane from only one side, and this pressure would force the membrane and the piston toward the fresh water if the piston were not locked in place. This pressure on a semipermeable membrane due to ions in solutions is called *osmotic pressure*.

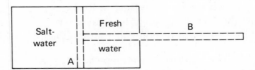

FIGURE 3 Piston B with semipermeable membrane mounted at A. Moving the piston to the left makes the saltwater more concentrated and increases the volume of fresh water.

Now suppose that the piston is unlocked and slowly pushed by an external force *toward* the seawater. In the process water molecules will pass through the membrane, permitting the piston to move, and the salt ions will be concentrated in a smaller volume of water. A volume v of seawater will be desalinated when the piston is moved so that the fresh-water compartment is larger and the seawater compartment is smaller by that amount. To move the piston against the osmotic pressure, the external force must do work on (that is, add energy to) the system.

The work done (W) by the force moving the piston is just the product of the osmotic pressure (P) and the change in volume of the seawater compartment v.

$$W = P \cdot v \qquad (1)$$

We assume the fractional change in the volume of the seawater compartment is small enough that we can neglect the slight increase in P that occurs during the displacement of the piston. This is assured if v is much less than the initial volume of the seawater compartment, which we call V.

The value of the osmotic pressure is found in a quite accurate way by treating the ions in the seawater compartment exactly as if they were gas molecules moving in a vacuum. The equation that relates pressure (P), volume (V), and absolute temperature (T) for an *ideal* gas also relates osmotic pressure, volume, and absolute temperature for a *dilute* solution. This equation, which underlies the "gas laws" of chemistry, is

$$PV = nRT \qquad (2)$$

where n is the number of moles present (either moles of gas or moles of ions) and R is a fundamental constant of nature:

$$R = 1.99 \text{ calories per mole per Kelvin degree}$$

There are typically 1.11×10^{-3} moles of ions in a gram of seawater,[8] and its density is 1.025 grams per cubic centimeter. Assuming that we can treat seawater as a dilute solution, we can calculate the osmotic pressure at room temperature ($T = 20°C = 293°K$) from equation (2):

$$P = (1.11 \times 10^{-3}) \times (1.99) \times (293) \times (1.025)$$
$$= 0.66 \text{ calories per cubic centimeter}$$

or about 27 times atmospheric pressure.

To desalinate 1 gram of seawater requires moving the membrane so that the fresh-water compartment is enlarged by $v = 1.00$ cubic centimeter (since the density of fresh water is 1.00 gram per cubic centimeter). Thus, using equation (1), 0.66 calories must be expended to desalinate 1 gram of fresh water by this method.

We stipulated earlier that the piston be moved *slowly* because then the system always stays close to equilibrium and it is possible to retrace the states of the system by reversing the direction of motion of the piston (moving it slowly toward the fresh water). A process carried out in such a way that the states of the system can be retraced is said to be carried out *reversibly*. A starting point of thermodynamics is the statement that

[8] The reader can get this answer approximately by knowing that the salinity of seawater is generally within a few percent of 35 parts per thousand by weight and by assuming that all of the salt is sodium chloride. He can then look up the actual composition of seawater in Appendix 4 to work out the 5 to 10 percent corrections due to magnesium, sulfate, and other ions.

the amount of energy used to bring about a change from one state to another is smallest when that change is carried out reversibly and that this minimum amount of energy is independent of which reversible procedure is chosen to bring about that change. Thus thermodynamic arguments permit us to generalize the result we have just obtained: *no matter what desalination process is tried* 0.66 calories is the minimum amount of energy required to desalinate 1 gram of seawater, given that the initial and final temperatures are both 20° C.

17. ENERGY

John Harte and Robert H. Socolow

The sun is the primary source of energy that supports life on the earth. Some of this energy is spent or converted almost immediately into useful goods and some is stored in "energy banks"; the energy for man's industrial activity is obtained from the energy savings of the geological past. It is helpful to divide the subject and to look first at the earth's energy ledger in the absence of man and then at the magnitude of the impact of man's energy consumption on the environment.

I. Solar Energy and the Earth's Heat Balance

The sun radiates 3.8×10^{26} watts of power in the form of photons (electromagnetic radiation), which corresponds to the production of 9.1×10^{25} calories of energy every second radiating outward from the sun. A much smaller amount of energy is radiated as charged particles and neutrinos. At a distance, R, of 92 million miles from the sun, where the earth floats, the sun's radiation is spread out over a spherical shell of area $4\pi R^2$. At the earth the flow of energy per unit area is found by dividing the incoming power by the total area of this shell. This ratio is a quantity called the *solar flux* and usually denoted by the letter Ω.

$\Omega = 0.032$ calories per second per square centimeter

The value of Ω is easily remembered as roughly 2 calories per *minute* per square centimeter.

The earth as seen from the sun is a disk with area πr^2, where r, the radius of the earth, is about 4000 miles (6.4×10^8 centimeters), and so the total energy flow received by the earth is $\pi r^2 \Omega = 4.1 \times 10^{16}$ calories per second or 1.3×10^{24} calories per year.[1] (The numerical values of the quantities of energy discussed in this essay are assembled in Table 1.)

What actually happens to the 1.3×10^{24} calories of solar energy received by the earth every year? Approxi-

[1] The earth is a spinning sphere and not a disk, of course, so the solar energy is distributed over the surface area, $4\pi r^2$, of the earth. Therefore the energy flux received on the earth, averaged over night and day and over all latitudes, is one-quarter of the solar flux at the earth, or 8.0×10^{-3} calories per second per square centimeter.

276

mately 35 to 40 percent of this energy is reflected back into space and does not significantly affect the heat balance of the atmosphere or the surface of the earth. Most of this reflection occurs from clouds and atmospheric dust, but some occurs from such reflecting surfaces as the oceans, snow, and sand.

The fraction of the energy incident on an object which is then reflected is called the *albedo* of the object. The albedo of the earth *as a whole* is 35 to 40 percent, and the average albedos of the oceans, forested land, and snow are, respectively, 2 percent, 3 to 10 percent, and as high as 80 or 90 percent. The albedo of a water surface varies considerably, decreasing as the sun approaches high noon or the water gets choppy.

Another 10 to 15 percent of the solar energy incident on the earth is absorbed by gases in the atmosphere and thus directly affects the heat balance of the atmosphere. Approximately one-half of this radiation is absorbed by water vapor and carbon dioxide, and most of the remainder by ozone, a rare molecular form of oxygen (O_3) that is present in the upper stratosphere some 20 to 30 miles above the surface of the earth. This ozone layer absorbs most of the short wavelength (ultraviolet) radiation which otherwise would strike the earth in intensities that would be lethal to present forms of life.

The remaining portion of the incident energy (about 6.5×10^{23} calories per year) is absorbed at the earth's surface. This portion of the solar flux drives a variety of processes on earth, some of which will be described

TABLE 1. Quantities of Energy Discussed in This Essay

Energy Transfer	Rate in Calories per Year
Energy radiated by the sun into space	2.8×10^{33}
Solar energy incident on earth	1.3×10^{24}
Solar energy affecting earth's climate and biosphere	8.1×10^{23}
Energy used to evaporate water[a]	2.2×10^{23}
Solar energy used in photosynthesis	9.4×10^{20}
Energy stored in net primary productivity	7.2×10^{20}
Energy conducted from earth's interior to its surface	2.0×10^{20}
Total energy consumed externally by man (1969)	4.8×10^{19}
Energy content of food consumed by man (1969)	2.9×10^{18}
Total energy consumed in U.S. (1969)	1.6×10^{19}
Electrical energy produced in U.S. (1969)	1.2×10^{18}

[a] See the essay on Water, Comment 2, page 266.

later, and is subsequently reradiated at wavelengths that are longer than those of the incoming radiation. The incident radiation is largely visible and that reradiated is largely infrared. The reradiated energy, on its way up from the surface of the earth, interacts with the atmospheric gases. Carbon dioxide (CO_2) and water vapor, both present in the atmosphere in small quantities, are more effective absorbers of infrared radiation than of the incident short-wavelength radiation. Thus they trap a portion of the solar flux. This is the so-called greenhouse effect, and it contributes significantly to the heating of the lower atmosphere; since some of this heat is radiated back to the surface of the earth, the greenhouse effect influences the surface temperature of the earth. The carbon dioxide in the atmosphere thus affects the earth's surface temperature.

The portion of the solar energy which affects the heat balance of the earth and which is available to maintain life on earth is the 60 to 65 percent which is *not* reflected, roughly 8.1×10^{23} calories per year. This amount can change, however, if the albedo of the earth changes. Therefore the quantity of snow on earth and dust and water vapor in the atmosphere, and the size of the ripples of the ocean also influence the earth's temperature. (See Comment 1: *The earth's heat balance*, for further details.)

Comment 1: The earth's heat balance

If the earth continually receives energy from the sun, why does it not get hotter and hotter without limit? The answer is that it is continually radiating heat to outer space as well. To calculate the equilibrium temperature of the earth, we write a heat balance equation: the rate at which energy is coming into the system must equal the rate it is leaving.

As we saw in footnote 1, the average rate per unit area at which heat is incident upon the earth and atmosphere is one-quarter of the solar flux:

$$\text{incoming flux} = \frac{\Omega}{4}$$

The formula for the rate at which heat leaves each square centimeter of the system is complicated and cannot be written down exactly. A reasonable approximation is to take the following:

$$\text{outgoing flux} = a\frac{\Omega}{4} + sT^4$$

The first term on the right represents the rate at which the solar flux is reflected, a being the albedo of the earth. The second term gives the rate per unit area at which energy is radiated away from the earth into space. The temperature, T, is the absolute temperature, measured in degrees Kelvin (1 degree Kelvin = 1 degree centigrade + 273). If the system was the most efficient radiator (a glowing coal is a good approximation to one), then the constant, s, would be 1.43×10^{-12} calories per square centimeter per second per (degree)^4. The earth is not a perfect radiator, and therefore s is actually somewhat smaller, but this value is accurate enough for our purposes.

For the earth to be in thermal equilibrium, we must have the relation

$$\frac{\Omega}{4} = a\frac{\Omega}{4} + sT^4$$

and therefore the equilibrium temperature, T_0, is

$$T_0 = \left[\frac{\Omega(1-a)}{4s}\right]^{1/4}$$

Try to verify that T_0 is approximately $-30°$ C. The temperature you obtain in this very approximate manner is colder than the familiar average temperature of the earth's surface. What you have obtained instead is an estimate of the temperature of that portion of the earth's atmosphere which is responsible for most of the radiation loss of the earth. It corresponds roughly to the temperature 4 to 5 miles above the earth's surface (the middle of the troposphere).

Equation (1) suggests one of the ways in which the temperature of the earth may change. If the albedo of the earth increases, for example, by the addition into the atmosphere of new sources of water vapor (perhaps the jet trails of supersonic transport in the lower stratosphere) or by an increase in atmospheric particulate matter as a consequence of industrial pollution, then the equilibrium temperature will decrease.

To calculate the *surface* temperature you would have to take into account the rate of heat transfer between the surface and the atmosphere (which in turn is governed by the albedo of different layers in the atmosphere), the greenhouse effect, heat transfer effects associated with winds and evaporation, and other effects that at present are poorly understood. This is the vast, and as yet unsolved, problem of climate prediction; it is an exciting subject.

II. Energy in the Biosphere

Energy and the Green Plant

How is the energy that the earth receives from the sun actually utilized? Let us look first at green plants and the process of photosynthesis.

A green plant takes in carbon dioxide (CO_2) and water (H_2O) and produces glucose ($C_6H_{12}O_6$), oxygen (O_2), and water. The synthesis of glucose occurs in several stages. The overall reaction is summarized by the formula

$$6CO_2 + 12H_2O + \text{sun's energy} \rightarrow C_6H_{12}O_6 + 6O_2 + 6H_2O$$

Note that the equation "balances"; for example, 24 atoms of oxygen are taken in and given off. The amount of energy required for this process is customarily given as the amount needed to produce a mole of glucose [2]

[2] A mole of a substance is defined as 6.02×10^{23} molecules (Avogadro's number) of that substance and is just that amount such that the numerical value

(180 grams) and this amount is observed to be about 6.7×10^5 calories, or 3700 calories per gram of glucose.

Now we can calculate how much of the solar flux is used by green plants for photosynthesis. On the average throughout the whole world, roughly 320 dry grams of green plant matter are produced on every square meter every year, or a total of $4\pi r^2 \times 320 = 1.6 \times 10^{17}$ grams per year is produced on the whole earth.[3]

To be more precise, let us define this concept operationally. Pick yourself a square meter of earth and watch it. As the green plants growing there reach maturity, pluck them, remove the 90 percent or so of water they contain, and weigh the rest. Do this all year and add up the weight. If you picked an average square meter of the earth, you will accumulate a weight of 320 grams. This number, the number of dry grams of green plant matter produced per square meter per year, is called the *net primary productivity*. Because it does not include the weight of the green plant matter which was metabolized during the year, the word *net* is used.

There is considerable uncertainty surrounding the productivity of the earth. The relative distribution of biomass between the land and the sea, for example, is not well understood, though it is believed that more than half of the 1.6×10^{17} dry grams of green plant matter produced yearly is produced on the land. Considering that 70 percent of the surface area of the earth is ocean, the net primary productivity of the land exceeds that of the oceans. The largest part of production on land occurs in forests.

About one-half of the 1.6×10^{17} dry grams of green plant matter produced yearly consists of carbon. Thus each year approximately 0.8×10^{17} grams of carbon are "processed" from carbon dioxide into glucose by photosynthesis. The actual amount is about 20 to 30 percent higher; glucose production exceeds biomass production because plants metabolize one-fourth of the glucose they produce, giving off carbon dioxide and obtaining energy for their life processes. In addition, some living plants are consumed by herbivores. So let us round off the estimate to 10^{17} grams of carbon. Since carbon is 40 percent of glucose by weight, this much carbon corresponds to 2.5×10^{17} grams or 1.4×10^{15} moles of glucose ($C_6H_{12}O_6$). The production of 1 mole of glucose requires 6.7×10^5 calories, and thus we find that a total of 9.4×10^{20} calories per year is required for photosynthesis of plant life on earth. Comparing with the 8.1×10^{23} calories per year available (incident and not reflected) from the

---

of its weight in grams is numerically equal to its molecular weight. For example, a mole of molecular oxygen, O_2, would contain 6.02×10^{23} molecules and would weigh 32 grams, for the molecular weight of oxygen is 32. [A mole of any *gas* is observed to occupy a volume of 22.4 liters at standard temperature ($O° C$) and standard pressure (76 centimeters of mercury); this result is independent of which gas one chooses because relations between temperatures, pressures, and volumes of gases are nearly entirely determined by the *number* of molecules present.]

[3] This number is only an estimate; it is probably correct to within 50 percent, which will be accurate enough for our purposes. See R. H. Whittaker, *Communities and Ecosystems*. New York: Macmillan, 1970.

sun, we see that this is about 1/860 of the available energy from the sun. In areas of the world such as the Everglades where the net primary productivity exceeds the global average by as much as a factor of 10, the utilized fraction of available solar energy exceeds 1 percent.

The plant is an energy bank; over its lifetime a plant stores roughly 75 percent of the energy supplied by the sun for photosynthesis, and it metabolizes only 25 percent. The stored energy does not disappear when the plant dies because energy is conserved. We know we can burn a dried plant and get heat in the process; in fact, we can extract about 4.5×10^3 calories per gram from typical dried plant material, roughly the same amount of energy as was required to make 1 gram of glucose. The energy per gram which can be extracted varies with the plant, reflecting the fact that the living plant carries out many chemical processes other than photosynthesis. For example, certain kinds of algae yield values almost twice as high. The stored energy in plant material provides calories for the organisms that eat it; in the process of decomposition, some heat is always emitted.

One might wonder whether the energy released upon decomposition could be utilized for further photosynthesis of new plants. In fact, this is not possible; whereas visible light is necessary to trigger the photosynthesis reaction, the heat of decomposition is radiated in the infrared region.

Cycles in the Atmosphere

To support life on earth, more energy is required than just the amount used for photosynthesis. Additional energy, for example, is required to power the hydrocycle and the gas cycles upon which life is dependent. In the essay on Water in this book, we estimate the energy needed to evaporate the water that rains each year. The amount is huge—about 250 times as much as is used in photosynthesis.

Carbon dioxide is a necessary ingredient for photosynthesis and also affects our climate by its influence on the atmospheric heat balance. Atmospheric oxygen, of course, is essential for animal life. Before we discuss the effects of man's activities on these vital resources, let us first estimate a few critical numbers describing quantities of these gases in the absence of man. For example, a natural question to ask is what fraction of the atmospheric storehouse of carbon dioxide and oxygen is utilized or produced each year by plants?

We have seen that each year roughly 10^{17} grams of carbon are processed from atmospheric carbon dioxide by plants. Since the molecular weight of carbon is 12, this corresponds to an intake of $\frac{1}{12} \times 10^{17}$ moles of carbon dioxide, which occupy a volume, at standard temperature and pressure, of

$$(\frac{1}{12} \times 10^{17}) \times 22.4 \text{ liters} = 1.9 \times 10^{20} \text{ cubic centimeters}$$

Now, the effective volume of the atmosphere is approximately 4.0×10^{24} cubic centimeters.[4] It is observed that 0.034 percent of the atmosphere by

---

[4] The effective volume is the volume the atmosphere would have if it all were at standard temperature and pressure. (Actually, the pressure decreases with

volume consists of carbon dioxide; thus the carbon dioxide alone would occupy 1.3×10^{21} cubic centimeters at standard temperature and pressure. The ratio of the amount of carbon dioxide used by plants each year to the amount of carbon dioxide in the atmosphere is therefore roughly $1.9 \times 10^{20}/1.3 \times 10^{21}$ or one-seventh. Photosynthesis has been going on for more than seven years, of course, and the reason that the carbon dioxide is still in the atmosphere is that plants respire some carbon dioxide and, more important, most of the remaining amount is returned to the atmosphere when the plants decompose.

In addition, the oceans play an important role in the carbon dioxide cycle. Every year a quantity of carbon dioxide roughly comparable to that used in photosynthesis is cycled back and forth between the atmosphere and the oceans (which store 50 times as much of this gas as does the atmosphere). The solubility of carbon dioxide in water increases as the water temperature decreases, and thus seasonal temperature variations play a role in this transfer process. In fact, a positive feedback could be set in motion because of this temperature-solubility relation. As air warms, more carbon dioxide will be released from the warming oceans. The additional carbon dioxide will, through the greenhouse effect, warm the air still further, and so it goes. Such a process has been suggested to explain glacial retreats.

In a similar way we can estimate the amount of oxygen given off by plants each year. For every mole of carbon dioxide consumed in photosynthesis, a mole of free molecular oxygen is produced, and therefore the volume of oxygen produced each year will equal the volume of carbon dioxide taken in by photosynthesis. Whereas 0.034 percent of the atmosphere, we saw, was carbon dioxide (by volume), 21 percent of the atmosphere (by volume) consists of molecular oxygen; that is, there are 620 oxygen molecules for every carbon dioxide molecule in the atmosphere. Thus plants return each year to the atmosphere approximately $1/7 \times 1/620 = 0.02$ percent of the atmospheric supply of oxygen.

It is sometimes stated that the oxygen that plant photosynthesis puts into the atmosphere each year is vital for animal life on earth. This is true, but a little misleading; for when plants decompose and return carbon dioxide to the atmosphere, or when they respire, they are consuming oxygen. The process of respiration and of decomposition of dead biomass (the latter process occurs primarily through the action of bacteria and fungi) takes oxygen from the atmosphere at a rate that is very nearly equal to the rate at which living plants produce oxygen. It appears, then, that plants do not contribute a significant *net* amount of oxygen each year for the support of

increasing altitude.) This atmosphere would be of constant density equal to the density at sea level (1.29×10^{-3} grams per cubic centimeter) and would be roughly 5 miles high. The effective volume is computed by dividing the actual mass of the atmosphere, 5.14×10^{21} grams, by the sea level density. It is also useful to work out the total number of moles of all gases in the atmosphere: since 1 mole of any gas at standard temperature and pressure occupies 22,400 cubic centimeters, the atmosphere contains $4.0 \times 10^{24}/22,400 = 1.8 \times 10^{20}$ moles of gases.

animal life. However, that is not quite true either, for some decomposition of plant matter occurs inside the animals that feed on plants. The oxygen that is used to "burn" that plant matter was inhaled by an animal and could be considered part of the oxygen needed to support that animal.

From the viewpoint of man and his energy needs, a critical component of the oxygen cycle is a process that has been occurring since life began on earth. At a very small rate some plants are buried before decomposition takes place. Every year a certain amount of biomass will sink to the bottom of the sea, get buried under landslides or in mud, or in some other way end its life in an environment where not enough oxygen is available to bring about decomposition.[5] The chemical energy stored in the plant when solar energy had earlier been absorbed thus finds its way underground.

One result of such burial is a net gain of oxygen in the atmosphere equal to the difference between the amount of oxygen given off during photosynthesis and the amount used for respiration during the plant's life. This is the amount of oxygen that would have been required to decompose the buried plant. It has been speculated that our atmospheric supply of oxygen accumulated in this way over the long history of life on earth.

But another result of plant burial particularly concerns us here, because of its effect on the recent history of man: natural forces have processed some of this buried biomass into coal, natural gas, and oil—the fossil fuels. Now man is excavating these energy sources which are the result of millions of years of photosynthesis, and he is burning the available fuels in a few centuries. What might the consequences of this activity be?

III. Energy and Man

Magnitude of Energy Consumption

The energy consumed by man is of two forms: internal (or nutritional) energy and energy for external requirements, including heat and light for our homes and power for our factories and automobiles. The energy consumed internally by man, expressed in calories per day per person, has remained relatively constant throughout the past. The average daily per capita consumption of energy from food is roughly 2.2×10^6 calories today,[6] averaged over the whole world.[7]

There are 3.6×10^9 people in the world today (1970), implying a total calorie intake of $365 \times (3.6 \times 10^9) \times (2.2 \times 10^6) = 2.9 \times 10^{18}$ calories per year. If all our food consisted of vegetation, this number of calories would be

---

[5] It is estimated that at present less than one part in a thousand of plant biomass escapes decomposition each year. See Wallace S. Broecker, "Man's Oxygen Reserves," *Science*, Vol. 168, No. 3939, p. 1537 (1970).

[6] Many discussions of food and nutrition use a unit, called the large calorie, equal to 10^3 calories.

[7] The United Nations estimates 2.4×10^6 calories per person is needed to meet daily internal energy needs, a figure that more than half of the people of the world do not achieve.

An eruption of the Myojin Reef volcano, 170 miles south of Tokyo. Energy flows from the interior of the earth to the surface primarily by conduction through the rocks in the earth's crust; about 2.0×10^{20} calories per year is transported in this fashion (five times the energy man currently releases at the surface of the earth when he burns fossil fuels). A much smaller amount of energy is transported by convection in volcanos and geysers.

Man only knows how to exploit this energy flow in situations in which heated water is either being vented naturally or is trapped underground. Although hydrothermal sources are now being tapped for energy, primarily in Italy, New Zealand, and Iceland, this is a limited source of energy for man. The total natural heat flow from all hydrothermal sources is estimated to be less than 2 percent of the rate of man's current energy consumption. (U.S. Navy Photo)

provided by about 1/250 of the total amount of plant matter produced each year. In actuality we consume even more than this amount of biomass in order to obtain our calories because of our penchant for meat. Suppose our diet consisted entirely of cows or other herbivores. Then, since a herbivore retains on the average only 10 percent of the biomass it eats,[8] we would, in effect, be consuming ten times more vegetation, or about 1/25 of the annual world production of all forms of life, to support our "habit." Any serious attempt to cope with world hunger must take this sobering concept into account. To summarize, the internal per capita energy needs of man have remained relatively constant, but both our increasing numbers and the increasing number of us meeting our energy needs by meat consumption are increasing the total demand on our land resources.

On the other hand, the external energy needs of man, and even the per capita needs, have skyrocketed. Data on world and United States yearly energy consumption are shown in Figure 1. During the last few decades, the doubling time for world energy consumption has been approximately

[8] For a discussion of consumption and retention of food at successive levels of the food chain see the Comment accompanying Essay 7 (pp. 93–95).

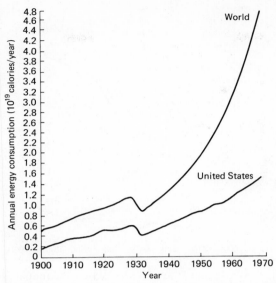

FIGURE 1 World and United States total annual energy consumption.

20 years, corresponding to an annual growth rate of 3.6 percent. The world population, on the other hand, has been growing at the rate of 2 percent a year, so that the per capita increase is about 1.6 percent a year. In the United States, energy consumption is growing at the rate of 2.8 percent a year, while the population is growing at the rate of 0.9 percent a year. The rate of per capita increase in the United States (1.9 percent a year) is about the same as for the world as a whole, but the figure that is being compounded is about six times as large. (See Figure 2.) The United States accounted for about one-third of the world's energy consumption in 1969.

In 1969 roughly 4.8×10^{19} calories of energy were consumed in the world, almost all of it from fossil fuels. Taking the approximate energy value of 10^4 calories per gram [9] of oil, this is the energy obtained from 4.8×10^{15} grams or 3.5×10^{10} barrels of oil.[10] Equivalently, since coal yields roughly 0.7×10^4 calories per gram, this is the energy produced from about 0.75×10^{10} tons of coal. A third way to obtain this energy is from natural gas. A cubic foot of methane gas, CH_4, at standard temperature and pressure contains 28.32 liters or 28.32/22.4 moles or $(28.32/22.4) \cdot (12+4)$ grams. This gas yields about 1.2×10^4 calories per gram, and so 2×10^{14} cubic feet of the gas would be required to meet the world's energy needs in

[9] In part II it was stated that 3.7×10^3 calories of solar energy are required to make one gram of glucose, and roughly 4.5×10^3 calories are obtained from the combustion of dried plant material. Underground, further chemical processes occur, yielding coal, gas, and oil, but the energy content of the fossil fuels is not very different from (roughly twice as large as) the energy content of the original glucose.

[10] A barrel of oil equals 42 United States gallons. If the barrel contains average crude oil, it weighs about 310 pounds.

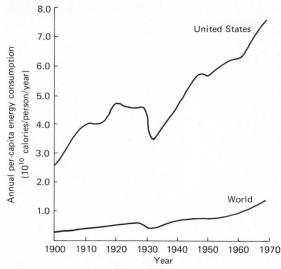

FIGURE 2 World and United States per-capita annual energy consumption.

1969. Actually coal and oil each accounted for 40 percent of the fuel consumed in 1969, and natural gas accounted for most of the remainder. The energy consumed by man which has been generated by nuclear fission, water power, wood and solar power, all combined, is less than 5 percent of man's total energy consumption. Estimates are that about 5 percent of the available supply of fossil fuels have already been consumed, although such an estimate involves an uncertain prediction about the size of any as-yet-undiscovered underground reserves.[11]

The sources of energy consumed in the United States have been changing. Figure 3 shows, by percentage, the sources of energy in the United States since 1850. It is clear that fossil fuels are the primary source today. Because of the impending shortages of these fuels and the growing energy demand, it is quite likely that nuclear sources, which today produce less than 1 percent of the United States energy, will dominate the graph 50 years from now.[12]

Table 2 shows the distribution by economic sector of energy consumption in the United States in 1947, 1955, and 1965. The fraction of the total energy consumption that can be attributed to the production of electricity has been growing steadily, although it is still only about one-fifth of the total energy consumption in the country. Energy that has not first been transformed into electricity is consumed in industrial processes (for exam-

[11] An excellent assessment of resource depletion is found in *Resources and Man*, National Academy of Sciences and National Resources Council, San Francisco: W. H. Freeman, 1969.

[12] It is because of this impending change and the unfamiliarity of much of the subject matter that we have devoted an entire essay to the subject of radiation.

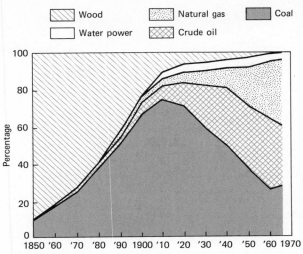

FIGURE 3 Percentage of total energy consumed in the United States that is attributable to five primary energy sources annually since 1850. The contribution of nuclear energy sources is too small to appear on this graph. (From Brian Skinner, *Earth Resources*. Englewood Cliffs, N.J.: Prentice-Hall, 1969)

ple, burning coal with iron to make steel), in homes and commercial installations (mainly for the purpose of heating), and in transportation (almost exclusively to power the internal combustion engine). Each of these three sectors of the economy still consumes more energy than the electrical industry. It is clear from these data, for example, that for our society to change over to electric-powered automobiles would necessitate a dramatic redistribution of energy consumption among sectors of the economy.

Electrical energy consumption, as distinct from total energy consumption, is growing at 7 percent a year in the United States, doubling approximately every ten years. Since population is growing at about 1 percent per year, per capita electrical energy consumption is growing at an annual rate of 6 percent. This is one of the most important and clear examples of an economic indicator whose growth is only slightly attributable to population growth and is primarily a result of growing consumption by the individual. From Table 2 we see that, in 1965, 20.7 percent of 1.35×10^{19} calories, or 2.7×10^{18} calories, were used to produce electric power. Almost all of this energy was derived from fossil fuels. According to the *World Almanac*, the electrical energy produced in 1965 was 0.9×10^{18} calories. The difference between the fuel energy input and the electrical energy output is the waste heat at the power plant. The ratio of the electrical energy output to the fuel energy input is called the efficiency of the power plant. From our data, we can infer that the average efficiency of fossil-fuel electrical power plants is approximately 0.9/2.7, or 33 percent.

The production of 0.9×10^{18} calories of electrical energy in 1965 in the United States can be restated in a form that will be more familiar to any-

TABLE 2. Fuel Consumption in the United States by Consuming Sector

Consuming Sector	1947	1955	1965
Household and commercial	20.4%	21.6%	21.9%
Industrial	38.2%	34.7%	32.6%
Transportation	26.5%	24.6%	23.6%
Electrical utilities	13.3%	16.7%	20.7%
Other	1.6%	2.4%	1.2%
Total energy in calories per year	0.84×10^{19}	1.01×10^{19}	1.35×10^{19}

SOURCE: Adapted from *U.S. Energy Policies—An Agenda for Research*, Resources for the Future Staff Report, Johns Hopkins Press, Baltimore, Maryland, 1968.

one who has looked closely at his monthly electricity bill. A *watt* is a unit of power, or rate of energy consumption, equal to the consumption of 1 joule of energy every second, or 0.24 calorie every second. (Thus 1 watt-second is a joule, and 1 kilowatt-hour is $1000 \times 3600 = 3.6 \times 10^6$ joules of energy, since 1 kilowatt = 1000 watts and 1 hour = 3600 seconds.) Since there are 3.15×10^7 seconds in a year, the average rate at which electric power was consumed in the United States in 1965 was

$$0.9 \times 10^{18} \text{ calories}/3.15 \times 10^7 \text{ seconds}$$
$$= 1.2 \times 10^8 \text{ kilowatts}$$

Since the 1965 population was about 190 million, this means that the per capita power consumption was about 0.63 kilowatt, or equivalent to every man, woman, and child burning ten 60-watt bulbs continuously. By 1969 the per capita electrical energy consumption had reached 0.75 kilowatt.[13]

In the transmission of electricity through wires and in the utilization of electricity by motors, air conditioners, and so forth, there are further losses of useful energy to heat. The heat given off by a light bulb, for example, represents an inefficient use of energy, as does, in a slightly different sense, the air conditioning of a restaurant or movie theater to the point where customers need to wear sweaters and coats.

[13] An often quoted figure is the national electric power *capacity* installed per person, which was about 1.4 kilowatts in 1969. Power plants, evidently, run at about 50 percent of capacity on the average.

There has been a great deal of attention in the energy industries to efficient energy production; the efficiency of electric power plants, for example, has steadily increased. Society has not paid the same attention to achieving efficiency in energy utilization, however, partly because often the cost of energy is only a few percent of the total cost of an industrial or household process. Now the impact of energy consumption on the environment is becoming a more central concern, and also "brownouts" are occurring with increasing frequency, reflecting a consumer demand that exceeds the available energy supply. This suggests that there is a need to assess the general question of energy consumption and to assure that energy is being used efficiently; in other words, energy conservation must be understood in the naturalist's, not only the physicist's, meaning of the word "conservation."

To a certain extent, energy utilization can be made more efficient by technical innovation in devices. It may be possible, for example, to slow a subway by transferring much of its kinetic energy into the rotational energy of a flywheel (instead of into heat and noise) and then to start the subway up again by transferring the kinetic energy back from flywheel to train. In such a process, in principle, fewer calories will be required to drive the subway.

How efficiently energy is used may also be affected by regulations governing the cost of electric power. At present, bulk rates for electricity are much lower than the rates small customers are charged, and this may decrease the incentive for the large user to use energy efficiently; on the other hand, bulk rates may bring about economies of scale. Whether pricing policies could be designed to affect the efficiency of energy use is an interesting problem for research.

The total consumption of energy may also have to be regulated by society. There will probably be laws governing the use of air conditioners and automobiles, especially in situations where energy use is affecting the microclimate of a city. In addition, energy-intensive solutions to environmental problems, like nuclear-powered recycling plants and water desalination plants, will have to be examined carefully to discern the trade-offs between the burdens they impose and the burdens they eliminate.

Impact on the Environment

In estimating the impact of man's energy production on the environment, an important concept to keep in mind is that the energy we produce —no matter whether it is produced from fossil fuels or nuclear sources, whether it is lighting a light bulb or moving an automobile—ultimately is dissipated as waste heat energy. Heat energy is the energy of random motion, whereas other forms of energy express organized motion; the statement that the energy of a closed system becomes steadily more disorganized is a statement of the second law of thermodynamics.

If we compare the 4.8×10^{19} calories produced in 1969 by man with the solar flux, this will give us one indication of our effect on the thermal balance of the globe. Recalling that 8.1×10^{23} calories per year from the sun affects the heat balance at the earth's surface, we see that the contribution of global fossil fuel consumption (4.8×10^{19} calories) is at present 0.006 percent of that figure.[14]

An analysis of the earth's heat balance suggests that an increase by 1 percent in the amount of heat energy reaching the earth's surface would result, after a new equilibrium had been reached, in an average global temperature rise of roughly 1° C.[15] The 0.006 percent increase in energy at the earth's surface that now results from man's energy consumption is probably of negligible impact globally. If the current 3.6 percent annual growth rate of energy consumption were to continue, then the figure of 0.006 percent would double every 20 years. It is not understood at present how long this could continue before the climatic repercussions would become catastrophic.

However, this way of estimating the environmental impact of energy production is misleading for many reasons. First, power sources are not uniformly spread over the globe, and *localized* effects on weather of waste heat dumped into the atmosphere will certainly become apparent even before noticeable *global* climatic changes occur. Energy production in densely populated urban areas, for example, is probably already having an effect on the weather in cities. To get an idea of the magnitude of urban energy production, we can compare the energy output in New York City with the solar flux directly on that city.

Approximately 7.8 million people live within an area of 365 square miles in New York City. If we assume that New Yorkers consume energy at the national per capita average rate of 7.5×10^{10} calories per year (see Figure 2), then every year 5.8×10^{17} calories of energy are consumed within 365 square miles. In other words, energy production in New York City is roughly 1.6×10^{15} calories per square mile per year. The reader can check that this figure is about 7000 times larger than the global average.

The solar flux affecting the earth's climate and biosphere (see Table 1), averaged over the surface of the earth, is, in these units, 4×10^{15} calories per square mile per year. This is only about 2½ times larger than our estimate for the energy consumption by man in New York City. Modern cities would experience catastrophic climatological repercussions from the high density of waste heat production, except for the fact that they are still small enough for winds to dissipate their heat adequately.

Second, not all of the waste heat that results from man-made energy is dissipated into the atmosphere. In fact, we have learned from Professor Eipper's essay in this book that the cooling of electric power plants by water causes harmful biological effects. Some power plants are cooled in

[14] If the entire world population were to enjoy the current per capita energy consumption of the United States, then this ratio would increase to nearly 0.04 percent.

[15] See Reid A. Bryson, "All Other Factors Being Constant. . . ," *Weatherwise*, April 1968, p. 56.

the ocean, some in lakes, and others in rivers. In order to estimate the effect of man's waste heat on the hydrosphere we have to make some assumptions. If we assume that the entire quantity of water on earth, estimated to be 1.4×10^{24} cubic centimeters, absorbs all the energy consumed by man each year, then the 4.8×10^{19} calories produced in 1969 would be enough heat to raise all of that water a mere 1/30,000 degree centigrade.

The assumption that *all* the water on earth can be used to cool power plants is, of course, unrealistic. In the essay on Water, a more detailed discussion of the impact of power plant cooling on water resources is presented. (See especially Comment 1, pp. 263–264.)

Finally, we turn to what may be the most serious effect of energy consumption on earth, the changing composition of the atmosphere. When fossil fuels are consumed, not only is heat produced but oxygen is taken from and carbon dioxide is returned to the atmosphere. Let us estimate the magnitudes involved. We have stated that man consumed 4.8×10^{19} calories of energy from fuels in 1969. Since, as we saw in part II, the heat content of fuel is about 10^4 calories per gram, this corresponds to 4.8×10^{15} grams of fuel. Most of this fuel is in the form of hydrocarbons which are molecules consisting of carbon and hydrogen. In combustion, atmospheric oxygen combines with the fuel and energy is produced. The simplest possible fuel is pure carbon and the formula for its combustion is [16] $C + O_2 \rightarrow CO_2$

Since coal, the most widely used fuel on earth, consists primarily of carbon, we shall assume that this simple combustion formula describes *all* fuel consumption and we shall calculate the amount of oxygen consumed and carbon dioxide produced each year by fossil fuel consumption. Since 1 mole of carbon weighs 12 grams, the 4.8×10^{15} grams of carbon correspond to 4×10^{14} moles of carbon. For every mole of carbon that is burned, 1 mole of oxygen is consumed and 1 mole of carbon dioxide is produced.

Thus in 1969 approximately 4×10^{14} moles of molecular oxygen were taken from the atmosphere and approximately 4×10^{14} moles of carbon dioxide were returned to the atmosphere as a result of fossil fuel consumption. (The reader may want to see how the complications mentioned in footnote 16 would affect these results.)

The atmosphere, we saw in part II, contains 1.8×10^{20} moles of gases, of which 21 percent, or 4×10^{19} moles, are molecular oxygen, and 0.034 percent or 6×10^{16} moles, are carbon dioxide. (The photosynthesis process, we also saw, involves annually the ingestion of 8×10^{15} moles of carbon dioxide and the release of 8×10^{15} moles of molecular oxygen.) The combustion of fossil fuels at current rates, therefore, removes only

[16] Hydrocarbons, such as methane (CH_4) or ethane (C_2H_6), which are two principal ingredients of natural gas, produce water (H_2O) as well as carbon dioxide when they burn. The reader may enjoy "balancing" the chemical formulas for the combustion of these gases and determining the ratio of the number of molecules of oxygen consumed to the number of molecules of carbon dioxide produced.

$$4 \times 10^{14} / 4 \times 10^{19} = 1.0 \times 10^{-5}$$

of the atmospheric oxygen each year, but adds

$$4 \times 10^{14} / 6 \times 10^{16} = 1/150$$

or almost 1 percent to the atmospheric reservoir of carbon dioxide. The oxygen reservoir in the atmosphere is so large that neither plant cycles nor fossil fuel consumption is affecting it substantially; the carbon dioxide reservoir, on the other hand, is much smaller and therefore man's influence on it is much greater.

From the graph of world energy consumption in Figure 1 the reader can estimate the total quantity of carbon dioxide added by man to the atmosphere since 1900.[17] Such an estimate reveals that a quantity of carbon dioxide equal to roughly 15 percent of our present atmospheric supply of carbon dioxide has been added to the atmosphere by man as a result of fossil fuel consumption. Has a corresponding increase in the amount of carbon dioxide in the atmosphere been observed? Although there are experimental uncertainties, measurements [18] indicate an 11 percent increase in atmospheric carbon dioxide between 1900 and the present (1969); this is somewhat smaller than the amount of carbon dioxide man has added to the atmosphere by fuel consumption. (See Comment 2: *The partition of new carbon dioxide between ocean and atmosphere*, for further discussion.)

You might expect that an 11 percent increase in atmospheric carbon dioxide would increase the surface temperature of the earth because the greater amount of carbon dioxide would increase the "greenhouse" effect. And this too has occurred. Between 1900 and 1940 the mean world temperature increased by 0.35° C. Meteorologists estimate that half of this temperature increase may be attributable to the increase of CO_2 in the atmosphere.

But in 1940, the world mean temperature reached a maximum; by 1965 it had dropped nearly 0.2° C, despite the fact that the CO_2 concentration was still increasing, in fact faster than ever. No one is sure of the reason for this reversal in what appeared to be a trend. A popular explanation attributes the reduction in the earth's average surface temperature to an observed increase in the amount of particulate matter ("dust") in the atmosphere. Particulate matter in the atmosphere acts to increase the albedo in at least two ways: by directly reflecting the incoming sunlight and by providing condensation centers that increase the formation of fog and clouds, both of which reflect the sunlight better than land or water. Thus the increasing amounts of particulates in the atmosphere probably act to lower the amount of *direct* solar radiation reaching

---

[17] Assume that the average heat content of all the energy sources consumed since 1900 is roughly 10,000 calories *per gram of carbon.*

[18] J. Murray Mitchell, Jr. "A Preliminary Evaluation of Atmospheric Pollution as a Cause of the Global Temperature Fluctuation of the Past Century," p. 141, in S. F. Singer (ed.), *Global Effects of Environmental Pollution.* AAAS Symposium, Dallas, Texas. Berlin: Springer-Verlag, 1970.

the earth's surface and this, in turn, might be expected to result in a decrease in the average surface temperature of the earth. It is estimated [19] that a 3 to 4 percent increase in the concentration of particulate matter over the entire earth could give rise to a 0.4°C decrease in the earth's average surface temperature.

Two reasons have been suggested to account for the observed increase of atmospheric particulates: fossil fuel consumption and increasing volcanic activity. Particulates released from fuel consumption generally do not rise above the lower atmosphere where their residence time (time interval before they settle to the ground) typically varies from several days to several weeks depending upon the rainfall. Thus widespread dispersal of these particulates is inhibited and their climatic effects are likely to be greatest near urban areas. Observations made over Washington, D.C. indicate a 57 percent increase in atmospheric turbidity (a measure of the density of dust particles) since the beginning of the century.[20]

Volcanic activity also releases particulates to the atmosphere and much of this matter will end up in the stratosphere, where the residence time varies from a month to several years. In general, the longer the residence time, the greater the dispersal of the particulates. The observed variations in volcanic activity are consistent with the idea that this natural phenomenon is responsible for global temperature changes; activity was relatively low from 1870 to 1940, but several major eruptions have occurred during the past three decades.

Nevertheless, it is by no means certain that the increase in particulate matter is the cause of the global cooling trend since 1940. In order to substantiate the hypothesis, far better global measurements will have to be made of the concentration of particulates in the atmosphere. In addition, a better theoretical understanding of the role of particulate matter in determining global climate will have to be achieved. For example, it has been pointed out [21] that if most particulate matter is concentrated at very low altitudes, the temperature of these regions could actually *increase*, because the particulates would not only reflect some solar radiation but would also *absorb* radiation, and would transfer the absorbed energy to the rest of the atmospheric constituents near ground level, thus warming the surface.

There can be other effects of particulates as well. Particulates that settle out of the atmosphere in polar latitudes will reduce the albedo of the ice caps and increase their rate of melting. Thus particulates can affect climate in several contradictory ways and the observed increase in atmospheric particulates is a subject deserving further analysis.

If, indeed, the popular explanation turns out to be correct, and fossil fuel consumption is responsible for both an effect that raises the earth's surface temperature (increases in CO_2 concentration) and one that lowers

[19] Reid A. Bryson, *op. cit.*

[20] *Ibid.*

[21] H. E. Landsberg, "Man-Made Climatic Change," *Science*, Vol. 170, p. 1265 (1970).

it (increases in albedo due to particulate concentrations) this does not warrant complacency. It is ominous that our capacity to change our global climate has outrun our understanding of what is happening. If, at present, a variety of large effects are canceling, we should not feel sanguine. Environmental stability, like stability among the nations, may not be enhanced by the opposition of strong forces.

Comment 2: The partition of new carbon dioxide between ocean and atmosphere

It should not be surprising that some of the carbon dioxide released by man is no longer in the atmosphere. Carbon dioxide emitted into the air can subsequently pass into solution in the oceans. It is presumed that the "missing" carbon dioxide has been stored in the oceans, although this has not been checked directly. In fact, eventually most of the carbon dioxide which has been emitted will find its way into solution in the oceans, but this process takes centuries to occur. The reason that the rate is so slow is that the process is critically dependent on the slow rate at which the deep layers of the ocean mix with the surface layer. Estimates are that one year after some carbon dioxide is emitted into the atmosphere, roughly nine-tenths of it is still there, and only one-tenth is in the oceans.[22]

Recalling that there is 50 times as much carbon dioxide in the oceans as in the atmosphere, one might guess that this ratio would be retrieved when a new equilibrium is finally reached; that is, that 98 percent of the carbon dioxide emitted into the atmosphere would eventually end up in the oceans. But this is not true. In the process of absorbing the carbon dioxide which it now contains, the oceans have become more acidic, and this makes them less able to absorb further carbon dioxide. Estimates [23] are that not more than 90 percent of the carbon dioxide which has been emitted into the atmosphere will eventually end up in the oceans. The remaining 10 percent would be our legacy to the air indefinitely far into the future.

It is possible, however, that a proper accounting for carbon dioxide is more complicated: a net transfer of carbon dioxide from the atmosphere to *plant matter* could be occurring at the present time—that is, averaged over a year's time plant growth could be exceeding plant decay. The total amount of plant matter on earth (the standing biomass) is not well known at present; and thus, necessarily, small percentage changes in this total cannot be detected. Small percentage changes in the standing biomass are significant, however, for there is roughly as much carbon dioxide locked up in plant matter as there is carbon dioxide in the entire atmosphere.[24] Our legacies to the air and to the forests are linked in ways that we do not now understand.

[22] G. Skirrow, "The Dissolved Gases—Carbon Dioxide," p. 284, in *Chemical Oceanography*, J. P. Riley and G. Skirrow (eds.). London and New York: Academic Press, 1965.

[23] *Ibid.*

[24] One estimate of the total biomass on earth is 1.8×10^{18} dry grams. (R. H. Whittaker, *Communities and Ecosystems*. New York: Macmillan, 1970, p. 83).

18. RADIATION

John Harte, Robert H. Socolow,
and Joseph N. Ginocchio *

There are three steps to any radiation event, the emission
of radiation by a source, the passage of radiation through
a medium like air, and the absorption of radiation by an
absorber. The sources of radiation that will concern us
include the sun, uranium in brick walls, X-ray machines,
and nuclear power plants; we shall take the absorber to
be the human body.

The radiation itself is in the form of high energy particles. Part I of this essay describes the most important
of these particles and how they interact with matter.
Part II considers the biological effects of radiation, emphasizing the almost completely intractable problem involved in passing from a description of microscopic
damage to an assessment of biological damage. Part III
describes the sources of radiation, both those nature has
supplied and those our twentieth-century technology
has generated. Included in Part III is a brief discussion
of some of the government standards for radiation protection.

I. The Particles of Radiation and
Their Interactions with Matter

What Radiation Is Made Of

When radioactivity was first being investigated
at the end of the previous century, the fact that several
different kinds of particles were involved was recognized, but their full identity was unknown. The names
expressed the prevailing ignorance: alpha rays, beta rays,
and gamma rays. It was known that alpha rays and beta
rays consisted of charged particles, that beta rays could
pass through greater thicknesses of matter than alpha
rays, and that gamma rays were electrically neutral and
the most penetrating of the three.

Today one knows that gamma rays are a form of
electromagnetic radiation, differing from X rays, visible

* The authors wish to thank Frank Firk, Jack Gibbons,
George Holeman, Marvin Kalkstein, Paul Howard-Flanders, Barry
L. Nichols, Haig Papazian, and Kenneth Thomson for helpful com-
ments on earlier drafts of this essay.

light, radio waves, and other forms of electromagnetic radiation in that the particles of gamma radiation are of higher energy. The general word for any particle of electromagnetic radiation is *photon,* and a gamma ray is a photon whose energy is roughly 100,000 electron volts or larger. (The electron volt, or eV, is the common unit of energy in the atomic domain. Larger units derived from the electron volt are 1 KeV = 10^3 eV, and 1 MeV = 10^6 eV. One electron volt is the energy acquired by an electron after it is accelerated by a voltage of 1 volt. One electron volt = 1.6×10^{-12} erg.)

Photons may be specified by their wavelength instead of their energy; either property determines the other—the longer the wavelength the lower the energy. Visible light ranges from red light (wavelength of 7×10^{-5} centimeter, photon energy of 1.7 eV) to violet light (wavelength of 4×10^{-5} centimeter, photon energy of 3.1 eV). Photons are called X rays when their energy lies roughly between 1 KeV and 100 KeV.

Photons are not the only kind of electrically neutral radiation known today. In this essay we shall also say a little about two other neutral particles, neutrons and neutrinos.

The alpha rays are now known to be the nuclei of helium. The beta rays are now known to be electrons, carrying negative electric charge, and positrons (the antiparticles of electrons) carrying positive electric charge. Of the many other electrically charged forms of radiation, we shall discuss only protons here.

Radioactive Decay

In considering the emission of radiation, the most important distinction we need to make is between radiation that arises from the spontaneous decay of a radioactive isotope and radiation that is generated in a man-made device, like a cyclotron or an X-ray machine, which artificially accelerates and then releases high energy particles. In the first case, the radiation cannot be turned off.

There are more than a thousand different radioactive isotopes (sometimes called radioactive nuclei, and sometimes called radionuclides). An isotope of an element is specified by the symbol associated with the element and a superscript giving the total number of neutrons and protons in the nucleus. Thus Sr ^{90} (read as "strontium ninety") is an isotope of strontium with 38 protons (one has to look that number up, but it is the same for all the isotopes of strontium) and $90 - 38 = 52$ neutrons. Associated with every radioactive isotope is a characteristic time, which may range from less than a second to millions of years; after this time interval has elapsed, precisely half of a very large number of nuclei of that isotope present at the beginning of the time interval will have decayed. This length of time, called the *half-life,* is a property of a large collection of identical nuclei, but it cannot be usefully associated with a *single* radioactive nucleus because it is a fundamental fact of nature that one cannot tell how long a *particular* nucleus will take to decay. The half-life of Sr ^{90} is 28.1 years, but a particular nucleus of Sr ^{90} might decay tomorrow morning or a million years from now.

The rate of emission of radiation by bulk quantities of an isotope is measured with a special unit, called *curie,* such that one curie corresponds to 3.7×10^{10} disintegrations per second.[1] (A smaller unit is the microcurie, one-millionth of a curie, or 3.7×10^4 disintegrations per second.) By a "bulk sample," we mean enough atoms so that the characteristic *average* behavior of the large number of atoms dominates the random behavior of the individual atoms.

Using a rating measured in curies to describe a sample is evidently not adequate unless one also specifies the time at which the measurement was made. At one instant we can have a curie of iodine-131 (half-life of 8.1 days) and a curie of plutonium-239 (half-life of 24,000 years); both will emit particles at the same rate to start with, but 80 days later the iodine sample is down to one-thousandth of a curie (since 80 days is 10 half-lives and 2^{10} is about 1000), whereas the plutonium sample is virtually as potent as before. The reader should note that at a given time, a curie of a long-lived isotope contains more radioactive nuclei than a curie of one that is short lived.

Paths of Energetic Particles

Whether the radiation results from a radioactive isotope or from a man-made device, the particles of radiation lose energy by collisions with the particles of matter that lie in their way. Eventually, the particles are so slowed down that they can be said to have "stopped." In actuality, all matter is in thermal motion, but particle kinetic energies associated with thermal motion are much less than 1 electron volt, whereas particles emitted by radioactive nuclei generally carry more than a million electron volts of energy when they are emitted.

The manner in which particles are slowed down depends on whether the particles are electrically charged or electrically neutral. A charged particle loses energy much as a bullet does when it enters a thick piece of wood: the slowing down is the result of a very large number of collisions, in each of which the fast particle is slowed only a little bit. Most of these collisions are with atomic electrons in the matter being traversed, and almost always the electron will bounce away with enough energy to be driven out of the atom in which it was previously held, leaving an ionized atom behind. Quantitatively, a charged particle will lose roughly 30 eV on the average in the process of ionizing each atom in its path. So a 6 MeV alpha particle will ionize 200,000 atoms before slowing down. The ionized atom will have chemical properties different from the neutral atom, and if the atom happens to lie in a crucial molecule of a cell it is conceivable that even a single ionization will affect the whole organism.

We are also interested in knowing the depth of penetration of charged particles in the absorber. We certainly expect that on the average, the larger the incident energy, the greater the penetration depth. Less obvious

---

[1] The definition is historical: Pierre and Marie Curie discovered radium in 1898, and 1 gram of naturally occurring radium has an activity of 1 curie.

is the fact that a large number of charged particles of a given type and with the same energy all travel roughly the same distance (called the *range*) in the same material. The explanation lies in the fact that the slowing-down process for fast-moving charged particles consists of a very large number of separate collisions (200,000 for a 6 MeV alpha particle), and so variations in the energy loss per event and in the distance between successive events tend to average out.

The range depends on the incident particle and the absorber as well as on the incident energy. Consider 6 MeV particles incident on water. At 6 MeV the range of an alpha particle is $1/200$ centimeter, the range of a proton is $1/20$ centimeter, and the range of an electron is 3 centimeters. Ranges in water and in biological tissue are comparable. Since the alpha particle is heavier than the proton, which in turn is heavier than the electron, one observes that the heavier the particle at a given energy, the shorter the range.

In all three cases nearly the same amount of energy, roughly 30 eV, is lost in an average collision of the incident particle with a water molecule, and thus roughly 200,000 ionizations occur before the incident particle is brought to rest. Moreover, 30 eV is also the average energy loss per collision when any of these particles is slowed down by materials other than water as long as the incident energy is a few MeV.

Evidently, the 200,000 ionized atoms that form the wake of a 6 MeV alpha particle lie closer together than do the 200,000 ionized atoms that mark where a 6 MeV electron has passed. Using the value of the range just given, on the average, the former lie $0.005/200,000 = 2.5 \times 10^{-8}$ centimeters apart. This distance is roughly twice the size of an atom, so this means that a 6 MeV alpha particle ionizes roughly half of the atoms in its path. A 6 MeV electron, which has a range 600 times as large, will ionize only one atom out of 1200 in its path. The path the alpha particle has followed, although shorter, is left more devastated.

When X rays and gamma rays traverse matter, they tend to lose large fractions of their energy in individual collisions, rather than losing their energy a little bit at a time like charged particles. The atomic electrons again are responsible for absorbing most of the energy, but in this case they frequently carry away enough energy from the collision with the photon to travel distances which on the atomic scale are substantial. They can then, in turn, create a path of ionization branching off from the photon's path. Particles, such as these electrons, whose large kinetic energy originates in an earlier collision, are called *secondaries,* and the relative fraction of the ionization caused by secondaries rather than by the primary beam increases with the photon energy.

Not only do gamma rays lose their energy in big chunks as they traverse matter, but also, once they have lost enough energy to drop into the X-ray energy region, they tend to lose almost all their energy all at once. So there are not a large number of sites where the photon has interacted, and hence there is not a well-defined range associated with a photon of given energy incident on a given material. Instead, one speaks of a "median path length"—that length beyond which just *half* of the photons will have their original energy. For a 6 MeV gamma ray incident on water, the median path length is 20 centimeters, and this means that 60 centimeters (3 path lengths) into water $(\frac{1}{2})^3 = \frac{1}{8}$ of the original photons will still have their *original* energy.

Neutrons have almost no interaction with electrons, because neutrons are electrically neutral. (Photons, although electrically neutral also, still interact with electrons, because photons are the carriers of the electromagnetic field.) Neutrons are slowed down by colliding with atomic *nuclei*. These nuclei are consequently displaced and, being charged, can cause ionization; if the neutrons have very high energy and the nuclei are sufficiently light, the nuclei could be knocked clear out of the atoms. Fortunately, the only place where one is likely to encounter a significant amount of neutron radiation is right beside a nuclear reactor, where neutrons are emitted in nuclear fissions. The neutron is itself radioactive (its half-life is 12 minutes), and hence free neutrons (as opposed to neutrons bound with protons in nuclei) do not remain in the environment very long.

Describing the Amount of Radiation Present

Suppose radiation does not arise from the decay of a radioactive isotope, but instead is generated by a man-made device. In that case, the curie is no longer a useful unit because the radiation does not arise from a disintegrating isotope. Still, we want to be able to speak quantitatively of the amount of radiation that is found at any particular distance from a source, and the customary unit is the *roentgen*. The roentgen is defined and the concept is discussed in Comment 1: *Units of radiation: the roentgen and the rad.*

The basic unit of radiation absorption is the *rad:* 1 rad corresponds to the deposit of 100 ergs of energy per gram of absorbing material. One also defines the millirad, equal to 10^{-3} rad. Thus, for example, if a 60-kilogram person is subjected to a beam of 10^{10} particles, each carrying 6 MeV of energy, then, assuming that all these particles have stopped within him, he will have received a "whole body dose," that is, a dose averaged over the whole body, of 6×10^{10} MeV$/60$ kg $= 10^6$ MeV/gram $= 1.6$ ergs/gram $= 16$ millirads.

Comment 1: Units of radiation: the roentgen and the rad

Conceptually the roentgen and the rad are quite different. The roentgen unit is a unit of exposure; it is defined in terms of what radiation *would do*, were it to interact with some particular substance (dry air), chosen by convention. The rad unit is a unit of absorbed dose; it describes what radiation *has done* to any specified substance, while passing through.

A beam in transit through a medium, like air, that does not attenuate the beam significantly, is best described in roentgens. The effects of radiation on a system, usually a biological system, are best described in rads.

One roentgen is defined to be that intensity of radiation which, on passing through one cubic centimeter of dry air, would create 1 esu of positively charged ions and 1 esu of negatively charged ions. The esu is a unit of electric charge 2.08×10^9 times larger than the charge on a single electron or proton. As part of the definition, one also specifies that the air is at standard temperature (0°C) and standard pressure (760 millimeters of mercury). Experimental observations have established that this definition is essentially equivalent to another: one roentgen is that amount of radiation which would deposit 87.8 ergs of energy in 1 gram of dry air lying in its path.[2]

A dose of one rad is absorbed by a substance, by definition, when 100 ergs are absorbed per gram. Hence, comparing the definitions, when one roentgen is incident on air, the air absorbs a dose of 0.878 rads. Thus, in describing radiation incident on air, the numerical values in roentgens and in rads agree to within 12 percent.

When radiation is incident on water and on most biological tissue, it happens again to be true that one roentgen of radiation deposits a dose of about one rad. This is the reason why the two units are often treated as if they were interchangeable.

The reason why air, water, and most biological tissue absorb nearly the same amount of energy per gram is that they are made of atoms with similar atomic weights. Bone, on the other hand, contains 15 percent by weight of the relatively heavy element calcium. There are some ranges of energy for various incident particles where bone, as a consequence, responds to the same incident radiation differently from other body tissue. However, when the radiation consists of beta or gamma particles carrying from 100 keV to 5 MeV of energy, the absorption per gram in bone and in other tissues will be very similar, because in this energy region the determining characteristic of the absorbing material is the total number of electrons per gram and this is nearly the same.

II. Biological Effects of Radiation

In order to predict the biological damage to an organism upon exposure to a given source of radiation, it is useful to break down the problem into three stages: first, we want to know the properties of the source and its position relative to the organism; second, we want to know how the radiation from that source will ionize individual atoms or molecules in the organism; and finally, we want to be able to estimate the consequences of that ionization for the organism as a whole. One might hope that if we knew all the information at one stage, we could deduce the consequences

---

[2] Using the fact that the density of air at standard temperature and pressure is 1.29×10^{-3} grams per cubic centimeter, the interested reader can establish from the equivalence of these two definitions that when particles of radiation traverse dry air, on the average 34 eV of energy is deposited for each ion pair produced.

at the next stage. However, we shall see that this field of science is not so well advanced.

A source of radiation is specified by the kinds of particles emitted, the emission rate, and the energy distribution of these particles. The effect of the source on a receiver cannot be specified, however, unless one knows the amount and the kind of shielding between the source and the receiver and whether the source of radiation is concentrated or diffuse. For example, a source of alpha particles is toxic in the human lung because the decay site can be alongside sensitive tissue; even a few centimeters of air would bring the alpha particles to rest. If the alpha particles originate from a speck of inhaled solid material (such as plutonium oxide), the damage will be concentrated in a single part of the lung, whereas if the source is a gas (such as radon), the damage will be dispersed through the organ. The energy of the particles is important to know for two reasons. First, at very low energies there will be *thresholds* in the radiated organism, that is, energies below which certain kinds of significant atomic or molecular damage cannot be incurred. (There may also be energies below which certain kinds of biological damage cannot be *detected* by present experimental methods, even though they have occurred.) Second, the depth of penetration of the radiation from any source increases as the energy of the particles increases.

At the molecular level we wish to characterize the ionization damage from that source. The parameters we need to know are the number of ionized sites, their density, their location in the organism, and the time interval during which the damage takes place. In order to make the concept of damaged site density explicit, imagine a focused beam of particles incident on an organism. The damaged sites will extend over some region of the exposed tissue, and the number of damaged sites per unit volume in the tissue will vary with the depth of that volume within the organism and with the location of that volume relative to the focus of the beam.

Why is the density of damaged sites expected to be an important parameter in determining biological damage? In order to bring about a certain amount of biological damage to an organism (for example, shorten the life-span by a given amount, or induce a particular cancer growth, or increase the mutation rate by a given amount), a certain number of cells have to be killed or sufficiently damaged. Damage to the individual cell will increase as the number of atoms in that cell which have been disrupted by the incoming radiation increases. Moreover, a cell can survive more easily if the damage sites are spatially far apart than if they are close together.

Because cell damage increases when damage sites are close together, scientists working in the field of radiation protection have introduced another unit, called the *rem* (and the *millirem* $= 10^{-3}$ rem), very similar to the rad but somewhat more closely reflecting the biological damage that accompanies the absorption of a given amount of radiation. The rem is rather loosely defined by

$$1 \text{ rem } = 1 \text{ rad } \times \text{ (modifying factor)}$$

where the "modifying factor" depends on the nature of the radiation and its energy.[3] The modifying factor is nearly always taken to be 1 for electrons, positrons, and gamma rays, and it is generally assigned a value between 10 and 20 for alpha particles in the MeV energy range, with higher values for higher energies. In this way the increased toxicity of closely spaced damage sites, which alpha particles in the MeV energy range produce, is crudely taken into account.

That the rem dosage, which attempts to reflect biological damage, should increase with the rad dosage is reasonable; the rad dosage is a good measure of the number of damage sites and thus takes into account a portion of the information needed to characterize the damage at the cellular level. That the biological damage should be *directly proportional* to the rad dosage is a guess based upon observations with doses above 25 rads.

But the problem of assessing biological damage is more complex than this; not only does the distribution of damaged sites within a cell determine whether a given cell is destroyed, but the distribution of destroyed cells determines how severely the organism is injured. In many regions of the body the ability to survive damage to the cells is greatest when the damaged cells lie close together; as an extreme example, a person could function with one totally destroyed kidney, but not with two half-destroyed kidneys. Thus for a given radiation dosage, the competing effects of high ionization density within a cell and wide distribution of damaged cells determine the damage to the organism.

In addition, natural repair mechanisms within the cell and within the exposed organ play a large role in determining biological damage. However, these repair mechanisms can operate only at a certain maximum rate. In general, one expects that the longer the time interval over which a given radiation dose is applied, the less cumulative damage is incurred.

The definition of the rem unit does not take into account these factors nor others, such as the sensitivity of the radiated organ and the age of the individual being exposed, which are surely important to know before estimating biological damage. An attempt to take these factors into account is made when rem dose standards for particular organs and individuals are set.

There are short-term and long-term biological effects of radiation, and radiation standards, ideally, would take both into account. However, standards have been set almost exclusively on the basis of information about short-term effects, which can be quantitatively established by working with plants and animals in the laboratory. In addition, there are data gathered from observations of survivors of Hiroshima and Nagasaki and from victims of occasional accidents. It has become possible to identify the "acute radiation syndrome" associated with whole-body doses in excess of 25 rads

---

[3] The detailed definition of the rem is cloaked in greater obscurity than the definition of any other unit in physics with which we are familiar.

absorbed by human beings in a short period of time. The biological damage and medical prospects are summarized in Table 1.

We only know a little about how gross biological damage arises from radiation damage at the cellular level, and almost nothing about how it arises from radiation damage at the atomic level. It is not difficult to estimate the damage at the atomic level from a lethal dose of 500 rads (see Comment 2: *Quantitative estimate of damage at the atomic level....*) But a detailed quantitative understanding of the relations between ionized atoms on the one hand and fetal development, mutant genes, cancer growth, and repair mechanisms of cells and organs on the other has not been achieved.

If we could achieve this understanding, we might be able to extrapolate the data observed at high radiation doses to predict with some confidence what the damage from smaller radiation doses would be. We might also be able to predict the long-term biological effects of radiation. It is clear that some short-term biological damage is associated with even the smallest amount of radiation, but it is also possible that natural biological mecha-

Table 1. Effects of Acute Radiation Doses on Man

Gamma-Ray Whole-Body Dose (rad)	Effects	Remarks
25–100	Blood changes; person feels little or no effect.	Lymph nodes and spleen damaged; lymphocyte count drops. Bone marrow damaged; decrease in white blood cell, platelet, and red blood cell count.
100–300	Blood changes, vomiting, malaise, fatigue, loss of appetite.	Antibiotic treatment may be necessary. Recovery can be expected.
300–600	Above effects plus hemorrhaging, infection, diarrhea, epilation, and temporary sterility.	Antibiotics and blood transfusions administered. Expect recovery in about 50% of cases at 500 rad. Possible bone marrow transplant.
More than 600	Above symptoms in addition to damage to the central nervous system; incapacitation at doses in excess of about 1000 rad.	Death almost a certainty. Sedation. Possible bone marrow transplantation in lower portion of this range.

SOURCES: Adapted from G. S. Hurst and J. E. Turner, *Elementary Radiation Physics*, New York: Wiley, 1970.

nisms allow an organism to repair much of the damage from low doses. The way standards are in fact set is to take the smallest doses that can induce observable effects in the laboratory and to divide by a "safety factor," which is in the neighborhood of 100.

Nonetheless, there are long-term effects of radiation which belie these procedures. Somewhat mysteriously, the incidence of leukemia among Hiroshima survivors peaked seven years after the bomb was dropped. Radiation at low doses may be responsible for a substantial percentage of all cancers. In that case, to the extent that man's technology increases the background level of radiation, it will also be increasing the number of cancer victims. Moreover, the addition of small increments in the overall background radiation can still permit the appearance of large increments in the radiation in biological systems in some cases. There are biological concentration mechanisms by which iodine, for example, is concentrated in the thyroid gland. In addition, some radioactive isotopes can, like DDT, become concentrated from organism to organism as the isotopes move up the food chain; large and still unexplored concentrations of some long-lived isotopes will be reached at the end of the lifetime of a long-lived species like man. Although some isotopes, like cesium-137, are rejected by the human body after a period of a few years, strontium-90 and plutonium-239 are among the long-lived radioactive isotopes that man is believed to concentrate throughout his entire lifetime.[4]

Comment 2: Quantitative estimate of damage at the atomic level from a lethal dose of radiation

For large organisms, a typical lethal dose of radiation is 500 rads averaged over the entire body. Let us estimate how many ionizations per cell this corresponds to. Five hundred rads corresponds to the deposit of 5×10^4 ergs or 3.1×10^{16} eV of energy per gram of tissue. Since the body is primarily water, a gram of tissue occupies approximately 1 cubic centimeter. Because a typical cell is 10^{-3} centimeters across, there are roughly 10^9 cells per cubic centimeter. The lethal dose of 500 rads thus corresponds to the deposit of 3.1×10^7 eV per cell. Since one atom is ionized for every 30 eV of energy deposited in matter (roughly), we conclude that an average of 10^6 atoms (1 million atoms) are ionized per cell in every cell of the body by a lethal dose of radiation.

There is another way to look at this result. Because an atom is approximately 10^{-8} centimeters in diameter, there are 10^5 atoms along a line from one side of the cell to the other, and a particle that ionizes *every* atom in its path would ionize 10^5 atoms in traversing the cell. Thus a lethal dose corresponds on the average to ten such particles traversing every cell of the body.

We can gain an understanding of the importance of the location and shape of the source if, instead of assuming that the 500 rad dose is spread out over the entire body, we imagine that an emitter of 6 MeV

---

[4] Additional discussion of concentration mechanisms can be found in the essays by Professor Loucks and Professor Eipper in this book.

alpha particles is lodged in the middle of some tissue. Assume the emitter is spherical and is 1 centimeter in diameter. Since a 6 MeV alpha particle has a range of 0.005 centimeter (recall part I, page 298), the reader can easily show that the energy will be deposited over a spherical shell whose volume contains roughly 10^7 cells. For a 60-kilogram person, a whole-body dose of 500 rads corresponds to the deposit of 2×10^{21} eV. But now, instead of assuming that this radiation is averaged over the whole body, we shall assume that this entire radiation is absorbed by the 10^7 afflicted cells. Then 10^{13} atoms are ionized in each of the afflicted cells, or about 1 percent of the atoms in these cells.

Genetic Effects

Radiation can induce genetic mutations. When a sex cell of an organism is irradiated, a gene may undergo a mutation, that is, may be altered in such a way that the cell continues to function but the genetic message which it carries is altered. The altered message then becomes part of the gene pool of the species, and it is possible for the mutation to propagate through many generations. Will the mutation always remain in the gene pool of the species? In general, no.

Even if the original irradiated sex cell participates in the creation of progeny, the gene may be lost in later generations. This is quite likely to happen in a sexually reproducing organism like man, because, of all the genes a person passes on to his children, only half, on the average, come from each of his parents. Since only one parent, presumably, has contributed any particular mutation, that mutation should only be passed on to about half of his children.

The mutant gene can also disappear by natural selection, if carrying the gene has an adverse effect on the ability of the organism to reproduce. Finally, the possibility exists that a second mutation (called a "back mutation") restoring the original message can occur in one or more organisms of later generations.

A mutant gene can be debilitating and yet evolution may still not work very quickly against it. This can happen because the mutation is recessive or because, although dominant, it is not lethal prior to the age of reproduction. When the mutation responsible for hemophilia arises, for example, it propagates for many generations, because hemophilia, while hindering those who are stricken from leading full lives, still does not prevent reproduction. Nonlethal but debilitating mutations will add to the world's stock of misery far into the future.

Radiation can also produce mutations in the genes within ordinary cells. In that case the damage can only affect the irradiated individual, not his progeny. It is quite likely that some cancers are the result of a form of mutation in ordinary cells.

The experimental investigations of genetics have concentrated on those genes in an organism which can be associated with some particular physical trait. Traits which can be used range from blood group, to flower color, to the ability or inability of an organism to synthesize some amino acid. In each case, a large number of organisms in one generation is compared with a large number of organisms in a subsequent generation. The investigator, assuming the laws of Mendelian genetics are operating, can predict what should be observed from generation to generation in the absence of mutations, and (with uncertainties which make this a very challenging and subtle subject) he can attribute departures from his prediction to mutations.

Mutations can be artificially produced, but they also occur naturally. A useful rule of thumb is that, in man, one out of every 100,000 genes will naturally undergo a mutation in every generation. In simpler organisms like bacteria, the rate per generation appears to be considerably smaller. Some of these mutations may be the result of background radiation interacting with the sex cells of the organism prior to the time at which the organism reproduces. It is virtually certain, however, that not all the natural mutations are due to background radiation, because mutations can also be induced by a variety of chemicals (including mustard gas and formaldehyde), by heat, and, presumably, by simple chemical errors during cell division which are the manifestation of the random thermal motion of the molecules in a cell.

Information about radiation-induced mutations has been obtained primarily from laboratory studies of mice, fruit flies, and microorganisms. Quantitative and controlled studies of the radiation sensitivity of human genes have not been carried out, and therefore estimates of the dependence of the induced mutation rate in man on the applied radiation dose involve considerable guesswork. One essentially has to assume that what is true for lower animals is true for man.

In the laboratory one can vary the total dose received by the sex cells and the time interval during which the dose is applied. If the dose is absorbed within a few seconds, it is called an "acute" dose; if it is spread over hours, days, or longer, it is called a "chronic" dose.

For acute doses above 25 rads, there appears to be a linear relation (that is, a proportionality) in all test animals between the number of induced mutations and the size of the dose.[5] For mice the rate of induced mutation for specific genes is roughly 2.5 mutations per 10^7 genes per rad per generation. Thus an acute dose to mice sex cells of 1000 rads, for example, will induce 2500 mutations of any given kind in every 10^7 sex cells, these mutations being measured by their appearance in the next generation. For fruit flies the rate of induced mutations per rad per generation is 15 times smaller, but the relation between dose and number of induced mutations is still observed to be linear.

There is evidence from these studies that the rate of induced mutation depends on whether the damage is to the mature sex cell or to a

[5] The quantitative data in this paragraph are adapted from Curt Stern, *Principles of Human Genetics*, San Francisco: Freeman, 1960.

precursor of the sex cell. Genes of the mature sperm cells are generally more sensitive to radiation than are the genes in the precursors of the sperm cells, called spermatogonia. However, mature sperm cells are alive for a much shorter period of time than are spermatogonia, and therefore receive less natural radiation.

There are several difficulties involved in applying these experiments to make predictions about how changes in the background radiation level will affect the mutation rate in man. First, the dose received from background radiation is chronic, not acute. Studies on mice indicate that acute doses result in four times as large a rate of appearance of induced mutations as chronic doses spread out over a week or so. Thus some repair mechanisms appear to be operative. The effects of *very* long-term chronic doses are not known at present.

Second, if there is a fifteen-fold increase in the rate of mutations per generation induced by large acute doses in going from fruit flies to mice, how can one estimate the correction to be made in going from mice to men? One of the more difficult corrections arises from differences in lifespan. It becomes necessary, in particular, to know under what circumstances damage to a precursor of a sex cell can lead to a mature sex cell that carries a mutation. If a precursor can sustain this form of damage over the whole lifetime of the organism prior to reproduction, then the longer an organism lives before reproducing, the larger the chance that background radiation will cause a mutation. At the other extreme, some kinds of mutations may occur only when mature sex cells are struck by radiation; in that case the lifespan of the mature sex cell rather than the age of an organism at reproduction is the relevant time period to use in comparing different organisms.

Third, one can question whether the proportionality between size of dose and rate of production of mutations which is observed at doses above 25 rads can be extrapolated to the much smaller doses characteristic of background radiation. The average person, as we shall see, receives a gonadal dose of roughly 0.2 rad per year at present, two-thirds from natural sources and one-third during medical examinations. Thus in the 30 years of a human generation, he receives about 6 rads. To do experiments using such low doses requires following millions of organisms through several generations, and studying the effects of a 6-rad chronic dose is beyond present experimental technique.

Nonetheless, it is instructive to see what one obtains using the linear assumption and the mouse data in calculating the rate of production of mutations in human genes in 30 years. A dose of 6 rads in 30 years, multiplied by a mutation-induction rate of 2.5 mutations per 10^7 genes per rad per generation, gives 0.15 mutations per 10^5 genes. Dividing by 4 to make an attempt to correct for the chronic character of the dose, one finds a mutation rate of 0.04 mutations per 10^5 genes. This is 25 times less than the average natural mutation rate observed in man, 1 per 10^5 genes per generation, quoted earlier.

If the assumptions which underlie this calculation can be believed, then it would imply that the chronic doubling dose for man (that is, the chronic radiation dose which would *double* the currently observed mutation rate in man) is roughly $25 \times 6 = 150$ rads per generation. Moreover, it would support the belief that background radiation is not responsible for most of the observed mutations in man.

The entire evolutionary history of life has been dependent on mutations: mutations provide new variants on existing species, which then compete for survival and, if they are better suited, gradually replace the earlier strains. However, the overwhelming majority of mutations are not favorable, and the offspring of organisms where such mutations have occurred will die out. No one is able to predict which mutations will be favorable nor predict in any detail just what the risks are of an increased mutation rate in man. One must anticipate as well that any increase in the level of background radiation due to man's activities will cause additional mutations in other organisms in the biosphere, for example, in insects or bacteria. The possibility arises that such mutations could result in offspring which the natural environment is unable to control.

Far more research will have to be carried out before an accurate assessment can be made of the genetic risks to man arising from radiation.

III. Natural Radiation and Radiation from Man's Technology

What are the sources of radiation? Part of the radiation we receive originates from natural sources and part from man-made sources, as shown in Table 2.

Natural Radiation

The important natural sources of radiation are cosmic radiation from outer space (including, of course, the electromagnetic radiation from the sun), radioactive elements that exist in the rocks and water of the earth, and radioactive elements in organic matter, including our own bodies.

The earth's atmosphere protects us from getting the total onslaught of cosmic radiation, and hence the dose from this source varies with altitude. (There is also a variation with latitude because the earth's magnetic field deflects charged particles toward the magnetic poles.) The average amount of cosmic radiation received on the surface of the earth is approximately 30 millirads per year. In Table 3 we show the variation of the cosmic radiation with altitude.

Three abundant radioactive isotopes in the earth's crust are an isotope of uranium (U^{238}), an isotope of thorium (Th^{232}), and an isotope of potassium (K^{40}). All of these have half-lives of over 1 billion years, which is the reason why they are found at all. Had the half-life been much shorter, they would have vanished during the time (approximately 5 billion years) since the earth was formed. Both U^{238} and Th^{232} eventually decay into isotopes of lead, through a chain of decays which produces alpha particles, elec-

TABLE 2. Sources of Radiation Dose to Persons in the United States

Source	Genetically Significant Dose[a] (millirads per year)
Man-made	
Diagnostic X-ray[b] (1964)	55
Theraputic X-ray[b] (1964)	10
Radioactive fallout[c] (1964)	9
Nuclear industry[d] (1970)	1
Subtotal	75
Natural[b]	
Terrestrial radiation, external to body	60
Cosmic rays	30
Radioisotopes, internal to body	25
Subtotal	115
Total	190

[a]The genetically significant dose (GSD) is a weighted average of the gonadal doses received by members of a population. Each dose is weighted by the reproductive potential of the person being irradiated. This potential decreases with age and can be determined from demographic tables. For radiation which is uniformly distributed over the entire body and without regard to age, such as natural radiation, the GSD is equivalent to the average whole body dose or to the average gonadal dose.

SOURCES:
[b]National Council on Radiation Protection and Measurements, Report No. 39, *Basic Radiation Protection Criteria*, Washington, D.C.: NCRP Publications, 1971.
[c]A.M.F. Duhamel (ed.). *Health Physics*, Vol. 2, Part I. New York: Pergamon Press, 1969.
[d]K.Z. Morgan, "Never Do Harm," *Environment*, Vol. 13, No. 1, p. 28 (1971).

trons, and neutrinos. K^{40} can decay either into calcium-40, emitting an electron, or into argon-40, capturing an atomic electron and emitting a gamma ray.

Much of the radiation we receive from these natural sources is in the form of gamma rays that are produced as the charged particles emitted by the isotopes are slowed down by rocks, water, and air. The charged particles rarely reach our bodies, because their range, even in air, is quite short. The whole-body dosage resulting from these naturally occurring isotopes external to the human body averages about 60 millirads per year. There are sizable variations over the surface of the earth, however. In fact, different types of dwellings radiate differently, a concrete house giving off about one and a half times as much radiation per year as a brick house and about three times as much as a wooden house. In certain regions in India where a thorium sand called monazite is used as a construction material,

TABLE 3. Cosmic Radiation Exposure at Different Altitudes

Altitudes	Mean Dose (millirads / year)
Sea level	
Equator	23 to 33
Within 40° of the poles	26 to 41
5,000 ft (all latitudes)	40 to 60
10,000 ft	
Equator	56 to 89
Within 40° of the poles	66 to 128
40,000 ft	2.8×10^3
30 km–600 km	7.3×10^3
22,000 km	
(Van Allen Belt)	8.8×10^6

SOURCES: Adapted from Encyclopedia Britannica, 1966 ed., Vol. 18, p. 1028, and K. Z. Morgan and J. E. Turner, *Principles of Radiation Protection*. New York: Wiley, 1967, p. 10.

the average yearly background dose is over ten times the world average.

Since alpha particles travel such small distances, they are most toxic when they have been ingested. Most of the damage caused directly by alpha emitters in the natural background radiation arises from trace quantities of two isotopes of the *gas* radon (Rn^{220} and Rn^{222}, half-lives of 54 seconds and 3.8 days, respectively) which can pass directly into the lungs. These isotopes arise as by-products of the decays of Th^{232} and U^{238}, respectively; the latter are found in many building materials.

The third largest naturally occurring source of the radiation absorbed by the human body is the radioactive material within the body. This source is not negligible; it contributes about 25 millirads per year. The isotope contributing almost all of the dose is potassium-40, an isotope present everywhere in the organism where potassium is found.

This is quite remarkable. It means that there is no way to build a radiation-proof spaceship, because within the organism itself radioactive isotopes are stashed away. It is quite straightforward to calculate the dosage the body receives from K^{40} decay. (See Comment 3: *Radiation dose from K^{40} in the body's own potassium.*)

When all of these sources are added together, the average natural background radiation on the earth is about 115 millirads per year.

Comment 3: Radiation dose from K^{40} in the body's own potassium

To estimate the radiation dose from K^{40} decay in body tissues we need to know five things: (1) A 70-kg (or 150-lb) man has about 150 grams of potassium in him. (2) One out of 8500 potassium nuclei is a K^{40} isotope.

(3) The mean life of K^{40} is 1.8×10^9 years (half-life is 1.3×10^9 years), so that one out of every 1.8×10^9 K^{40} nuclei present in the body will decay each year. (4) The energy released in each K^{40} decay is about 1.3 MeV. (5) Not all of that energy is absorbed by the body, because about half of the energy is carried by neutrinos—particles that pass through the body without (almost ever) interacting. In addition, a small fraction of the photons emitted in K^{40} decay will pass out of the body without being completely slowed down.[6]

Combining (4) and (5), we *estimate* that 0.6 MeV is absorbed by the body for each K^{40} decay. One gram of tissue, using (1) and (2), contains 2×10^{-7} gram of K^{40}. Recalling that Avogadro's number of K^{40} atoms (6×10^{23} atoms) weighs 40 grams, we get that 1 gram of tissue contains 3×10^{15} nuclei of K^{40}. Using (3), this means 1.6×10^6 disintegrations per year, depositing 9.6×10^5 MeV or 1.5 ergs per gram of tissue, a dose of 15 millirads.

The dose due to C^{14} decay is much less—about 1 millirad per year. One might have supposed that the radioactive isotope of carbon, C^{14}, would contribute more radiation than K^{40}, because there is about ten times as much carbon in the body as potassium. Also the C^{14} half-life (5700 years) is about 200,000 times shorter than the half-life of K^{40}, so that a C^{14} nucleus is 200,000 times as likely to decay as a K^{40} nucleus in the same time period. However, the C^{14} nuclei in nature have been formed in the atmosphere by collisions of cosmic rays with atmospheric nitrogen, and this has resulted in a C^{14} abundance in the atmosphere and in living tissue [7] that is only 1.3×10^{-12} of the abundance of C^{12}. When these facts are combined, it turns out that in every gram of living matter there are about three C^{14} disintegrations and four K^{40} disintegrations every minute. What makes the K^{40} dose so much larger, finally, is that the average energy deposited in the body per decay is about twelve times larger for the K^{40} decay than for the C^{14} decay.

Radiation from Man's Technology

The radioactivity over which man has control is associated with medical care, industry, weaponry, and power production. Radiation from medical uses in the United States in recent years, averaged over the population, has resulted in a gonadal dose of about 65 millirads per year, which is

---

[6] This calculation has been done previously by Willard F. Libby, *Science*, Vol. 122, p. 57 (1955); some of the nuclear physics data which he uses have since been revised. Libby points out the following additional effect: in a dense crowd of people, many of the photons which escape from one person's body will be absorbed by another's. This adds 2 millirem per year to the total radiation received by each person in the center of a crowd. Thus we find, in unexpected corners, further arguments for population control!

[7] Because the rate of production of C^{14} in the atmosphere has been constant for many thousands of years, the fraction of C^{14} in living tissue has been the same through history. This has permitted the dating of archaeological remains, like bones or the ash from fires. If a tree was burned 5700 years ago, its ashes will have only one-half of the C^{14} they had at the time of the fire; if it was burned 11,400 years ago, one-fourth of the C^{14} would still be around, and so on.

about half as large as the radiation from natural sources. Diagnostic X rays account for about 85 percent of this, with the remainder resulting from the growing use of radiation for therapeutic purposes. The average chest X ray results in a dose of 80 millirads to the chest and approximately one millirad to the male gonads or 0.2 millirads to the female gonads. A standard dental X-ray series delivers a dose of 1 rad to the jaw and a gonadal dose which is about 0.1 millirads. Less common examinations, including those of the pelvic and gastrointestinal regions, account for most of the dose to the gonads, even for people in the reproductive years. The dose received per diagnosis has been steadily reduced through improvements in X-ray sources and films. The total annual dose received per person from all medical diagnoses may be increasing, however, as a result of the greater frequency of use of X-ray and isotope methods of diagnosis in the population as a whole.

In the United States, exposures to those working with X rays are monitored at the state level by the various Departments of Health. Generally, however, records are not kept of actual doses delivered, except by the doctors and hospitals involved. Policy-makers at several levels of government have deliberately chosen not to establish limits on the total radiation dose received from medical sources by the whole population on the grounds that such limits would constrain the individual doctor treating his patient. A society more concerned with the genetic effects of such radiation and more concerned with possible abuses in diagnosis and treatment would probably have chosen to monitor this total medical radiation dose far more closely.

Industrial doses, when averaged over the entire United States population, are quite low at present. Individuals who work with isotopes and X rays are permitted by law to receive larger doses than the rest of the population. X rays are used, for example, to examine the structure and properties of materials, and radioactive isotopes are being used in an increasing number of applications.

In the late 1930s it was discovered that by bombarding one particular naturally occurring isotope, U^{235}, with slow neutrons, one could split up the uranium nucleus and release large amounts of energy. The same phenomenon has since been found to occur when slow neutrons strike uranium-233 and plutonium-239, man-made isotopes. The breakup process is called fission, and the energy released is the source of both the destructive power of the atomic bomb and the energy produced in nuclear power plants.[8]

When the uranium or plutonium nucleus undergoes fission, most of the fragments produced are radioactive. When fission occurs in the atmosphere, radioactive fallout is inevitable. Of all the radioactive fragments produced, the isotopes that are most toxic to human beings are those that are produced in large concentrations, have an intermediate half-life (from

---

[8] Radiation doses from the Hiroshima and Nagasaki bombs were lethal at distances of nearly a mile, as shown in Table 4.

1 to 100 years), and are readily absorbed but slowly discharged by the human body. The intermediate half-life isotopes are the most dangerous, because (1) the shorter-lived isotopes may be able to be isolated until the great majority of the nuclei have decayed, and (2) the very long-lived isotopes are nearly inert, only a small fraction of the nuclei decaying in a single person's lifetime. The two radioactive isotopes fulfilling these criteria that are of greatest concern in atomic bomb testing and nuclear waste disposal are strontium-90 (Sr^{90}) and cesium-137 (Cs^{137}). The human organism concentrates strontium in bones, and hence the radioisotope Sr^{90}, with a half-life of 28 years, is found primarily there. The radioisotope Cs^{137}, with a half-life of 30 years, is distributed throughout the entire body.

Under certain circumstances radioactive isotopes that meet only some of the above conditions may be environmental problems. We have already pointed out the hazards of the *short-lived* radon gas found in building materials. Another example is the radioisotope I^{131}, which is released in the fission process and has a half-life of 8 days. It is dangerous in spite of its short half-life, because iodine gas is easily occluded on dust particles and thereby transported on wind currents, and iodine, once it passes into the food chain, is concentrated in the thyroid gland. The possible result of the subsequent decay is a damaged thyroid.

Hydrogen bombs release most of their energy when two light nuclei coalesce, a process called fusion. While a hydrogen bomb contains a fission trigger within it, and thus releases all of the isotopes associated with the atomic bomb, there are additional radiation dangers posed by the large amounts of radioactive tritium (H^3) and C^{14} that are produced in a hydrogen bomb blast. Because carbon and hydrogen are constituents of the most basic molecules of life, the genetic (and therefore long-range) damage associated with these isotopes can be especially severe.

When nuclear weapons are tested in the atmosphere, within a week radioactive nuclei emitted in the tests can be detected all over the world,

TABLE 4. Estimates of Absorbed Dose as a Function of Distance from the Hypocenter at Hiroshima and at Nagasaki

Distance (meters)	Hiroshima		Nagasaki	
	Neutron Dose (rad)	Gamma Dose (rad)	Neutron Dose (rad)	Gamma Dose (rad)
0	14,800	10,900	4180	26,400
500	3,220	2,887	737	7,290
1000	192	260	37.1	903
1500	10.1	22.3	1.8	121
2000	0.54	1.9	0.089	17.9

SOURCE: K. F. Morgan and J. E. Turner, *Principles of Radiation Protection*, New York: Wiley, 1967, p. 41.

so fast is the dispersal of debris in the lower atmosphere (troposphere). Much of the radiation is associated with isotopes having half-lives of a few days or less, and the ambient radiation rapidly returns to close to pretest levels. The longer-lived isotopes, while not contributing more than a few percent to ambient levels, become dangerous as they work their way into food chains. The longer-lived isotopes are deposited at ground level over many years, in part because isotopes which the test had ejected into the upper atmosphere (stratosphere) take many years to fall to earth.

Uranium fission is also the basic energy-generating process in today's nuclear power plants. The nuclear reactor at the core of the power plant contains the radioactive debris of the fissions that have occurred, as well as plutonium and uranium nuclei. This intense radioactivity must be carefully handled, not because of the danger of a nuclear explosion (that danger is nil), but because the radioactive material itself represents a serious contaminant if it should ever become part of the biological environment. The risks involved in the nuclear power industry come from conventional industrial accidents, like fires, metal fatigue, and accidents to vehicles transporting materials. Any of these could permit the radioactive material that one is carefully trying to insulate from the biological environment to enter that environment. To this list of "accidents," one could add sabotage and destruction of a plant in a war—even a conventional war. An estimate of the magnitudes of radioactive wastes generated in nuclear power plants is found in Comment 4: *Radioactivity produced by bombs and power plants.*

Comment 4: Radioactivity produced by bombs and power plants

When an individual uranium or plutonium nucleus undergoes fission, an average of 200 MeV of energy is generated. About 180 MeV of this energy are released instantaneously, primarily in the form of kinetic energy of the fission fragments but also in the form of "prompt" gamma rays. The remaining 20 MeV are released in the subsequent decays of the fission fragments, too late to contribute to the effective energy of a bomb. Still, most of these decays occur not long after the original fission event: one day after the fission event an average of less than 2 MeV of energy per fission is still left to be released. Roughly half of the 20 MeV of energy released in the decay of fission fragments is carried off by neutrinos, which have no subsequent interactions, and half is released in the form of beta and gamma rays that, when the original fission has occurred in a nuclear reactor, will slow down in the reactor and contribute to its total thermal output. Thus about 180 MeV per fission contributes to the explosive energy of a bomb and 190 MeV per fission contributes to the thermal output of a reactor. In a restricted sense, therefore, it is meaningful to compare the fission output of bombs and nuclear reactors: when the energy released is the same, even though altogether different periods of time are involved, almost the same number of nuclear fissions will have occurred and therefore roughly the same number of radioactive isotopes will have been produced.

The energy output of bombs is typically measured in kilotons of TNT, a precise unit of energy:

$$1 \text{ kiloton of TNT} = 10^{12} \text{ calories}$$

(One megaton is 1000 kilotons.) However, if one is to assess the yield of isotopes in a bomb explosion, it is necessary to know not only the total energy released but also what fraction of the energy was released by nuclear fusion and what fraction by nuclear fission. In most of the hydrogen bombs tested in the atmosphere, fission reactions occurred not only in the trigger of the bomb but also in a "jacket" of natural uranium; the bomb designers chose to increase the energy released by the bombs by taking advantage of the fact that the abundant isotope of uranium, U^{238}, can be fissioned by *fast* incident particles. Using a uranium jacket, of course, greatly increases the yield of radioactive isotopes.

Between 1945 and 1963 (when the Limited Nuclear Test Ban Treaty was signed), 511 megatons of nuclear weapons were tested in the atmosphere or at the surface of the earth, and fission reactions accounted for 193 megatons, or 38 percent of the energy released.[9] The fallout from these tests gave a dose to the gonads of 9 millirads in 1964, averaged over the whole population of the United States. Since 1964 only China and France have tested nuclear weapons above ground, and this dose has been dropping slowly.

The Bell Station nuclear reactor discussed in Professor Eipper's essay (Essay 8) was designed to convert the thermal energy provided by nuclear fission into electrical energy, at roughly 32 percent efficiency, and to provide an electrical power output of about 800 megawatts. Thus nuclear fissions would have to proceed at a rate sufficient to provide $800/0.32 = 2500$ megawatts of thermal power. Using the relation,

$$1 \text{ megaton of TNT} = 132 \text{ megawatt-years}$$

that connects two huge units of energy and that is straightforward to verify, the reader can calculate that the same number of nuclear fissions would occur at the Bell Station power plant in 10 years as have occurred in all of the nuclear testing above ground.

At 190 MeV per fission, there would be 2.6×10^{27} fissions per year at Bell Station, a plant whose size and efficiency are typical of the nuclear plants now under construction. The short-lived isotopes resulting from these fissions will decay within the reactor, while the long-lived isotopes present a waste disposal problem whose magnitude we can now estimate.

To find out how much of any particular radioisotope is produced in a reactor, one needs to know what fraction of time a given nuclear fission results in a particular isotope.[10] Thus, for example, 5.77 percent

[9] U.N. Scientific Committee on the Effects of Atomic Radiation, Report (General Assembly 17th session, Supplement No. 14A/5814). New York: United Nations, 1964. Cited in C. L. Comar, "Movement of Fallout Radionuclides through the Biosphere and Man," *Annual Reviews of Nuclear Science*, Vol. 15, p. 176 (1965).

[10] These fractions are found in Earl K. Hyde, *The Nuclear Properties of Heavy Elements, III. Fission Phenomena*. Englewood Cliffs, N.J.: Prentice-Hall, 1964, Section 4.2.

of the fissions of U^{235} induced by slow neutrons result in the production of strontium-90, and 6.15 percent of such fissions result in the production of cesium-137. When plutonium-239 is the fissioning nucleus, the percentages are 2.25 and 6.63, respectively. These percentages are facts of nature.[11] Accordingly, the 800 megawatt reactor discussed above, if its fuel is uranium-235, will produce 1.5×10^{26} strontium-90 nuclei and 1.6×10^{26} cesium-137 nuclei every year. Knowing the mean lives of the two isotopes (the mean life is the half-life divided by 0.693; see Appendix 1), one can calculate the resultant radioactivity. The radioactivity produced each year that is associated with the strontium-90 (whose mean life is 40 years, or 1.26×10^9 seconds) is

$$\frac{1.5 \times 10^{26}}{1.26 \times 10^9} = 1.2 \times 10^{17} \text{ disintegrations per second}$$
$$= 3.2 \times 10^6 \text{ curies}$$

The radioactivity produced each year that is associated with the cesium-137 (whose mean life is 44 years, or 1.4×10^9 seconds) is

$$\frac{1.6 \times 10^{26}}{1.4 \times 10^9} = 1.1 \times 10^{16} \text{ disintegrations per second}$$
$$= 3.1 \times 10^6 \text{ curies}$$

We ignore, in these rough estimates, the fact that a small fraction of the isotopes produced at the beginning of the year will have decayed by the end of the year.

If a reactor is operated in such a way that all of the fuel elements are replaced at the same time, at intervals of one year, then these estimates would give the amount of radioactivity due to these isotopes that is present in the reactor at the moment of replacement, whereas immediately after replacement there would be much less radioactivity within the reactor.

These are very large quantities of radioactivity. The Federal Radiation Council recommends that the maximum amount of strontium-90 in the body of an individual working with radioactive materials should be kept below 2 *millionths* of a curie, which is 1.6×10^{12} times smaller than the amount of strontium-90 present at the end of one year within the 800 megawatt reactor we have just considered. When an individual has ingested 2 millionths of a curie of strontium-90, 74,000 disintegrations are taking place in his body every second, a rate ten times as large (assuming he weighs 60 kg) as the rate—calculated in the previous Comment—at which potassium-40 and carbon-14 decays are occurring naturally in his body. The task of adequately isolating a reactor's radioactivity from the biosphere is evidently a formidable one.

---

[11] The percentage yields of strontium-90 and cesium-137 in bombs tested in the atmosphere (where most of the fissions were induced by fast particles) are quoted as 3 to 4 percent and 5 to 6 percent, respectively, in Samuel Glasstone (ed.). *The Effects of Nuclear Weapons*. Washington: The Atomic Energy Commission, p. 475 (1964).

Prospects and Procedures for the Nuclear Power Industry

During normal operation, how is the radioactivity handled? Almost all of the radioactive fragments associated with uranium-235 fission remain in the canisters (called fuel rods) in which the uranium was initially sealed. A small amount of the radioactivity turns up elsewhere in the power plant, either because a fission fragment passes through small leaks in the casings of the fuel rods or because a neutron penetrates the casing and then collides beyond the fuel rod with another nucleus, in a reaction which makes a radioactive end product.[12] Most of this stray radioactivity will be found in what is called the primary coolant: the water that passes in a closed circuit through the hot fuel rods. Very little radioactivity is directly transferred to the water that flows through the condensers that cool the spent steam, because the water passing through the condensers is completely independent of the primary coolant.[13]

The high-level radioactive sources in the fuel rods and the low-level radioactive sources elsewhere in the power plant are processed separately. Every year or so, when the plant is to be refueled, the reactor is shut down and the fuel rods are moved to a storage vault where they are left for several months while the shorter-lived radioactivity subsides. The rods are then shipped to a plant where they are chemically processed to recover the unused fuel and to remove the fission products. The fission products are usually stored underground, except for the gaseous wastes, including krypton-85 (half-life of 10.6 years), which are released to the atmosphere. Present procedures are regarded as temporary, and further processing and relocation of the stored wastes is currently under discussion in many countries.

As more power plants are built, a better way will have to be devised to dispose of the accumulating wastes. One proposal is to solidify these wastes and store them in abandoned salt mines. Salt mines have been suggested because they are embedded under rock, have had no contact with groundwater, have high heat conductivity so that the wastes can cool off, are relatively plastic so that they give with earthquakes, and are already equipped with shafts and corridors. It is believed that there are enough mines in the United States to store the radioactive wastes that will accumulate in the next few hundred years.

To control the low-level radioactive sources, all of the materials that have been in close proximity to the fuel rods (including the primary coolant and the water used to clean equipment) are passed through a purification system. All but a small fraction of the solid or liquid radioactive substances removed during the purification process are collected and concentrated. When enough of these wastes accumulate they are encased in barrels and buried.

[12] Yet another source of stray radioactivity is uranium that has been left on the outside surface of fuel elements during the manufacturing of these elements.

[13] See Figure 1 of Professor Eipper's essay, p. 113.

The remaining solid and liquid radioactive elements are discharged to the waterway which serves the power plant by cooling the condenser. The radioactive gases released during purification pass into the atmosphere through a small chimney.

All United States plants must be licensed by the AEC, and must adhere to standards set by that agency. In setting these standards, the AEC follows the recommendations of the Federal Radiation Council (FRC), the National Committee on Radiation Protection and Measurements (NCRP), and the International Commission on Radiological Protection (ICRP). The AEC also has the responsibility for seeing that radiation standards are followed by all users of radioisotopes produced in AEC reactors.

Two kinds of standards exist today. First, there are emission standards that place an upper limit at each power plant on the concentration of radioactivity in the effluent from the power plant that is permitted to be discharged into the surrounding air and water. At certain locations, additional emission standards are imposed to restrict the *total* amount of radioactivity released. For example, one regulation states that the gross quantity of radioactive material released into a sanitary sewage system by an AEC licensee cannot exceed 1 curie per year.

A second type of standard restricts the whole-body radiation dose, *exclusive of the dose from naturally occurring radiation and medical X rays,* that individuals are permitted to receive. For example, present standards restrict individuals in nuclear industries to whole-body doses of 5 rems per year; all other individuals in the population are restricted to doses of 0.5 rems per year; and the average dose to the United States population as a whole cannot exceed 0.17 rems per year. An additional dose of 0.17 rems, or 170 millirems per year will approximately double a person's total radiation exposure, since, as explained earlier in this essay, the average dose from naturally occurring background radiation is about 115 millirems per year, and an average medical dose is about 65 millirems per year.[14]

Standards that restrict the total dose received each year by the population are clearly desirable, but by themselves they are inadequate. Because they restrict the doses received from all sources, they are difficult to enforce. If total dose levels are ever exceeded, the emission standards will presumably have to be tightened, but tightened where?

Over the years all of the standards imposed by the AEC have been tightened, as awareness of the dangers of radiation has increased and as a larger population has been exposed. Criticism of the radiation standards is based largely upon the biological uncertainties described in part II. As we have said, the short-term damaging effects of large doses of radiation have been well documented, but there is considerable uncertainty about the effects of small doses applied over long periods of time. To put the argument succinctly, we do not know the biological impact of the natural background radiation (for example, how much genetic damage or how many cancer cases a year are due to it), and therefore we certainly do not know the bio-

[14] Doses in rads and rems are virtually equivalent in these cases.

logical risks associated with an additional yearly dose that would approximately double the total dose received. Standards cannot, perforce, incorporate information that does not now exist.

What seems certain is that there is some biological cost that cannot be escaped but must be "traded off" in exchange for the benefits of nuclear technology. Such trade-offs are implicit in the expansion of virtually every technology, not just nuclear technology. Excess deaths from cancer due to increases in ambient radiation have their non-nuclear analogs in excess deaths from emphysema due to air pollution. That some of the biological and psychological cost is borne by future generations is also not unique to nuclear technology: creating long-lived isotopes that will not decay for several generations is analogous to altering the level of carbon dioxide in the atmosphere in a way that may threaten injurious modifications in the future climate, and is also analogous to destroying the wilderness heritage by damming up river valleys.

But there is one unique biological cost associated with the decision to increase the amount of radioactivity in the environment, and that is the genetic cost, a cost borne *entirely* by future generations. These generations pay a double price: they carry mutant genes that we have passed on to them, and they are vulnerable to further mutations that can be caused by the long-lived radioactive isotopes we have introduced into their environment. Although these ethical considerations need not deter the development of nuclear technology, at least they ought to make somewhat harder the justification of each innovation.

The nuclear power industry is still in its infancy. In addition, using nuclear weapons for peaceful purposes, like digging canals and recovering natural gas, is scarcely under way. As a result, the actual total dose due to the radioactivity engendered by nuclear technology is now far below the AEC limit of 170 millirem per year. (It is probably less than 1 percent of that figure.) In fact, the existence of a ceiling that is far above current levels may actually be psychologically harmful, because those responsible for emissions from an individual power plant may be led to believe that *their* emissions are insignificant.

Soon the nuclear power industry will begin to grow rapidly; the number of curies of radioactive material produced in power plants is already rising steadily. Moreover, the amount of energy produced at nuclear installations relative to non-nuclear installations is also growing. It is possible that someday fusion power will be developed, and this is expected to permit fewer curies of radioactive materials to be produced per kilowatt-hour of electrical energy. But there is likely to be a period of time—the last two decades of this century and perhaps beyond—when industrialized societies are dependent on power from nuclear fission.

Each additional kilowatt-hour of electrical energy generated by nuclear fission increases the amount of radioactive waste which must be disposed of. Accordingly this increases both the small but finite medical damage resulting from routine operation and the risk of severe medical damage in the event of an accident during one of the many stages of processing of that waste.

Moreover, the more advanced fission reactors ("breeder" reactors) that are expected to be in operation in about ten years will be fueled by plutonium instead of or in addition to uranium.[15] Plutonium fuel, unlike uranium fuel, can be used in a nuclear weapon without elaborate processing to separate isotopes. It thus becomes easier to divert reactor material to military purposes, and problems of international monitoring to assure that this is not being done become more acute. In short, additional kilowatt-hours of energy production will be associated with additional political risks as well as additional medical risks.

It seems likely, therefore, that the period of dependency on nuclear fission reactors will be a time during which the environmental costs of energy production are widely perceived. Such a perception could unsettle a society committed to growth.

[15] Glenn T. Seaborg and Justin L. Bloom, "Fast Breeder Reactors," *Scientific American*, Vol. 23, No. 5, p. 13 (1970).

APPENDIX 1. HALVING AND DOUBLING

One often tries to predict how a quantity will grow or diminish with time. Few quantities change in a simple way. But two types of growth are so simple to discuss mathematically that this simplicity leads us to apply the mathematics widely.

The first type of growth is called *linear* growth: the *amount* of growth is independent of time. An infant gaining half a pound each week is growing in a linear fashion. A beach losing 100 tons of sand each year is shrinking in a linear fashion.

The second type of growth is called *exponential* growth: the *percent* of growth is independent of time. Your dollar getting 5 percent compound interest annually is appreciating in value exponentially. A beach losing 5 percent of its sand each year is shrinking exponentially.

The mathematics of linear growth is too simple to merit discussion; the mathematics of exponential growth is more interesting. The basic idea underlying exponential growth is that the amount of change that occurs *depends on how much of the substance is present,* and moreover depends

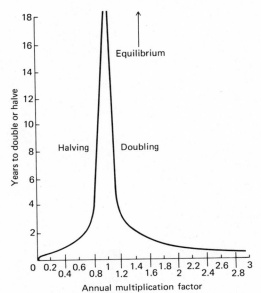

FIGURE 1 This figure shows how many years it takes for a population either to double its size or to shrink to half its size, assuming that the size is changing in an exponential fashion. The horizontal scale is the ratio of the population in one year to the population in the previous year. If this ratio is exactly 1, the population is in equilibrium, and the time to double or halve becomes infinitely long.

321

on the amount in a specific way such that the rate of change is always proportional to the amount present.

Suppose that the ratio of the amount of substance at any time to the amount a year earlier is K, and suppose that an amount, S, is present originally. We assume that K is *greater* than 1, so that the quantity is *increasing*. Then after one year the amount present will be KS, after two years it will be K^2S. The question can be asked: After how many years will the amount of the substance double? Mathematically we want the relation between K and N such that

$$K^N S = 2S \qquad \text{or} \qquad K^N = 2$$

TABLE 1

Annual Multiplication Factor (K)	Annual Growth Rate (percent)	Doubling Time (N) (in years)
1	0	Infinite
1.01	1	69.7
1.02	2	35.0
1.03	3	23.4
1.04	4	17.7
1.05	5	14.2
1.06	6	11.9
1.07	7	10.2
1.08	8	9.0
1.09	9	8.0
1.10	10	7.3
1.12	12	6.1
1.14	14	5.3
1.16	16	4.7
1.18	18	4.2
1.20	20	3.8
1.30	30	2.6
1.40	40	2.1
1.50	50	1.7
1.60	60	1.5
1.70	70	1.3
1.80	80	1.2
1.90	90	1.1
2.00	100	1.0
3.00	200	0.6
4.00	300	0.5
5.00	400	0.4
10.00	900	0.3
100.00	9900	0.15

discussions of growth that we display the relation here in Table 1. We also graph the relation between K and N in Figure 1. The reason that N is not an integer in the table is because the growth is proceeding continuously, and N gives the length of time it takes to double. We list in the table not only the annual growth factor, K, but also the percent growth. Thus if a substance is growing by a factor of 1.05 in a year, we speak of a "5 percent annual growth rate." From the table we see that doubling occurs in 14.2 years in that situation. (Compare this with *linear* growth at 5 percent a year, where 20 years are required for doubling.)

Exponential *shrinking* is just as simple an idea as exponential growing. Just let K be smaller than 1. This time we ask for the number of years for the amount of the substance to halve (a length of time called the *half-life*). The answer comes from solving the equation

$$K^M S = \tfrac{1}{2} S \qquad \text{or} \qquad K^M = \tfrac{1}{2}$$

where M is the half-life.

TABLE 2

Annual Multiplication Factor (K)	Annual Shrinkage Rate (percent)	Halving Time (M) (in years)
1.00	0	Infinite
0.99	1	69.0
0.98	2	34.3
0.97	3	22.8
0.96	4	17.0
0.95	5	13.5
0.94	6	11.2
0.93	7	9.6
0.92	8	8.3
0.91	9	7.3
0.90	10	6.6
0.88	12	5.4
0.86	14	4.6
0.84	16	4.0
0.82	18	3.5
0.80	20	3.1
0.75	25	2.4
0.70	30	1.9
0.65	35	1.6
0.60	40	1.4
0.55	45	1.2
0.50	50	1.0
0.40	60	0.8
0.30	70	0.6
0.20	80	0.4
0.10	90	0.3
0.01	99	0.15

The relation between K and M is tabulated in Table 2, and graphed on Figure 1. The two parts of Figure 1, one showing doubling-time and one showing halving-time, are related, because, for example, it takes the same number of years to double at a growth rate of 1.01 as it takes to halve at a growth rate of $1/1.01$.

To delve more deeply into exponential growth we must skip to a higher level of mathematics. But we already have the most important ideas in front of us.

Let $y(t)$ stand for the amount of quantity y at time t and let dy/dt stand for the rate at which the quantity is changing at time t. Then the equations that embody the assumption of exponential growth are the following:

$$\frac{dy}{dt} = \frac{1}{T}y(t) \tag{1a}$$

$$\frac{dy}{dt} = -\frac{1}{T}y(t) \tag{1b}$$

Here $1/T$ is a constant of proportionality, which we always assume to be positive. Then, since y is positive, the left-hand side of equation (1a) is positive, and the left-hand side of equation (1b) is negative. Hence equation (1a) describes exponential growth, and equation (1b) describes exponential shrinking. All of the information about the growth (or shrinking) is contained in the single constant T, whose value we must measure or guess for each special case. The constant T has units of time, and in situations of exponential shrinking T is called the *mean-life*.

Consider explicitly the case of a radioisotope, say carbon-14. Assume that a large amount is present. Equation (1b) describes how the amount present will change with time, and T, the mean-life, is a known constant for each isotope. For carbon-14, T is 8200 years, and (1b) becomes

$$\frac{dy}{dt} = -\frac{y(t)}{8200 \text{ years}} \tag{2}$$

Approximating the left-hand side by a ratio of small but not infinitesimal quantities, $\Delta y/\Delta t$, where Δy is the change in the amount of the substance which occurs in the time Δt, we can rearrange equation (2) to get

$$\frac{\Delta y}{y(t)} = -\frac{\Delta t}{8200 \text{ years}} \tag{3}$$

Equation (3) is an approximation to equation (2), with the approximation only being a good one if both sides of equation (3) are very much smaller than 1. In that limit, which, for example, would be valid if Δt were one year, we learn that the fractional change during a time Δt in the amount of isotope present, $\Delta y/y$, is given by the ratio $\Delta t/T$. Hence in one year, $1/8200$ of the amount of isotope which we start with will decay. This is why the mean-life is useful.

The connection between half-life and mean-life is obtained by solving the differential equation (1b). The answer is

$$y(t) = y(0)e^{-t/T} \tag{4}$$

where $y(0)$ is the amount present at time $t=0$, and the equation gives the amount present at time t later. (The number e, approximately equal to 2.718, is the most important constant of mathematics, with the possible exception of π.) Let M again be the length of time to halve, that is,

$$y(M) = \tfrac{1}{2}y(0) \tag{5}$$

From equations (4) and (5) we can obtain

$$2 = e^{M/T}$$

or

$$\ln_e 2 = \frac{M}{T}$$

where $\ln_e 2$ is the natural logarithm of 2 and equals 0.693. Thus

$$0.693T = M \tag{6}$$

That is, the half-life is about 69 percent of the mean life. The mean life is longer because it takes longer for all but $1/e=37$ percent of a substance to be left than for one-half to be left.

Finally, let us learn a trick that bankers know, which lets them memorize the first part of Table 1. We had found that K (the annual growth factor) and N (the doubling time) were related by

$$K^N = 2$$

but this can be written, using natural logarithms again, as

$$N \ln_e K = \ln_e 2 = 0.693$$

If K, the annual growth factor, is close to 1, a property of the natural logarithm allows us to simplify this equation in an approximate way. Writing $K=1+c$, with c much less than 1, it is true that $\ln(1+c)$ is closely approximated by c. Then

$$Nc = 0.693 \qquad \text{or} \qquad N = \frac{69.3}{100c}$$

We recognize that $100c$ is just the percent growth. Hence the rule of thumb: If you know the percent growth, you can get the doubling time by dividing the percent growth into 69.3. This is approximate and only valid when the percent growth is small. Compare the rule of thumb with the data in Table 1 to get a feel for what is going on.

APPENDIX 2. MATHEMATICS OF DEMOGRAPHY

Basic Concepts

We are interested in those concepts which help make precise any discussion of populations. These can be populations of light bulbs, or algae, or Eskimos. All that is essential is that each member of the population have a well-defined beginning (birth) and a well-defined end (death). When the members of the population are human beings, the analysis is called demography.

Much of demography concerns the *prediction* of how populations will change and the *explanation* of why changes have occurred in the past. All such analysis is beyond the scope of this appendix. An example of demographic analysis is the essay by Lincoln and Alice Day in this book. Here we confine ourselves to the *description* of populations.

The principal subject of interest is the number of members at each age at every moment of time, the number of 10-year-olds on January 1, 1968, for example. Let x be an integer labeling the ages from 0 to the age beyond which no member is alive (which we can take to be infinity) and let t label calendar time. The population is completely described by a function $N(x, t)$ of two variables, which is called the *age-specific population* at time t. For the same special case, suppose that when $x=10$ years and $t=$ January 1, 1968, $N(x, t)$ had the value 250. This would mean that there were 250 10-year-olds (population members between the ages of 10 and 11) on January 1, 1968.

The total population at time t, $T(t)$, is evidently

$$T(t) = \sum_{x=0}^{\infty} N(x,t) \tag{1}$$

Total pop = Sum of age specific populations 0 → ∞

where the symbol Σ ("capital sigma") means take the *sum* of all the terms enumerated, these being the terms in which x takes all possible integer values between 0 and ∞.

If no one died, then the equation

$$N(x + 1, t + 1) = N(x,t) \tag{2}$$

would describe how the population evolved in time. Here we measure calendar time in years, the same unit in which we measure age, to keep things simple. Equation (2) says, for example, that all 10-year-olds on January 1, 1968, are 11-year-olds on January 1, 1969, a statement that is correct because we have assumed that no 10-year-olds can die.

To take death into account, we assume (as do all life insurance companies) that it is possible to assign a probability $p(x, t)$ that any member of the population who is age x at time t will survive to be age $x+1$ at

326

time $t+1$.[1] Such survival probabilities are statistical averages, which can be obtained when large populations are studied. Like all probabilities, $p(x,t)$ is restricted to values between 0 and 1; 1 means the individual is sure to survive the year, 0 means he has no chance of making it.

Now we see how to modify equation (2): The number of 11-year-olds on January 1, 1969, equals the number of 10-year-olds on January 1, 1968, times the fraction of 10-year-olds on January 1, 1968, who survive one year. That fraction is just what is meant by the probability $p(10;$ January 1, 1968). So, in general, we see that equation (2) becomes

$$N(x + 1, t + 1) = N(x,t)p(x,t) \tag{3}$$

We permit $p(x,t)$ to depend on t in order to allow for innovations that increase or decrease the chance of survival. These innovations might be in medicine (new drugs), diplomacy (ending a war), technology (safer cars), and so on.

Note that t assumes discrete values one year apart in our treatment. An alternative treatment in which time and age are continuous variables is briefly mentioned in the Comment: *Treatment of population development with continuous variables.*

Comment: Treatment of population development with continuous variables

The mathematically inclined reader will realize that although we have discussed populations and conditional survival probabilities only for discrete time and age intervals, a limit exists in which these intervals become arbitrarily short and the population distribution becomes a function of two continuous variables (x,t), one for age, one for calendar time. In the same limit, it is convenient to replace $1 - p(x,t)$ by a function $d(x,t)$ representing the probability *per unit time* that a person age x at time t will die. In this limit, equation (3) then becomes a partial differential equation

$$\frac{\partial N(x,t)}{\partial x} + \frac{\partial N(x,t)}{\partial t} = -N(x,t)d(x,t)$$

In addition, there is a boundary condition relating $N(0,t)$ to the instantaneous birthrate. We will not discuss the continuum case again.

The probability that a person age x at time t survives to age $x+2$ at time $t+2$ is the product of the probability $p(x,t)$ that he survives from time t to time $t+1$ and the probability $p(x+1,t+1)$ that he survives from time $t+1$ to time $t+2$. It is useful to introduce the probability $S(x,t)$ that an in-

---

[1] We assume that $p(x,t)$ is independent of whether a person is exactly age x or somewhere between age x and age $(x + 1)$ at time t. The probability, $1 - p(x,t)$, that a person does *not* survive the year between age x and age $(x + 1)$ is essentially the quantity called the *age-specific death rate* by demographers.

fant between 0 and 1 at time t will survive to the age of x at time $(t+x)$.[2] Again, this is a product of probabilities, each describing the likelihood of survival in a one-year interval, and we can write

$$\prod_{r=0}^{x-1} [p(r,t+r)] = S(x,t) \tag{4}$$

The symbol $\prod$ (capital pi) means take the *product* of all the terms enumerated. The significance of $S(x,t)$ lies in the fact that the population age x at time $t+x$ is equal to the population age 0 at time t multiplied by the probability that they will have survived the x years beginning at time t:[3]

$$N(x,t+x) = N(0,t)S(x,t) \tag{5}$$

To take births into account, let us introduce the function $B(t)$ representing the number of live births occurring during the year preceding time, t.[4] The *crude birthrate* is defined by demographers to be $B(t)/T(t-\frac{1}{2})$, the number of births in a year divided by the total population in the middle of that year.

Special Populations: The Stationary and the Stable Distributions

The essay by Lincoln and Alice Day in this book distinguishes three special ways in which a population distribution can evolve in time.

1. A _stable distribution_ is a population distribution in which the *fraction* of the population at any age is the same every year. Mathematically, this means that $N(x,t)$ can be written as a product of two functions, one depending only on age, the other depending only on calendar time:

$$N(x,t) = F(x)T(t) \tag{6}$$

In this expression $F(x)$ gives the fraction of the total population at age x, and $T(t)$, as defined in equation (1), is the total population at time t.

The fractions $F(x)$ explicitly do not depend on time: the same fraction of the population is age ten year after year. The same fraction is also age zero year after year; a stable age distribution implies a constant birthrate.[5]

---

[2] Because of the assumption stated in the previous footnote, $S(x,t)$ is the probability that a newborn survives to his x-th birthday and also the probability that a six-month-old survives to age x and a half.

[3] We ignore immigration and emigration. The demographer who studies the population of a city cannot do this. How is equation (5) modified when immigration and emigration are taken into account?

[4] Note that if no children died between the ages of zero and 1, the number of children between zero and 1 at time t, $N(0, t)$, would equal $B(t)$. It occasionally simplifies the discussion below to assume a world in which $B(t) = N(0,t)$; we warn the reader whenever this assumption is used.

[5] This last statement is almost, but not quite, precise. It can be made precise only by taking into account the distinction between $N(0,t)$ and $B(t)$ discussed in footnote 4.

A stable age distribution cannot arise unless the survival probability function $p(x,t)$ has a very restricted form. Using (3) and (6), we see that

$$\frac{F(x+1)}{F(x)} \cdot \frac{T(t+1)}{T(t)} = p(x,t) \qquad (7)$$

Thus the survival probability function $p(x,t)$ itself, just like the population function $N(x,t)$, must have the form of a *product:* a function only of age times a function only of calendar time.

The simplest assumption about $p(x,t)$ compatible with equation (7) is that it is independent of time (the case of "no innovation"). Then we may write $T(t+1) = T(t) \cdot q$, where q is a constant; q can be greater or less than 1. Also, since $p(x,t+1) = p(x,t)$ when $p(x,t)$ is independent of time, it follows by simple algebra that

$$T(t+2) = T(t+1)q = T(t)q^2$$

and similarly

$$T(t+3) = T(t)q^3, \quad \text{etc.}$$

Thus the entire population increases or decreases by a constant multiplicative factor every year while keeping a constant age distribution. Increase or decrease by a constant multiplicative factor every year is called *exponential growth.*

We have just proved the following theorem: If a stable age distribution is to persist without any change in the survival probabilities, the only way the total population can change with time is by growing (larger or smaller) exponentially.

2. A *constant total population* hardly needs discussion. The function $T(t)$ described in equation (1) will be independent of time, whereas no constraints whatsoever are imposed on the age *distribution* within the total population. The number dying each year is exactly equal to the number being born, that number being permitted to vary arbitrarily from year to year. Suppose, with a population like the present distribution in the United States, which is predominantly young, we insisted on immediately obtaining a constant population. This would mean that very few babies could be born next year, and for several years thereafter, because relatively few people would die. After a while, we would have proportionately not very many children and an abundance of adults. When these adults reached the end of their life-span, about 70 years from now, it would be time for the birth rate to boom again to replace the large numbers dying. Thus a peak in the population distribution would travel up through the ages and then start over again.[6]

> [6] It is interesting that this traveling bump in the age distribution will actually damp out, and the distribution will approach the stationary distribution discussed in the paragraphs immediately below. Graphs of the behavior of this "bump," as well as numerous other interesting calculations based on the United States population are found in Tomas Frejka, "Reflections on the Demographic Conditions Needed to Establish a U.S. Stationary Population Growth," *Population Studies,* Vol. 22, No. 3, p. 389 (1968).

3. A *stationary population* is at the same time a stable age distribution *and* a constant total population. This is the demographers' ideal case: neither total numbers nor fractions at each age change with time. Mathematically, this means that the entire function $N(x,t)$ is independent of time.

Looking at equation (5), which is valid no matter what we assume about the population distribution $N(x,t)$, we see that if $N(x,t)$ and $N(0,t)$ are independent of time, the probability of living to age x, $S(x,t)$, must also be independent of time. Dropping the now irrelevant index t, equation (5) becomes

$$N(x) = N(0)S(x) \tag{8}$$

In other words, the stationary age distribution is exactly proportional to the distribution of probabilities for living to age x. The ratio of population age 30 to population age 60 in the stationary population is exactly the same number as the ratio of the probability of surviving from age 0 to age 30 to the probability of surviving from age 0 to age 60. The function $S(x)$ is known for our society for every year in the recent past (in fact, known separately for men and women); this is what permitted the Days, in Figure 3, page 213, to draw the stationary population distribution for the United States in 1967 and to compare it with the actual 1967 population.

Births Revisited: GRR and NRR

We conclude this Appendix by returning to the subject of births in order to introduce some simple concepts which demographers use. We define $b(x,t)$ to be the probability that a woman who was age x at time t bore a child during the following year. (The demographers call this the *age-specific birthrate*.) The total number of births between t and $t+1$ we have already called $B(t+1)$. It is related to the function $N_f(x,t)$, the number of *females* of age x at time t, by

$$B(t+1) = \sum_{x=0}^{\infty} b(x,t)N_f(x,t) \tag{9}$$

since each term in the sum corresponds to the total number of babies born during the year between t and $t+1$ to mothers of age x at time t, and we count all the babies born in that year when we sum over all possible mothers' ages.

The age-specific birthrate, $b(x,t)$, contains all the information about the propensity to bear children; the crude birthrate, $B(t)/T(t-1/2)$, which we defined previously, does not contain such information in a simple way because it is dependent on the total population. For example, one can have two populations with exactly the same age-specific birthrates and exactly the same populations of childbearing age and above, but with different populations of children below childbearing age. Initially, the population with the larger number of children will have a lower crude birthrate because its total population is larger; in a few years, if the age-specific birth-rates stay the same, this same population will grow more quickly.

Although the function $b(x,t)$ contains all the detailed information about childbearing, we may want to be able to summarize this information by making reference to some *average* quantity. With this in mind, demographers have defined two concepts which make use only of the age-specific birthrates and which are averages of a sort. The first, called the *gross reproduction rate* (GRR), is the total number of *female* children a woman would have if she were born at time t and were to survive to the end of her childbearing years while the age-specific birthrates in effect at time t remained unchanged. Once we define a fraction f, which is the fraction of the babies born which are female babies (best estimates today give $f = 100/205 = 0.487$), we can write

$$\text{GRR}(t) = f \sum_{x=0}^{\infty} b(x,t) \tag{10}$$

The GRR gives a snapshot picture averaging over the childbearing propensities of all the women at a single moment in time.

The second, and even more useful, concept is the *net reproduction rate* (NRR), which differs from the gross reproduction rate only by making allowance for the fact that not all women will survive to the end of their reproductive years. Thus the age-specific birthrate is weighted by the probability that the woman will survive to age x:

$$\text{NRR}(t) = f \sum_{x=0}^{\infty} b(x,t) S_f(x,t) \tag{11}$$

(In this expression, $S_f(x,t)$, the probability that a woman will survive exactly to age x, is conventionally computed assuming that the age-specific death rates for women remained forever at the values that they had at time t.) Because $S_f(x,t)$ is never more than 1, NRR is never more than GRR.[7]

The significance of NRR is that it gives the average number of female children which a woman can expect to have, if age-specific birth and death rates remain unchanged indefinitely. When the NRR equals 1, each woman exactly replaces herself. The demographers refer to *births at the replacement level* when NRR = 1.

The reader can verify, after a few steps of algebra, that if one has a stationary distribution, one must have NRR = 1.[8] But the converse is not

---

[7] The data in Figures 2 and 3 of Essay 13 (pp. 210 and 213) suffice to calculate the NRR and GRR in the United States in 1967. The reader in search of a better grasp of these abstract concepts is urged to try these computations.

[8] One must also assume $B(t) = N(0,t)$. Moreover, equation (11) has not corrected for the possibility that a woman would have given birth in the second half of a year but died during the first half of that year. Both of these problems disappear in the continuum case. They cannot be avoided in the discrete case unless

true, of course: one can have NRR=1 with any initial population distribution whatsoever. No matter what the initial distribution, you might guess, it will approach the stationary distribution over a period of time, if the age-specific birth and death rates are always such that the NRR=1. This conjecture appears to be true for a wide class of circumstances. The reader is encouraged to invent a simple population (for example, do not let anyone live more than four years) and to work out how the population evolves from various initial distributions under various assumptions about the basic functions $b(x)$ and $p(x)$.[9]

one assumes that births and deaths occur only at the beginning of the smallest discrete interval (year, day) being considered.

[9] For a more extensive development of the mathematical methods of demography, consult the Suggestions for Further Reading for Essay 13 (pp. 347–348).

APPENDIX 3. RELATIONS AMONG PHYSICAL UNITS

TIME

	Seconds	Minutes	Hours	Days	Years
1 second =	1				
1 minute =	60	1			
1 hour =	3,600	60	1		
1 day =	86,400	1,440	24	1	
1 year =	31,536,000	525,600	8,760	365	1

LENGTH

A. Metric Units

	Fermis	Angstroms	Microns	Centimeters	Meters	Kilometers
1 fermi =	1					
1 angstrom =	10^5	1				
1 micron =	10^9	10^4	1			
1 centimeter =	10^{13}	10^8	10^4	1		
1 meter =	10^{15}	10^{10}	10^6	10^2	1	
1 kilometer =	10^{18}	10^{13}	10^9	10^5	10^3	1

B. Conversion from Other Units

	Centimeters	Inches	Feet	Meters	Kilometers	Miles
1 centimeter =	1					
1 inch =	2.540	1				
1 foot =	30.48	12	1			
1 meter =	100	39.37	3.281	1		
1 kilometer =	1.000×10^5	3.937×10^4	3281	1000	1	
1 mile (statute) =	1.609×10^5	6.336×10^4	5280	1609	1.609	1

TEMPERATURE

One centigrade degree = 1.8 Fahrenheit degees.

The zero of centigrade is at 273.15 degrees above absolute zero.

Conversion formulas between a temperature in Fahrenheit (F), Centigrade (C), and Kelvin (K):

$$F = 1.8C + 32$$
$$C = K - 273.15$$
$$F = 1.8K - 459.67$$

The Kelvin temperature scale is sometimes called the absolute temperature scale.

WATER QUANTITY[a]

	Grams	Pounds	Liters	Gallons	Cubic Feet	Cubic Meters	Acre-Feet	Cubic Miles
1 gram = 1 cm³ =	1							
1 pound =	453.6	1						
1 liter = 10³ cm³ =	1.000×10^3	2.205	1					
1 gallon (U.S.) =	3.785×10^3	8.347	3.785	1				
1 cubic foot =	2.832×10^4	62.45	28.32	7.481	1			
1 cubic meter =	1.000×10^6	2.205×10^3	1.000×10^3	264.2	35.31	1		
1 acre-foot =	1.234×10^9	2.720×10^6	1.234×10^6	3.259×10^5	4.356×10^4	1.234×10^3	1	
1 cubic mile =	4.170×10^{15}	9.191×10^{12}	4.170×10^{12}	1.101×10^{12}	1.472×10^{11}	4.170×10^9	3.379×10^6	1

[a] Both volumes and masses are listed together. A density of exactly 1 gram per cubic centimeter is assumed. (It is useful to be aware of the existence of two "ounces" and three "tons." One fluid ounce = 1.043 ounces avoirdupoid; one gallon = 128 fluid ounces; and one pound = 16 ounces avoirdupoid. One long ton = 1.016 metric tons = 1.120 short tons; one long ton = 2240 pounds; one metric ton = 1000 kilograms; and one short ton = 2000 pounds.)

ENERGY

	Electron Volts	Ergs	Joules	Foot-Pounds	Calories	Btus	Kilowatt Hours
1 electron volt =	1						
1 erg =	6.242×10^{11}	1					
1 joule = 1 watt × 1 sec =	6.242×10^{18}	10^7	1				
1 foot-pound =	8.464×10^{18}	1.356×10^7	1.356	1			
1 calorie =	2.613×10^{19}	4.186×10^7	4.186	3.087	1		
1 Btu =	6.585×10^{21}	1.055×10^{10}	1.055×10^3	777.9	252.0	1	
1 kilowatt-hour =	2.247×10^{25}	3.600×10^{13}	3.600×10^6	2.655×10^6	8.601×10^5	3413	1

Note that 1 calorie is the amount of energy needed to raise 1 gram of water 1° C, and 1 Btu is the amount of energy needed to raise 1 pound of water 1° F.

Note also that the "calorie" discussed by weight watchers is 1000 times larger than the calorie described here.

APPENDIX 4. SOME FACTS ABOUT OUR PLANET

EARTH

Mass	5.983×10^{27} grams
Mean density	5.522 grams per cubic centimeter
Polar radius	6.357×10^8 centimeters
Equatorial radius	6.378×10^8 centimeters
Total area	5.101×10^{18} square centimeters
Area of continents	1.488×10^{18} square centimeters
Area of oceans	3.613×10^{18} square centimeters
Acceleration of gravity near earth's surface	978 centimeters per square second at equator
	983 centimeters per square second at poles
Magnitude of magnetic field near earth's surface	0.3 gauss at equator
	0.6 gauss at poles
Average distance from sun	1.495×10^{13} centimeters
Average distance from center of earth to center of moon	3.844×10^{10} centimeters

The solar flux (the rate at which solar energy falls on a surface perpendicular to the sun, lying outside the earth's atmosphere and at the average distance of the earth from the sun) is 1.94 calories per square centimeter per minute.

The albedo (the fraction of solar energy reflected by the earth and returned to outer space at the same wavelength as the incident radiation) is approximately 0.36.

The length of the earth day is increasing at a rate of 0.001 second per century as the earth dissipates its rotational kinetic energy by tidal friction.

ATMOSPHERE

Mass (including water vapor)	5.14×10^{21} grams
Average mass of water vapor	1.3×10^{19} grams
Average density at sea level at 0° C	0.00129 grams per cubic centimeter
Number of moles (1 mole = 6.02×10^{23} molecules)	1.8×10^{20}
Standard air pressure at sea level (this pressure is called 1 atmosphere)	1.013×10^{6} ergs per cubic centimeter
Height above sea level at which pressure drops to one-half atmosphere (pressure decreases approximately exponentially with height)	5.5×10^{5} centimeters (18,000 feet)

Composition at sea level (excluding water vapor)	Fraction by volume or by number of molecules	Fraction by weight
Molecular nitrogen (N_2)	0.7809	0.7553
Molecular oxygen (O_2)	0.2095	0.2314
Argon (A)	0.0093	0.0128
Carbon dioxide (CO_2)	0.00034	0.00046
Neon (Ne)	1.8×10^{-5}	1.25×10^{-5}
Helium (He)	5.2×10^{-6}	7.24×10^{-7}
Methane (CH_4)	1.4×10^{-6}	7.25×10^{-7}
Krypton (Kr)	1.14×10^{-6}	3.30×10^{-6}
Nitrous oxide (N_2O)	5×10^{-7}	7.6×10^{-7}
Molecular hydrogen (H_2)	5×10^{-7}	3.48×10^{-8}

OCEANS

Mass	1.43×10^{24} grams
Typical density at surface in mid-ocean	1.028 grams per cubic centimeter
Average depth	3.80×10^5 cm
Density at average depth	1.045 grams per cubic centimeter
Average temperature	4° C (39° F)
Typical salinity in mid-ocean	3.48 percent by weight

Composition (excluding dissolved gases)	Percent by weight
Oxygen in water	85.72
Hydrogen in water	10.80
Total water	96.52
Chlorine (Cl^-)	1.92
Sodium (Na^+)	1.07
Oxygen—primarily in sulfate (SO_4^{--}) and carbonate (CO_3^{--}) ions	0.19
Magnesium (Mg^{++})	0.13
Sulfur—primarily in sulfate (SO_4^{--}) ions	0.09
Calcium (Ca^{++})	0.04
Potassium (K^+)	0.04
Total dissolved salts	3.48

LIVING MATTER

Standing biomass (estimate)	1.8×10^{18} dry grams
On the continents	1.8×10^{18} dry grams
In the oceans	3.3×10^{15} dry grams
Net primary production (estimate)	1.6×10^{17} dry grams per year
On the continents	1.0×10^{17} dry grams per year
In the oceans	0.6×10^{17} dry grams per year

The standing biomass is the average dry weight of the standing crop of green plant matter at any instant. The net primary productivity is the dry weight of the green plant matter produced annually. If these concepts included animal matter, the quantities above would be only slightly larger.

SUGGESTIONS FOR FURTHER READING

Essay 1. An Empire of Dust

Marsh, George Perkins. *Man and Nature*, 1864. Reprinted by Harvard University Press, 1965. (David Lowenthal, ed.)

Sears, Paul B. *Deserts on the March*. Norman, Okla.: University of Oklahoma Press, 1935 (rev. ed., 1959).

Smith, Frank Ellis. *The Politics of Conservation*. New York: Pantheon, 1966.

Steinbeck, John. *The Grapes of Wrath*. New York: Viking Press, 1939.

Thornthwaite, C. Warren. "Climate and Settlement in the Great Plains." U.S.D.A. Yearbook, *Climate and Man*, 1941, pp. 177–187.

Essay 2. The Collapse of Classic Maya Civilization

General

Adams, Robert McC. "Archaeological Research Strategies: Past and Present," *Science*, Vol. 160, No. 3833, pp. 1187–1192 (1968). A description of recent trends in archaeological thinking including ecological approaches to culture history.

Hole, Frank, and Robert F. Heizer. *An Introduction to Prehistoric Archaeology*, 2nd ed. New York: Holt, Rinehart and Winston, 1969. A general introduction to the science of archaeology; chapter 17, on ecology and systems theory, is especially relevant to the concerns of this book.

Sanders, William T., and Barbara Price. *Mesoamerica: The Evolution of a Civilization*. New York: Random House, 1968. A general view of development of civilization in Mesoamerica which emphasizes the ecological approach and places the Classic Maya in a broad environmental and historical perspective.

Vayda, Andrew P., and Roy Rappoport. "Ecology, Cultural and Non-cultural," in *Introduction to Cultural Anthropology: Essays in the Scope and Methods of the Science of Man*, edited by J. Clifton, pp. 477–497. Boston: Houghton Mifflin, 1968. The best short introduction to the ecological approach in anthropology, emphasizing the importance of a holistic view of man and his environment.

The Maya

Coe, Michael D. *The Maya*. New York: Praeger, 1966. A very good introduction to the culture history of the Classic Maya.

Morley, Sylvanus G., and George W. Brainerd. *The Ancient Maya*, 3rd ed. Stanford, Calif.: Stanford University Press, 1956. A classic in the field which summarizes various views of the collapse.

339

Stephens, John L. *Incidents of Travel in Central America, Chiapas* and *Yu-catán*, 2 vols. New York: Dover, 1969.

——. *Incidents of Travel in Yucatán*, 2 vols. New York: Dover, 1963. Ste-phens called attention to the Maya civilization in these classic works describing two journeys to Central America in the 1840s. The Dover edition reprints the illustrations by his travel compan-ion, F. Catherwood, that appeared in the original books.

Thompson, J. Eric S. *The Rise and Fall of Maya Civilization*. Norman, Okla.: University of Oklahoma Press, 1966, rev. ed. Another good introduction to the Maya which includes a description of Thomp-son's peasant revolt hypothesis.

The Collapse

Cowgill, George L. "The End of Classic Maya Culture: A Review of Re-cent Evidence," *Southwestern Journal of Anthropology*, Vol. 20, pp. 145–159 (1964). A good review of hypotheses of the collapse.

Jiménez Moreno, Wigberto. "Síntesis de la Historia Pretolteca de Mesoa-merica," in *Esplendor del México Antiguo* (C. Cook de Leonard, coordinator), Vol. 2, pp. 1019–1108 (1959). Centro de Investi-gaciones Antropologicas de México. (English translation appears in *Ancient Oaxaca*, edited by J. Paddock, Stanford University Press.) Presents an invasion hypothesis related to general develop-ments in Mexican and Guatemalan culture history.

Meggers, B. J. "Environmental Limitation on the Development of Culture," *American Anthropologist*, Vol. 56, pp. 801–824 (1954). Includes an important discussion of the environmental limitations of the Southern Lowlands, although it is mistaken in assuming a nonin-digenous origin for Classic Maya civilization.

Sabloff, Jeremy A., and Gordon R. Willey. "The Collapse of Maya Civiliza-tion in the Southern Lowlands: A Consideration of History and Process," *Southwestern Journal of Anthropology*, Vol. 23, pp. 311–336 (1967). Includes an extensive discussion of previous collapse hypotheses with many references for further reading and presents an invasion hypothesis based on excavations at the site of Seibal.

Essay 3. Window to the City: The Emergency Room

Brown, Claude. *Manchild in the Promised Land*. New York: Macmillan, 1965.

Harrington, Michael. *The Other America—Poverty in the United States*. New York: Macmillan, 1962.

Duff, Raymond S., and August B. Hollingshead. *Sickness and Society*. New York: Harper and Row, 1968.

Jacobs, Jane. *The Death and Life of Great American Cities*. New York: Random House, 1969.

Lewis, Oscar. *La Vida*. New York: Random House, 1966.

McHarg, Ian. *Design with Nature.* Natural History Press, 1969.
Mumford, Lewis. *Urban Prospects.* New York: Harcourt, Brace and World, 1968.

Essay 4. Low-Sulfur Fuels for New York City

"Air Pollution." Workbook prepared by the Scientists' Institute for Public Information, 30 East 68th Street, New York, N.Y. 10021.
"Air Pollution Control," *Law and Contemporary Problems.* Durham, N.C.: Duke University, School of Law (Spring 1968).
American Chemical Society Report, *Cleaning Our Environment: The Chemical Basis for Action.* American Chemical Society, 1155 Sixteenth Street, N.W., Washington, D.C. 20036.
Journal of the Air Pollution Control Association, 4400 Fifth Avenue, Pittsburgh, Pa. 15213.
National Air Pollution Administration, U.S. Department of Health, Education, and Welfare. *Air Quality Criteria,* Control Techniques.
Robinson, Elmer, and Robert C. Robbins. "Gaseous Sulfur Pollutants from Urban and Natural Sources," *Journal of the Air Pollution Control Association,* Vol. 20, p. 233 (April 1970).
Stern, Arthur C. (ed.). *Air Pollution,* 2nd ed., 3 vols. New York: Academic Press, 1968–1969.

Essay 5. Abortion: A Case Study in Legislative Reform

Abortion

Af Geijerstam, Gunnar K. (ed.). *An Annotated Bibliography of Induced Abortion.* Ann Arbor, Center of Population Planning, University of Michigan, 1969.
American Friends Service Committee, *Who Shall Live? Man's Control over Birth and Death.* New York: Hill and Wang, February, 1970.
Lader, Lawrence. *Abortion.* Indianapolis: Bobbs-Merrill, 1966.
Means, Cyril C., Jr. "The Law of New York Concerning Abortion and the Status of the Foetus, 1664–1968: A Case of Cessation of Constitutionality," *New York Law Forum,* pp. 411–515 (Fall 1968).
Newman, Sydney H., Mildred B. Beck, and Sarah Lewit. "Abortion Obtained and Denied: Research Approaches," New York, The Population Council, *Studies in Family Planning,* No. 53 (May 1970).
Pilpel, Harriet. "The Right of Abortion," *Atlantic Monthly,* Vol. 223, p. 69 (June 1969).
Rosen, H. (ed.). *Abortion in America.* Boston: Beacon Press, 1954.
Smith, David T. (ed.). *Abortion and the Law.* Cleveland: The Press of Western Reserve University, 1967.
Tietze, Christopher, and Sarah Lewit. "Abortion," *Scientific American,* Vol. 220, p. 21 (January 1969).

Local Government

Dahl, Robert A. *Who Governs?* New Haven: Yale University Press, 1961.
——. *Pluralistic Democracy in the United States: Conflict and Consent.* New York: Rand McNally, 1967.
Key, V. O., Jr. *American State Politics: An Introduction.* New York: Knopf, 1956.
——. *Politics, Parties, and Pressure Groups,* 5th ed. New York: Thomas Crowell, 1964.

Essay 6. The Helium Conservation Program of the Department of the Interior

Helium and the Conservation Program

Cook, Gerhard A. (ed.). *Argon, Helium and the Rare Gases: The Elements of the Helium Group,* 2 vols. New York: Interscience Publishers, 1961.
"Helium: Costs Jeopardize Future of Government Conservation Program," *Science,* Vol. 167, No. 3925, pp. 1593–1596 (1970).
Lipper, Harold W. *Helium Symposia Proceedings in 1968—A Hundred Years of Helium.* Washington, D.C.: U.S. Government Printing Office, 1969.
Seibel, Clifford W. *Helium, Child of the Sun.* Lawrence, Kansas: University of Kansas, 1968.
U.S. Bureau of Mines. "Helium," in *Mineral Facts and Problems.* Washington, D.C.: U.S. Government Printing Office, published every five years.
——. "Helium," in *Minerals Yearbook.* Washington, D.C.: U.S. Government Printing Office, published annually.

Economics of Conservation

Barnett, Harold J., and Chandler Morse. *Scarcity and Growth: The Economics of Natural Resource Availability.* Baltimore: Johns Hopkins Press, 1963 (for Resources for the Future).
Brown, Harrison, James Bonner, and John Weir. *The Next Hundred Years.* New York: Viking Press, 1963.
Burton, Ian, and Robert W. Kates (eds.). *Readings in Resource Management and Conservation.* Chicago: University of Chicago Press, 1965.
Scott, Anthony. *Natural Resources: The Economics of Conservation.* Toronto: University of Toronto Press, 1955.
Spengler, Joseph J. "The Economist and the Population Question," *American Economic Review,* Vol. LVI, pp. 1–24 (March 1966).

Essay 7. The Trial of DDT in Wisconsin

Carter, L. J. "Environmental Pollution: Scientists Go to Court," *Science,*
    Vol. 158, pp. 1552–1557 (1967).

——. "DDT: The Critics Attempt to Ban Its Use in Wisconsin," *Science,*
    Vol. 163, pp. 548–551 (1969).

Carson, Rachel. *Silent Spring.* Boston: Houghton Mifflin, 1962.

Graham, F., Jr. *Since Silent Spring.* Boston: Houghton Mifflin, 1970.

Gray, Oscar. *Environmental Law: Cases and Materials.* Washington, D.C.:
    Bureau of National Affairs, 1970.

Harrison, H. L., O. L. Loucks, J. W. Mitchell, D. F. Parkhurst, C. R. Tracy,
    D. G. Watts, and V. J. Yannacone, Jr. "Systems Studies of DDT
    Transports," *Science,* Vol. 170, pp. 503–508 (1970).

Hickey, J. J. *Peregrine Falcon Populations: Their Biology and Decline.*
    Madison: University of Wisconsin Press, 1969.

Hickey, J. J., and D. W. Anderson. "Chlorinated Hydrocarbons and Egg-
    shell Changes in Raptorial and Fish-eating Birds," *Science,* Vol.
    162, pp. 271–273 (1968).

Nash, R. G., and E. A. Woolson. "Persistence of Chlorinated Hydrocarbon
    Insecticides in Soils," *Science,* Vol. 157, pp. 924–926 (1967).

Peakall, D. B. "Pesticide-induced Enzyme Breakdown of Steroids in Birds,"
    Natura, Vol. 216, pp. 505–506 (1967).

Rudd, R. L. *Pesticides and the Living Landscape.* Madison: University of
    Wisconsin Press, 1964.

U. S. Department of Health, Education and Welfare. *Report of the Secre-
    taries Commission on Pesticides and Their Relationship to Environ-
    mental Health,* 1969.

Welch, R. M., W. Levin, and A. H. Conney. "Estrogenic Action of DDT
    and Its Analogs," *Toxicol. Appl. Pharmacol.,* Vol. 14, pp. 358–367
    (1969).

Woodwell, G. M., C. F. Wurster, and P. A. Isaacson. "DDT Residues in an
    East Coast Estuary: A Case of Biological Concentration of a Persis-
    tent Insecticide," *Science,* Vol. 156, pp. 821–824 (1967).

Wurster, C. F. "Chlorinated Hydrocarbon Insecticides and the World Eco-
    system," *Biological Conservation,* Vol. 1, pp. 123–129 (1969).

——. "DDT Goes to Trial in Madison," *Bioscience,* Vol. 19, p. 809 (1969).

Essay 8. Nuclear Power on Cayuga Lake

Description of the Cayuga Lake Story

Carlson, C. A., *et al. Radioactivity and a Proposed Power Plant on Cayuga
    Lake.* Ithaca, N.Y.: Authors, Fernow Hall, Cornell University, 1968.

Child, D., and R. Oglesby. *Annotated Bibliography of Limnological and
    Related Literature Dealing with the Finger Lakes Region.* Ithaca,
    N.Y.: Publication Number 29 of the Cornell University Water Re-
    sources and Marine Sciences Center, February 1970.

Comey, D. D. (ed.). *The Cayuga Lake Handbook,* Vol. I: *Thermal and Radioactive Pollution by a Proposed Nuclear Power Plant.* The Citizens Committee to Save Cayuga Lake, Box 237, Ithaca, N.Y., 1969.

Eipper, A. W., *et al. Thermal Pollution of Cayuga Lake by a Proposed Power Plant.* Ithaca, N.Y.: Authors, Fernow Hall, Cornell University, 1968.

Nelkin, Dorothy. "Nuclear Power and Its Critics: The Cayuga Lake Controversy." Ithaca, N.Y.: Cornell University Press, 1971.

Oglesby, R. T., and D. J. Alee (eds.). *Ecology of Cayuga Lake and the Proposed Bell Station (Nuclear Powered).* Ithaca, N.Y.: Cornell University Water Resources and Marine Sciences Center, 1969.

Additional References

Abrahamson, D. E. *Environmental Cost of Electric Power.* Workbook. New York: Scientists' Institute for Public Information, 1970.

Clark, J. R. "Thermal Pollution and Aquatic Life," *Scientific American,* Vol. 220, No. 3, pp. 18–27 (1969).

Coutant, C. C. *Thermal Pollution—Biological Effects and a Review of the Literature of 1969.* Bulletin BNWL–SA–3255. Richland, Washington: Battelle Mem. Inst., 1970.

Eipper, A. W. "Pollution Problems, Resource Policy, and the Scientist," *Science,* Vol. 169, pp. 11–15 (1970).

Federal Water Pollution Control Administration. "Industrial Waste Guide on Thermal Pollution." Corvallis, Oregon: Pacific Northwest Laboratory, 1968.

Federal Water Quality Administration. "Feasibility of Alternative Means of Cooling for Thermal Power Plants near Lake Michigan." Washington, D.C.: U.S. Department of the Interior, August 1970.

Hogerton, J.F. "The Arrival of Nuclear Power," *Scientific American,* Vol. 218, No. 2, pp. 21–31 (1968).

Mihursky, J. A., and V. S. Kennedy. *Bibliography on Effects of Temperature in the Aquatic Environment* (1220 entries). Nat. Res. Inst., Contribution No. 326. College Park: University of Maryland, 1967.

National Technical Advisory Committee. *Water Quality Criteria.* Washington, D.C.: Federal Water Pollution Control Administration, 1968.

Raney, E. C., and B. W. Menzel. "A Bibliography: Heated Effluents and Effects on Aquatic Life with Emphasis on Fishes" (1217 entries). Mimeo, 2nd ed., Dr. Edward C. Raney. Ithaca, N.Y.: Fernow Hall, Cornell University, 1967.

Thermal Pollution—1968, Part 1. Hearings before the Subcommittee on Air and Water Pollution of the Committee on Public Works, U.S. Senate, 90th Congress, 2nd Session, February 1968. Washington, D.C.: U.S. Government Printing Office, 1968.

U.S. Water Resources Council. *The Nation's Water Resources.* Washington, D.C.: U.S. Government Printing Office, 1968.

Essay 9. Warfare with Herbicides in Vietnam

Chemical and Biological Warfare. A special issue of *Scientist and Citizen* (now called *Environment*), Vol. 9, No. 7, pp. 113–172 (Aug.–Sept. 1967).

Galston, G. W. *Life of the Green Plant,* 2nd ed. Englewood Cliffs, N.J.: Prentice-Hall, 1964.

Janick, J. *Horticultural Science.* San Francisco: W. H. Freeman, 1963.

Rose, Steven (ed.). *CBW: Chemical and Biological Warfare.* Boston: Beacon Press, 1969.

Whiteside, T. *Defoliation.* New York: Ballantine/Friends of the Earth, 1970.

Essay 10. The Fight against Project Sanguine

Brown, K. V. "How We'll Broadcast With Mystery Waves," *Popular Science,* September 1969, p. 104.

Laycock, G. "Not All is Sanguine in Wisconsin," *Audubon,* January 1970, p. 105.

Libber, L. M. "Extremely Low Frequency Electromagnetic Radiation Biological Research," *Bioscience,* Vol. 20, p. 1169 (1970).

Sanguine System Environmental Compatibility Assurance Program (ECAP) Status Report, December, 1970, with four appendices. Available from the Sanguine Project Office, Naval Electronic Systems Command Headquarters, Washington, D. C. 20360.

"What is Project Sanguine?" Published by and obtainable from the State Committee to Stop Sanguine, P.O. Box No. 7, Ashland, Wisconsin 54806.

Essay 11. Mineral King: Wilderness versus Mass Recreation in the Sierra

Mineral King

"After 700 Road Curves, There's Old Mineral King . . . ," *Sunset Magazine,* pp. 42–52 (June 1967).

Anderson, Dewey. "Mineral King—A Fresh Look," *National Parks and Conservation Magazine,* Vol. 44., No. 272, pp. 8–10 (May 1970).

Hano, Arnold. "The Battle of Mineral King," *New York Times Magazine,* August 17, 1969.

Hope, Jack. "The King Besieged," *Natural History Magazine,* November 1968.

McCloskey, Michael. "Why the Sierra Club Opposes the Development of Mineral King," *Sierra Club Bulletin,* Vol. 52, No. 10, pp. 6–10 (November 1967).

Sierra Club

Brower, David (ed.). *Gentle Wilderness: The Sierra Nevada.* San Francisco: Sierra Club, 1967.

Colby, William E. (ed.). *John Muir's Studies in the Sierra.* San Francisco: Sierra Club, 1960.

Jones, Holway R. *John Muir and the Sierra Club.* San Francisco: Sierra Club, 1965.

Sierra Club Handbook. San Francisco: Sierra Club, 1969.

Man and the Land

Abbey, Edward. *Desert Solitaire.* New York: McGraw-Hill, 1968.

Dasmann, Raymond F. *The Destruction of California.* New York: Macmillan, 1965.

Douglas, William O. *A Wilderness Bill of Rights.* Boston: Little, Brown, 1965.

Leopold, Aldo. *A Sand County Almanac.* New York: Oxford University Press, 1949. In paperback, San Francisco: Sierra Club; New York: Ballantine Books, 1970.

McCloskey, Maxine E. (ed.). *Wilderness and the Quality of Life.* San Francisco: Sierra Club, 1969.

Thoreau, Henry David. *Walden.* New York: Dodd, Mead, 1946.

Essay 12. The Everglades: Wilderness versus Rampant Land Development in South Florida

Bent, Arthur Cleveland. *Life Histories of North American Marsh Birds.* New York: Dover Publications, 1963. Best source of information about the marsh birds inhabiting Everglades National Park and the Big Cypress Swamp.

Browder, Joe. "In Retrospect: The Everglades, the Jetport, and the Future," *Audubon Magazine*, March 1970. An excellent article, by the man who spearheaded the Audubon Society's efforts to stop the jetport, on the inadequacy of the decision at the federal level to relocate the jetport.

Egler, Frank E. "Southeast Saline Everglades Vegetation, Florida, and Its Management," *Vegetatio*, Vol. III, p. 213 (1952). A technical but readable discussion of Everglades vegetation, with especially interesting sections on hammock formation and evolution.

Environmental Problems in South Florida. Report of the Environmental Study Group of the National Academy of Sciences and the National Academy of Engineering. Spotlights some of the harmful effects on the air and water in South Florida which will result from a jetport and from uncontrolled land development.

May, Julian. *Alligator Hole.* Illustrated by Rod Ruth. Chicago and New York: Follett, 1969. A delightful book for children on the wildlife in an alligator hole in the Everglades.

Robertson, William B., Jr. *Everglades—The Park Story.* Coral Gables, Fla.: University of Miami Press, 1969. Best general source of information about the park flora and fauna. The photographs are excellent, as is the brief history of early Indian settlement in the Everglades.

Udall, Stewart. *Beyond the Impasse: The Dade Jetport and the South Florida Environment,* December 1969. Obtainable from the Overview Corporation, New York. A new kind of jetport for South Florida, one that might coexist with the park, is proposed in this provocative study.

United States Department of the Interior, *The Environmental Impact of the Big Cypress Swamp Jetport,* September 1969, obtainable from the Department of the Interior, Washington, D.C. A thorough analysis of the damaging consequences to the park of the proposed commercial and training jetports.

Essay 13. Toward an Equilibrium Population

Barclay, George W. *Techniques of Population Analysis.* New York: Wiley, 1958. Good, not overly technical discussion of demographic methods.

Berelson, Bernard. "Beyond Family Planning," in *Studies in Family Planning,* February, 1969. A comprehensive review of the various proposals made for dealing with the population problem beyond the current efforts of natural programs of voluntary family planning.

Cox, Peter R. *Demography,* 3rd ed. New York: Cambridge University Press, 1959. An actuarial introduction to demography.

Current Population Reports. United States Bureau of the Census. The best source for United States population statistics.

Davis, Kingsley. "Population Policy: Will Current Programs Succeed?" *Science,* November 10, 1967. A discussion of policies for population control.

Day, Alice Taylor. "Population Control and Personal Freedom: Are They Compatible?" *The Humanist* (Nov.–Dec., 1968).

Day, Lincoln H., and Alice Taylor Day. "Family Size in Industrialized Countries: An Inquiry into the Socio-Cultural Determinants of Levels of Childbearing," *Journal of Marriage and the Family* (May 1969).

———. *Too Many Americans.* Boston: Houghton Mifflin, 1964. Surveys the implications of population growth in the United States and refutes the arguments in support of continued population increase.

Demographic Yearbook. United Nations, Statistical Office. The best source for world-wide population statistics.

Flugel, T. C. *Population, Psychology and Peace.* London: Watts, 1944. A perceptive discussion of the psychological aspects of contraception and population control.

Freedman, Ronald (ed.). *Population: The Vital Revolution.* Garden City, N.Y. Doubleday, 1964.

Jaffe, A. J. *Handbook of Statistical Methods for Demographers.* U.S. Bureau of the Census, 1951.

Keyfitz, Nathan. *Introduction to the Mathematics of Population.* Reading, Mass.: Addison-Wesley, 1968. For the mathematically inclined.

Nam, Charles B. (ed.). *Population and Society.* Boston: Houghton Mifflin, 1968.

Peterson, William. *Population,* 2nd ed. New York: Macmillan, 1969. A good general introduction.

Poffenberger, Thomas. "Motivational Aspects of Resistance to Family Planning in an Indian Village," *Demography* (1968).

Population Index. Office of Population Research, Princeton University. A good current bibliography of demographic literature.

Essay 14. Toward a Stationary—State Economy

Boulding, Kenneth. "The Economics of the Coming Spaceship Earth," *Environmental Quality in a Growing Economy,* ed. Henry Jarrett. Baltimore: Johns Hopkins Press, 1966. An excellent and perceptive discussion of the transition from the "cowboy economy" to the "spaceman economy." All of Professor Boulding's writings are highly recommended.

Daly, Herman E. "On Economics as a Life Science," *Journal of Political Economy,* July 1968.

Keynes, J. M. "Economic Possibilities for Our Grandchildren," *Essays in Persuasion.* New York: Norton, 1963 (originally 1931).

Linder, Staffan B. *The Harried Leisure Class.* New York: Columbia University Press, 1970. A delightful socioeconomic analysis of the causes and consequences of the increasing price of time.

Lotka, A. J. *Elements of Mathematical Biology.* New York: Dover, 1956 (republication).

Mill, J. S. *Principles of Political Economy,* Vol. II. London: John W. Parker and Son, 1857.

Mishan, E. J. *The Costs of Economic Growth.* New York: Praeger, 1967. A broadside attack on the prejudices of growthmania.

Odum, Eugene P. "The Strategy of Ecosystem Development," *Science,* April 18, 1969.

Ohlin, Goran. *Population Control and Economic Development.* Paris: Development Centre of the Organization for Economic Cooperation and Development, 1967. A thorough survey of the literature on the population barrier to economic growth in underdeveloped countries.

United Nations. *Population Bulletin of the United Nations—No. 7, 1963.* New York: UN, 1965. An excellent summary of conditions and trends of fertility in the world, and correlations of fertility with diverse socioeconomic indices.

Essay 15. Adapting World Order to the Global Ecosystem

The Study of World Order

Aron, Raymond. *Peace and War: A Theory of International Relations.* Garden City, N.Y.: Doubleday, 1966. A comprehensive depiction of the workings of the state system on a world level.

Bozeman, Adda B. *Politics and Culture in International History.* Princeton, N.J.: Princeton University Press, 1960. A study that emphasizes the relevance of diverse cultural perspectives to the conduct of international relations.

Etzioni, Amitai. *The Active Society: A Theory of Societal and Political Processes.* New York: The Free Press, 1968. Difficult to read, but very creative and suggestive study of the prospects for drastic change at all levels of social organization.

Falk, Richard A., and Saul H. Mendlovitz (eds.). *The Strategy of World Order,* 4 vols. New York: World Law Fund, 1966–1967. A series of volumes designed to provide students of world order problems with a basic set of materials and interpretations.

Green, Philip. *Deadly Logic: The Theory of Nuclear Deterrence.* Columbus, Ohio: Ohio State University Press, 1966. The best critique of the false logic relied upon to sustain the war system in the nuclear age.

McDougal, Myres S., and Florentino P. Feliciano. *Law and Minimum World Public Order.* New Haven: Yale University Press, 1961. The most important recent effort to relate law to the issues of war and peace.

Sakharov, Andrei D. *Progress, Coexistence and Intellectual Freedom.* New York: Norton, 1968. Proposals for world order reform by a leading Soviet physicist.

Wagar, W. Warren. *The City of Man.* Baltimore: Penguin, 1967. A critical survey of the efforts to establish a more centrally organized world order system.

Implications of an Ecological Perspective

Brown, Harrison. *The Challenge of Man's Future.* New York: Viking, 1954. The first adequate account of the dangerous interrelationships between population pressure, resource depletion, and nuclear war.

Brown, Lester. *Seeds of Change: The Green Revolution and Development in the 1970's.* New York: Praeger, 1970. The best study of technological progress toward avoiding famine in the underdeveloped countries.

Commoner, Barry. *Science and Survival.* New York: Viking, 1966. A lucid account of why technological developments cause environmental peril.

Dubos, Rene. *Man Adapting.* New Haven: Yale University Press, 1965. A convincing exposition of the ways in which possibilities for human health and fulfillment are determined by the physical environment.

Edberg, Rolf. *On the Shred of a Cloud* (translated by Sven Ahmån). Tuscaloosa: University of Alabama Press, 1969. A lyrical yet serious depiction of the ecological crisis.

Falk, Richard A. *This Endangered Planet: Prospects and Proposals for Human Survival.* New York: Random House, 1971. A study of world order written from an ecological perspective, arguing for the necessity of a new world order and advancing a series of proposals to bring it about.

Feinberg, Gerald. *Prometheus Project: Mankind's Search for Long Range Goals.* New York: Doubleday, 1969. A provocative analysis of the need for the long-range goals of society to be broadly discussed in order to bring about changes in consciousness.

Park, Charles. *Affluence in Jeopardy: Minerals and the Political Economy.* San Francisco: W. H. Freeman, 1968. A discussion of resource depletion.

Reich, Charles. *Greening of America: The Coming of a New Consciousness and the Rebirth of a Future.* New York: Random House, 1970. An imaginative argument to the effect that a revolutionary consciousness is embodied in American youth.

Shepard, Paul, and Daniel McKinley (eds.). *Subversive Science: Essays Toward an Ecology of Man.* Boston: Houghton Mifflin, 1968. A fine collection of essays on the implications of ecology.

Essay 16. Water

A general view of limits on resources, including water, is found in:

Borgstrom, Georg. *Too Many, A Story of Earth's Limitations.* New York: Macmillan, 1969.

Skinner, Brian J. *Earth Resources,* Foundation of Earth Science Series. Englewood Cliffs, N.J.: Prentice-Hall, 1969.

At a somewhat more advanced level water treatment is discussed in:

Cleaning Our Environment—The Chemical Basis for Action. A report by the Subcommittee on Environmental Improvement, Committee on Chemistry and Public Affairs, American Chemical Society, Washington, D.C., 1969.

Eckenfelder, W. Wesley, Jr. *Water Quality Engineering for Practicing Engineers.* New York: Barnes and Noble, 1970.

Fair, Gordon and Charles Geyer. *Water Supply and Waste Water Disposal.* New York: Wiley, 1967.

A technical discussion of desalination can be found in:

Clawson, M. and H. H. Landsbery. *Desalting Seawater.* New York: Gordon and Breach, 1970.

Dodge, Barnett F. "Fresh Water from Saline Waters. An Engineering Research Problem." *American Scientist,* Vol. 48, p. 476 (1960).

Essay 17. Energy

Energy, Climate, and Atmosphere

Bryson, Reid A. "All Other Factors Being Constant . . ." *Weatherwise*, April 1968. Excellent discussion of factors that may be affecting the earth's climate.

Dobson, G. M. B. *Exploring the Atmosphere*. New York: Oxford University Press, 1968. Good, nonmathematical treatment of the role of the atmosphere in determining climate.

Hess, S. L. *Introduction to Theoretical Meteorology*. New York: Holt, Rinehart and Winston, 1959. For the mathematically inclined reader.

Lamb, H. H. *The Changing Climate*. London: Methuen and Co., 1966. Documents the meteorological parameters that have changed over the past 50 years.

Neuberger, H., and J. Cahir. *Principles of Climatology*. New York: Holt, Rinehart and Winston, 1969. Contains good explanations of how geographic features determine climatic conditions at the earth's surface. Many interesting questions are provided for the reader.

Ratcliffe, J. (ed.). *Physics of the Upper Atmosphere*. New York: Academic Press, 1960. Good source book at an advanced level.

Robinson, N. (ed.). *Solar Radiation*. New York: Elsevier Publishing Co., 1966. Contains many graphs and tables of data on the solar spectrum, heat exchange, absorption curves, etc.

Singer, S. F. (ed.), *Global Effects of Environmental Pollution*. AAAS Symposium, Dallas, Texas. Berlin: Springer-Verlag, 1970.

Energy and the Biosphere

"The Biosphere," *Scientific American*. Special Issue. September 1970. The entire issue is worth reading.

Kormondy, Edward J. *Concepts of Ecology*. Englewood Cliffs, N. J.: Prentice-Hall. Biological Science Series. An especially fine book on ecology and the energy flow in ecosystems.

Energy and Man's Technology

Broecker, Wallace S. "Man's Oxygen Reserves," *Science*, Vol. 168, No. 3939.

Man's Impact on the Global Environment—Assessment and Recommendations for Action; Report of the Study of Critical Environmental Problems (SCEP). Cambridge, Mass.: MIT Press, 1970.

Resources and Man. A Study and Recommendation by the Committee on Resources and Man. National Academy of Sciences, National Research Council, 1969.

Skinner, Brian J. *Earth Resources*. Englewood Cliffs, N. J.: Prentice-Hall, 1969.

Thirring, Hans. *Energy for Man*. New York: Harper and Row, 1962.

U.S. Energy Policies—An Agenda for Research. Resources for the Future staff report. Baltimore, Maryland: Johns Hopkins Press, 1969.

Essay 18. Radiation

Nuclear Physics (introductory level)

Howard, R. A. *Nuclear Physics.* Belmont, Cal.: Wadsworth, 1963.

Paul, E. B. *Nuclear and Particle Physics.* New York: Wiley, 1969.

Health Physics

Eisenbud, Merrill. *Environmental Radioactivity.* New York: McGraw-Hill, 1963.

Henry, Hugh F. *Fundamentals of Radiation Protection.* New York: Wiley-Interscience, 1969.

Hurst, G. S., and J. E. Turner. *Elementary Radiation Physics.* New York: Wiley, 1970.

Johns, H. E., and J. R. Cunningham. *The Physics of Radiology,* 3rd edition. Springfield, Ill.: Thomas, 1969.

Morgan, K. Z., and J. E. Turner. *Principles of Radiation Protection.* New York: Wiley, 1967.

Upton, Arthur C. *Radiation Injury—Effects, Principles and Perspectives.* Chicago: University of Chicago Press, 1969.

Genetics

Lerner, I. Michael. *Heredity, Evolution and Society.* San Francisco: Freeman, 1968.

Papazian, Haig P. *Modern Genetics.* New York: Norton, 1967.

Stern, Curt. *Principles of Human Genetics.* San Francisco: Freeman, 1960.

Nuclear Technology

Abrahamson, D. E. "Environmental Cost of Electric Power," Workbook prepared by the Scientists' Institute for Public Information, 30 East 68th Street, New York, N.Y. 10021.

Glasstone, Samuel (ed.). *The Effects of Nuclear Weapons.* Washington: U.S. Atomic Energy Commission, 1962.

———. *Sourcebook on Atomic Energy.* Princeton: Van Nostrand, 1958.

Hogerton, J. F. "The Arrival of Nuclear Power," *Scientific American,* Vol. 218, No. 2, p. 21 (1968).

———. *The Atomic Energy Deskbook.* New York: Reinhold, 1963.

———. *Atomic Power Safety.* Washington: Atomic Energy Commission, 1964.

Novick, Sheldon. *The Careless Atom.* New York: Dell, 1969.

"Nuclear Explosives in Peacetime," Workbook prepared by the Scientists' Institute for Public Information, 30 East 68th Street, New York, N.Y. 10021.

Seaborg, Glenn T., and Justin L. Bloom. "Fast Breeder Reactors," *Scientific American,* Vol. 23, No. 5, p. 13 (1970).

Tamplin, Arthur R., and John W. Gofman. *"Population Control" through Nuclear Pollution.* Chicago: Nelson-Hall, 1970.

NOTES ON CONTRIBUTORS

JOHN HARTE is an assistant professor of physics at Yale University. He graduated from Harvard College, class of 1961. Four years later he obtained his Ph.D. in physics from the University of Wisconsin. He spent the following year on a post-doctoral fellowship at the European Center for Nuclear Research (CERN) in Geneva and then two years as a postdoctoral fellow in physics at the University of California, Berkeley, before coming to Yale. Specializing in theoretical physics, Professor Harte's research has focused on the composite nature of "elementary" particles and the interactions of these particles at high energies. A long-standing and active interest in conservation and birdwatching led him to participate in the National Academy of Sciences study of the Everglades during the summer of 1969. He has since been involved in undergraduate education in environmental studies at Yale, where he has developed a new course in environmental science which uses the case-study approach.

ROBERT H. SOCOLOW is an assistant professor of physics at Yale University. He was educated at the Fieldston School in New York City and at Harvard University (B.A. 1959, M.A. 1961, Ph.D. 1964). Immediately following college, he had a fellowship for a year of travel in the Soviet Union, Asia, and Africa, which instilled a commitment to working on the relation of science to social problems. As a graduate student and at Yale he has been deeply involved in teaching science to nonscientists. A parallel commitment to advancing physics has involved him in teaching at the graduate level and in investigations in high energy physics concerned with the role of symmetry in subnuclear phenomena; some of the research was carried out in postdoctoral years at the Berkeley campus of the University of California and at CERN in Geneva, Switzerland. He has been a member of two National Academy of Sciences summer studies, which were focused on the environmental problems of the Everglades (1969) and of Kennedy Airport and Jamaica Bay (1970). He was awarded a Junior Faculty Fellowship from Yale for 1970–1971.

HERMAN E. DALY is associate professor of economics at Louisiana State University where he teaches courses in Latin American economic development, comparative economic systems, and a new course on the economics of population and environment. He has also taught at the University of Ceará (Brazil), and Vanderbilt University. In 1969–1970, he was a research associate at the Economic Growth Center, Yale University, during which time he somewhat perversely and ungraciously became disenchanted with orthodox growth economics and took enough time from his other research to write the article included in this volume. A long-standing interest in economic development led him to an interest in population problems, which in turn led to an interest in the general interrelations of economic growth and ecological limits.

ALICE TAYLOR DAY attended the Brearley School in her native New York City and received a B.A. from Smith College and an M.A. in sociology from Columbia. She has taught sociology at Mt. Holyoke College, the University of Massachusetts, and Albertus Magnus College. She has done research at the Bureau of Applied Social Research, Columbia University, and at the American Institute of Public Opinion (Gallup Poll). The Days have two children.

LINCOLN H. DAY is the chief of the Demographic and Social Statistics Branch of the United Nations. Until recently he held a joint appointment in the department of sociology and the department of epidemiology and public health at Yale University. A native of Iowa, he attended the public schools of Denver, Colorado, and received his B.A. degree at Yale and his M.A. and Ph. D. degrees at Columbia. He has taught sociology at Mt. Holyoke College, Princeton, and Columbia, and has been a Research Associate at both the School of Public Health at Harvard and the Bureau of Applied Social Research at Columbia. On two occasions, for a total of 31 months, he was a Visiting Fellow in Demography at the Australian National University.

ALFRED W. EIPPER is a fishery biologist of the U.S. Bureau of Sport Fisheries and Wildlife and leader of the New York Cooperative Fishery Unit at Cornell University. For the last 22 years he and his family have lived in Ithaca, New York. He received a B.A. in educational psychology from Reed College in 1942. After four years of Coast Guard sea duty in World War II, he obtained a B.S. in wildlife conservation at the University of Maine and a Ph.D. in fishery biology (with a minor in limnology) at Cornell in 1953. He conducts research on the effects of water qualities (such as temperature and oxygen) on larval fishes, and teaches a course in Fishery Resource Management. He is especially interested in objectives, principles, and policies for dealing with pollution problems.

RICHARD A. FALK is Milbank Professor of International Law at Princeton University. He is Director of the North American Section of the World Order Models Project and Vice-President of the American Society of International Law. He is the author of several books including *This Endangered Planet: Prospects and Proposals for Human Survival, Legal Order in a Violent World,* and *The Status of Law in International Society.* He has edited, in collaboration with Saul Mendlovitz, a multi-volume series entitled *The Strategy of World Order,* and with Cyril E. Black, the series *The Future of the International Legal Order,* and by himself, *The Vietnam War and International Law.* He is a member of a National Academy of Sciences panel, International Aspects of The Human Environment, and also a member of the Committee on the Human Environment of the Democratic Party Policy Council.

EDWARD FERRAND is Director of Technical Services in the New York City Department of Air Resources. He serves as principal advisor on science and technology to the Commissioner in the planning, direction, and evaluation of the air pollution control program and is in charge of special studies and research projects. He received his Ph.D. in chemistry from the Pennsylvania State University and was associate professor of chemistry at the Cooper Union School of Engineering and Science before joining the department. In 1963, while at Cooper Union, he was awarded a Science Faculty Fellowship by the National Science Foundation. Dr. Ferrand taught courses in chemistry and nuclear science and engineering at Cooper Union.

ARTHUR W. GALSTON is professor of biology, lecturer in forestry, and director of the Marsh Botanical Gardens at Yale University. He received his undergraduate training at Cornell University, his graduate training at the University of Illinois, and early postdoctoral and faculty experience at the California Institute of Technology. During World War II, he served for some time as Agriculture Officer in Naval Military Government on Okinawa. He has received a Guggenheim Fellowship for work in Stockholm and Paris, a Fulbright Fellowship for work in Canberra, Australia, and a National Science Foundation Science Faculty Fellowship for work in London. He has served as president of both the American Society of Plant Physiologists and the Botanical Society of America.

JOSEPH N. GINOCCHIO is an assistant professor of physics at Yale University working in the field of theoretical nuclear physics. He received his Ph.D. from the University of Rochester in 1964 and did postdoctoral research at Rutgers University and the Massachusetts Institute of Technology.

AUSTIN HELLER is the Secretary of the Department of Natural Resources and Environmental Control of the State of Delaware. From 1966 to 1970 he was the Commissioner of the New York City Department of Air Resources. He received an M.S. in sanitary bacteriology and chemical engineering from Iowa State University in 1941. He is a professional engineer and Diplomate of the Academy of Environmental Engineers.

ALBERT HILL has been a resident of the State of California for most of his life, and until recently of the Los Angeles area. He holds an A.B. and an M.A. (1968) degree in botany from the University of California, Los Angeles. He has been a co-author of articles in scientific journals on aspects of the cytology and ecology of California plants. Currently employed by the Botanical Garden of the University of California, Berkeley, he is involved in a portion of the education programs of the Garden. For the past several years, he has been active in the Mineral King Task Force, an informal committee of members of the Sierra Club interested in the preservation of wilderness values in general and specifically in Mineral King.

RICHARD D. LAMM, a practicing lawyer, has been a member of the Colorado House of Representatives (Dem.) since 1966. He was chief sponsor of Colorado's Therapeutic Abortion Law, of battered child legislation and of a revision of medical licensing. He is an associate professor of law at the University of Denver. He has devoted considerable time to environmental litigation, including actions to preserve Florisson National Monument and to review safety standards on Project Rulison, one of the Atomic Energy Commission's Plowshare programs for peaceful applications of nuclear weapons. He is the attorney for Planned Parenthood of Colorado.

ORIE L. LOUCKS is Professor of Botany in the Institute for Environmental Studies at the University of Wisconsin. He has worked in forest ecology and related environmental sciences in Canada and the United States for the past 15 years. He is currently utilizing the techniques of systems studies to identify gaps in our understanding of environmental deterioration and to assist court actions to protect the long-range public interest.

MICHAEL MCCLOSKEY is the Executive Director of the Sierra Club. He has worked for the club since 1961 serving in various capacities, most recently—until his appointment as Executive Director in 1969—as the club's Conservation Director. He has published widely both in the club's journals and in national magazines, producing both scholarly and popular articles. He did most of the staff work in the club's successful effort to have a Redwood National Park established. Mr. McCloskey earned his B.A. degree from Harvard in 1956 and earned a law degree from the University of Oregon in 1961. He has been in charge of the club's Mineral King campaign since 1965.

CHARLOTTE ALBER PRICE is a monetary economist, who has been concerned with problems of developing countries, and recently with conservation questions exemplified by her interest in helium. She has worked in the research department of the Federal Reserve Bank of New York, and has taught at Columbia College, Barnard College, and Vassar College. She is an alumna of Denison University, and did her graduate work at Duke and Columbia Universities.

JEREMY A. SABLOFF is an assistant professor of anthropology and an assistant curator of middle American archaeology at the Peabody Museum, Harvard University. His research in New World archaeology has been principally concerned with the ancient Maya. He is the co-author (with G. R. Willey) of *A History of American Archaeology* (to be published by Harcourt, Brace, and Jovanovich). He spent four seasons (1965–1968) as a member of the Peabody Museum archaeological project at the Maya site of Seibal, Guatemala.

PAUL SEARS was Professor of Conservation and Chairman of the Yale Conservation Program from 1950 until his retirement in June, 1960. Past president of the American Association for the Advancement of Science, the Ecological Society of America and the American Society of Naturalists and former Chairman of the Board of the National Audubon Society, his special field of interest is ecology, its application in the management of natural resources, and the history of vegetation.

Born in Bucyrus, Ohio in 1891, he holds a Bachelor's degree from Ohio Wesleyan, a Master's from Nebraska, and a doctorate from Chicago. He has taught at Ohio State, Nebraska, Oklahoma, Oberlin and Yale and is the author of several books and numerous articles, both technical and of general interest. He holds the honorary degrees of D.Sc., LL.D. and Litt.D., and until recently was a member of the National Science Board. He is currently a member of the Plowshare Advisory Committee of the AEC, a trustee of the Pacific Tropical Botanical Garden, and an honorary president of the National Audubon Society. He is now living in Taos, New Mexico. His books include *Deserts on the March, Charles Darwin, The Living Landscape,* and *Lands Beyond the Forest.*

KENT SHIFFERD was chairman of the Wisconsin State Committee to Stop Sanguine from its founding until April 1970. He teaches European history and Black Studies at Northland College in Ashland, Wisconsin. He is the holder of two Uhrig awards for excellence in teaching. He is a graduate of Northern Illinois University.

DR. H. LYLE STOTTS has been a full-time emergency room physician at Bridgeport Hospital, Bridgeport, Connecticut, since 1968. From 1953 to 1968 he had a private practice in Monroe, Connecticut. He graduated from the medical school at the University of Washington, Seattle, in 1952.

INDEX